Nonclinical Drug Administration

Nonclinical Drug Administration

Formulations, Routes and Regimens for Solving Drug Delivery Problems in Animal Model Systems

By

Shayne C. Gad
Principal
Gad Consulting Services

and

Charles B. Spainhour
Executive Vice President and Chief Scientific Officer
Calvert Holding & Calvert Laboratories
and
President and Chief Scientific Officer
Spainhour & Associates LLC

CRC Press
Taylor & Francis Group
Boca Raton London New York

CRC Press is an imprint of the
Taylor & Francis Group, an **informa** business

CRC Press
Taylor & Francis Group
6000 Broken Sound Parkway NW, Suite 300
Boca Raton, FL 33487-2742

Printed on acid-free paper

International Standard Book Number-13: 978-1-4665-0253-6 (Hardback)

Visit the Taylor & Francis Web site at
http://www.taylorandfrancis.com

and the CRC Press Web site at
http://www.crcpress.com

This work is dedicated to my wife, Novie Beth.

For your loving partnership and the joy, laughter, and smiles it brings.

Contents

Preface

The identification of promising therapeutic entities, the selection of the best from among these candidates to move into the expensive process of development, the evaluation of the efficacy and safety of the selected lead compounds, and prognostication of the relationship between administered doses of drugs and effects that are to be expected in humans are steps in drug development that have in common an evaluation in selected nonclinical (mammalian) models in which it is required that sufficient concentrations of drug ultimately be delivered to a therapeutic target organ site or tissue. Also required is that some multiple of the total systemic exposure necessary to be efficacious be shown to be safe. This process is subsequently replicated in humans relying heavily on methods of delivery (route, regimen, delivery devices, and formulation) that have been previously optimized in nonclinical models.

Paul Ehrlich's concept of the perfect medicine as a "magic bullet" – a molecule which goes with high selectivity to the therapeutic target site, does what it needs to do, and is subsequently cleared from the body with no (or minimal) action at any other site/tissue (as we now call it, no "off-target hits") – is especially relevant now. Much of the current cost of developing and bringing to market effective therapeutic agents and much of the post-market safety concerns arise from the inability to achieve adequate concentrations and selectivity in the due course of actually delivering the active drug moiety to where we would like it to be. And yet, in general, the means of achieving this aim, particularly in nonclinical models, largely remains the same as those employed at least a half century ago.

This volume will examine and present approaches of achieving the desired adequate and selective delivery using the currently available technology in four tool sets:

- *Route* – the entry points for introducing a drug into the body as efficiently and effectively at the desired target site as possible and addressing the species-specific differences in these approaches.
- *Regimen* – the use of chronological variability in the administration of therapeutic agents to optimize the attainment and maintenance of a consistent therapeutic concentration while concurrently maintaining the greatest possible safety margin (or separation from the tissue concentration threshold of toxicity).
- *Devices* – these provide means of controlling, to various degrees, the location, timing, and extent of therapeutic administration.

- *Formulation* – optimization of the selection of vehicles and excipients to maximize the bioavailability and selectivity at therapeutic targets while minimizing toxicity attributable to the nontherapeutic components of administered drug products. Included will be an appendix of available formulation components with their species-specific limitations and advantages in the cardinal nonclinical model species (mice, rats, dogs, rabbits, guinea pigs, non-human primates, and minipigs).

The authors bring with them more than 70 years of experience in working on optimizing these tool sets, and have previously published multiple papers, chapters, and books in these tool set areas. No integrated approach has previously been made available to nonclinical drug developers. This text will seek to bridge that gap and incorporate, update, and integrate the knowledge and experience in all four areas.

About the Author

Shayne C. Gad, B.S. (Whittier College, Chemistry and Biology, 1971) and after active duty service in U.S. Navy, Ph.D. in Pharmacology/Toxicology (Texas, 1977) DABT, Principal of Gad Consulting Services, a twenty-four year old consulting firm with nine employees and more than 500 clients worldwide in the pharmaceutical and medical device industries. Past President of the American College of Toxicology, the Roundtable of Toxicology Consultants and three of SOT's specialty sections, and recipient of the American College of Toxicology Lifetime Contribution Award in 2008. Served on Council, Membership, Program, and Animals in Research committees for ACT, multiple committees in SOT, and currently on the regulatory opinions committee of the Society of Toxicologic Pathologists. Previously at CMIR Chemical Hygiene Fellowship, Allied Chemical, Searle, Becton Dickinson and Synergen. He has authored or edited 49 published books and more than 350 chapters, articles and abstracts in the fields of toxicology, statistics, pharmacology, drug and device development and safety assessment. He has more than 39 years of broad based experience in regulatory toxicology, drug and device development, statistics and risk assessment, and has specific expertise in neurotoxicology, *in vitro* methods, cardiovascular toxicology, inhalation toxicology, immunotoxicology, risk assessment and genotoxicology. Grant reviewer for EPA, CAAT, NIH, and Canadian Health. He has direct involvement in the preparation of INDs (111 successfully to date), NDA, PLA, ANDA, 510(k), IDE, CTD, clinical data bases for phase 1 and 2 studies, and PMAs. Served as the COO of two pharmaceutical companies while a consultant.

Charles B. Spainhour has amassed approximately forty-seven years of experience in the pharmaceutical industry. For his first fifteen years Dr. Spainhour worked at a company known then as Smith Kline & French Laboratories and now as GlaxoSmithKline Corp., where he held positions in both the research and the development segments of the drug development process. Specifically, on the research side Dr. Spainhour's activities involved the areas of phosphodiesterase inhibition, anti-diuretic hormone agonism and antagonism, dopamine agonism and antagonism, neuroleptic biochemistry, anti- infective mechanisms and structure-activity-relationships. On the development, side Dr. Spainhour worked in pharmacokinetics & metabolism, radiosynthesis, process chemistry and drug safety evaluation. Specific areas of endeavor included H2-antagonists, leukotrienes, anti-infectives, anti-hypertensives, anti-hyperglycemic agents and anti-arrhythmics. Dr. Spainhour played a key role in the development of the first H2-antagonist, Cimetidine® and worked closely with Sir James Black.

Dr. Spainhour has a great interest and copious amount of experience in the research and development of natural products. Dr. Spainhour has also worked as an independent consultant in the areas of safety assessment and forensic toxicology, held a position on the faculty at the veterinary school of Texas A&M University and worked as a Toxicologist in the CRO Industry previously for Pharmakon Research International, Chrysalis International, Phoenix International Life Sciences and MDS Pharma Services. Currently Dr. Spainhour serves as the Executive Vice-President and Chief Scientific Officer for Calvert Laboratories and Calvert Holdings and is the President and Chief Scientific Officer of Spainhour & Associates LLC. Dr. Spainhour holds a B.S. from Michigan State University in Biochemistry, a veterinary degree from the University of Pennsylvania and a Ph.D. from Texas A&M University (Developmental Toxicology), and is board certified in forensic science, toxicology and forensic medicine.

1

Introduction

This book seeks to address a trio of objectives that, though separate, are linked and central to biomedical science and, ultimately, medicine. For nonclinical safety assessment, failure to view these three in an integrated manner (rather than the current practice as separate "silos") and to regularly and thoughtfully incorporate new information and technology has largely frozen in time, for decades, the practice of safety assessment.

The first objective is to present how to deliver more of a drug product systemically to facilitate the regulatory need for evaluating safety and efficacy in animal species (at elevated exposure levels) prior to advancing the drug to human testing, with the goal of translating the same lessons into humans and optimizing clinical use.

The second objective is to achieve better tolerance of therapeutics administration in test animals and humans which achieves the first and third objectives. Tolerance should be either local (at the site or organ of administration) or systemic (across the entire organism, especially at distal "off target" sites). Local tolerance is usually determined by evaluating either inflammation (erythema and edema) and tissue histopathology signs at the site of intravenous administration or changes in measures of the formed cell elements in the blood. Determining systemic tolerance (general toxicology) is most often what is entailed in general (or "systems") toxicology (Gad, 2016).

The third objective is to explore ways to improve therapeutic target receptor delivery performance, therefore improving both clinical pharmacodynamics bioavailability and specificity. In so doing, we also decrease off-target hits (in terms not just of receptors, but also of regionally delimited receptor sets) and both local tissue and systemic tolerance. The issue of cardiovascular and immune toxicity due to receptor effects distal to the therapeutically intended target tissues grows increasingly visible (Moslehi, 2016)

Our current model in pharmacodynamics is that we have a molecule which we want to adequately occupy a localized population of receptors so as to achieve our desired therapeutic effect (Keiser et al., 2009; Brunton et al., 2011). In this model, therapeutic efficacy is achieved by our drug molecules occupying target receptors for an optimal residence time period (Copeland, 2016). We have gotten very good at creating and identifying molecules to occupy specific protein receptors on a molecular level – but we must use animal models as surrogates to adequately understand how to do this on an intact higher-level organism and then be able to translate the findings to as close to an approval of human use methodology as possible before actually proceeding to humans. The common operating hypothesis is that if we get just enough test article to a target region, regardless of route, at sufficient plasma levels (minimum effective concentration, or MEC) while concurrently keeping C_{max} levels of the test article in the plasma below a minimum adverse level (minimum toxic concentration, or MTC) in order to achieve ideal therapeutic efficacy and safety (and therefore, at unintended targets below MTC), we can achieve therapeutic perfection [Figure 1.1]). There is a balance that needs to be maintained with plasma levels: the presence of sufficient test article to achieve the maximum effect or efficacy coupled with consideration being given to not have plasma or desirable and undesirable tissue levels that are so high that the incidence and severity of adverse off-target side reactions are minimized. At the same time, there is a recognition that what are desirable therapeutic (and undesirable off-target) adverse – "toxic" – thresholds vary among individuals, particularly in the heterogenous human population (illustrated in Figure 1.2). One should note here an implication that the optimum concentration will usually *not* be at a level that requires occupancy of all target receptors.

As an essential part of progressing a potential drug to the point where its clinical attributes may be evaluated, the practicing toxicologist is faced with multiple challenges in developing formulations and optimizing routes and regimens for the evaluation of candidate drug formulations, which are not shared elsewhere in drug development. These challenges apply (with minor exceptions) across all routes of administration, for which intended clinical

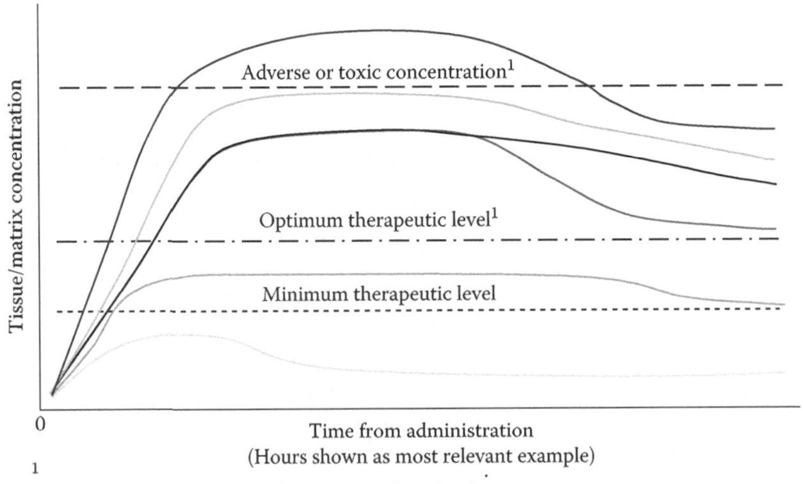

FIGURE 1.1
Optimum therapeutic levels for tissue or minimum levels for effective and safe therapy.

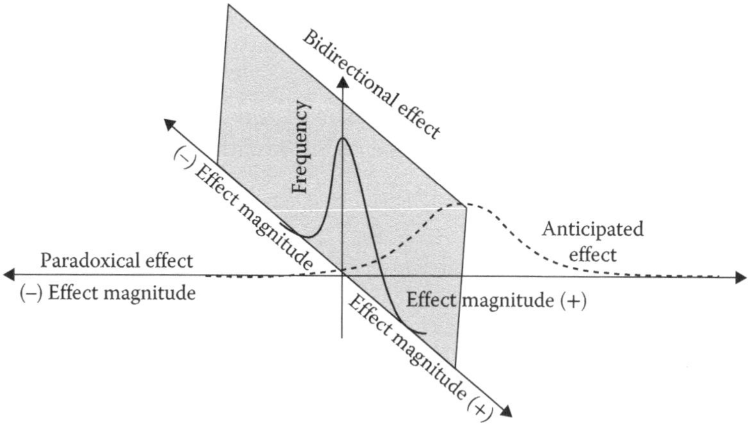

FIGURE 1.2
Few drugs have a single effect profile - rather, there are usually both desired (therapeutic target tissue) and unanticipated (toxicity target tissue) effect profiles which overlap - or not.

routes are preferred, if at all possible, to reduce the dimensions of extrapolation of results to the clinical use case:

- The components of formulations used to deliver a drug moiety must be tolerated – in multiple species (or duration of treatment) – while at the same time delivering plasma levels of candidate drug(s) that are of sufficient magnitude to please regulators. More than one formulation may be required over the course of development, each for a different species (or route).
- Nonclinical formulations are *not* required to be those used in clinical studies. Recent FDA expectations for dermal products have changed on this point, as there is also a desire for dermal systemic toxicity studies to imitate dermal transport metabolism and tolerance of a drug as it would be in use.
- Target doses that typically must be delivered are high – while volumes for delivery are limited to avoid inducing local tissue trauma in animals ("tolerance").
- Tolerances of specific species to different vehicle components must be considered, and there is only limited literature available to help with this (Gad et al., 2016).
- Study durations may allow uses in early (short-term) studies that cannot be extended to longer studies.
- For most routes and studies, maximizing levels of solubility and systemic bioavailability and concurrently maintaining drug stability are primary objectives. In our experience, the implications of this may actually complicate or preclude successful development.

- With the general exception of the dermal route, rates of administration matter – and the regimens can be one of the tools of manipulation to achieve study goals.
- There are "points of art" which can be applied to most routes to help optimize the meeting of all these requirements mentioned above.

Receptor Theory of Drug Efficacy and Drug Toxicity

In 1906, Paul Ehrlich conceptualized the "magic bullet" approach to a perfect therapeutic, as introduced in Figure 1.3 (Drews, 2000). A perfect therapeutic would be specific to the target receptor and go in unmodified form to a target site in a patient, do what it needs to do, and subsequently leave or be eliminated from the body. Ehrlich was an immunologist seeking to treat infectious diseases and there was at that time no knowledge of proteins as receptors, but his vision has remained the ideal for pharmacotherapy. It turns out that things are just not that simple.

Our current working model in pharmacodynamics is still underpinned by Ehrlich's model, being that we have one or more molecules which we want to adequately occupy a localized population of receptors so as to achieve our desired therapeutic effect (Keiser et al., 2009). The occupation may achieve its goal by either activating or blocking activation of receptors, or stabilizing them.

We have gotten very good at identifying molecules which occupy specific protein receptors on a molecular level – but we must use tissue and animal

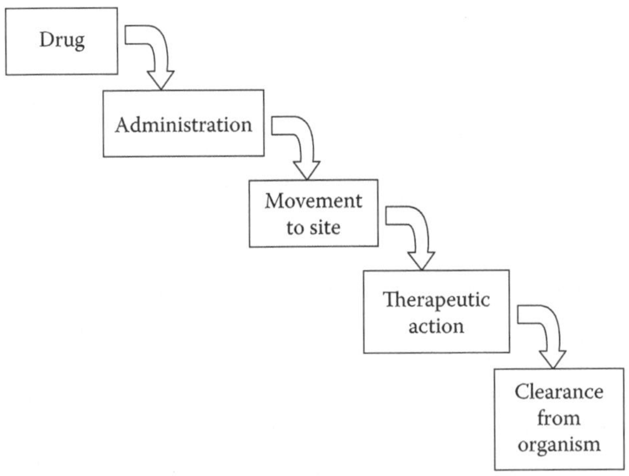

FIGURE 1.3
The magic bullet concept (Ehrlich hypothesis). Target sites are almost universally taken to be one or more forms of a protein or protein-complex receptor.

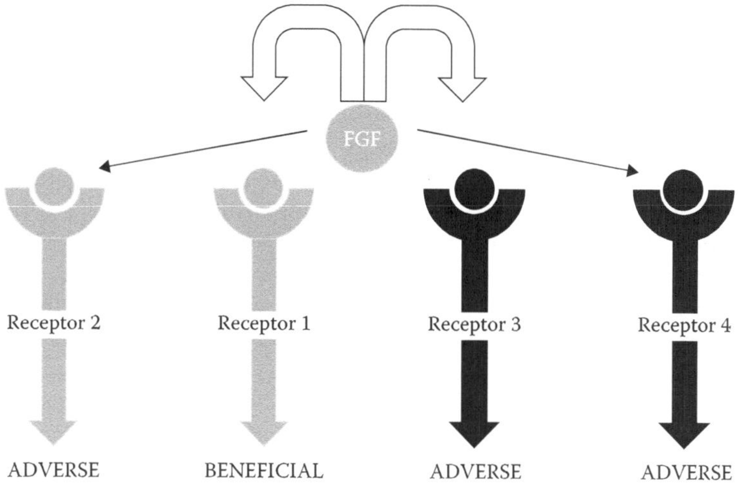

FIGURE 1.4
On- and Off-target effects.

models to wholly understand how to do this adequately, with sufficient receptor and regional specificity to avoid adverse or even dangerous effects at an organism level and be able to translate the animal model to at least an approximation of human-use methodology before actually proceeding to humans. Figure 1.4 illustrates the concept of "off-target hits" for drugs – therapeutic molecules reaching unintended receptors at levels sufficient to activate them.

As an essential part of nonclinical safety assessment, the practicing toxicologist is presented with multiple challenges in developing formulations for the *in vivo* evaluation of candidate drugs which are not shared elsewhere in drug development. These challenges apply, with minor exceptions, across all routes of administration. One of the ultimate goals is to determine, if at all possible, which use of the intended clinical routes is preferred to reduce the degree of uncertainty in extrapolation of results to the clinical use case.

Note that, however, where it is not possible to achieve a systemic exposure level by the intended route (not a rare case for topical routes), regulators may "recommend" use of a "systemic" route in an additional study to ensure that potential systemic adverse effects are identified and evaluated.

In the nine chapters that follow, we will seek to first assemble the basic concepts, principles, and hypotheses involved in quantitative receptor and chronological organism interaction dynamics central to the successful development of new therapeutics which depend on systemic administration to achieve desired therapeutic goals and in so doing avoid outcomes which limit, marginalize, or preclude the therapeutic use of so many molecules.

To do this we will first have to consider concepts central to current pharmacotherapy and toxicodynamics – and how current regulatory concepts and guidances hinder development but also could help improve the efficiency and success rate of development.

We will seek to understand the assumptions and concepts (and misconceptions) of (mostly nonclinical) pharmacokinetics. We both depend broadly on routes of administration to different species to give us accurate models by which to transfer therapies to humans (which largely disregard the differences of these species from humans, from one another, and between ages and genders in the model species).

Our technology and science have gifted us with many new tools and approaches to be able to more effectively and with specificity deliver therapeutic agents to desired targets, to guide them to targets, to activate and deactivate them when on target, and to quantitatively titrate their delivery to the target. Such delivery tools and approaches are overviewed in Chapter 5.

The concept of hormesis is a current version of the belief that dose responses (therapeutic and adverse) are not simply about "enough is better" but rather worse. There are, as illustrated in Figure 1.1, optimized levels.

Chapter 2 discusses routes of administration – that is, the varied ways of achieving access of a therapeutic to either a specific region of the body or to systemic distribution of the body. It should be noted that the minor routes (which do tend to be specific to regional or tissue-specific administration) are not specifically addressed here. These include intraurethral (into the urethra), epidural (into the spinal cord), intrathecal (into the brain), intracardial, intraventricular, and intraperitoneal (IP, used historically in nonclinical safety assessment of nontherapeutics but not generally with therapeutics or clinically). All of these (with the possible exception of IP) seek to achieve tissue- or organ-specific drug exposure.

Chapter 3 provides a primer on pharmacokinetics, with emphasis on the application (and limitations) of this discipline to nonclinical safety assessment.

Chapter 4 explores the different routes of administration for therapeutics, with their advantages and limitations and special consideration for the principal model species in toxicology.

Chapter 5 introduces the concept of delivery systems (and devices) for both small- and large-molecule therapeutics. These have applications in both nonclinical and clinical use.

Chapter 6 addresses the optimization of blood and tissue levels *along with duration* of treatment. How one achieves these optima can be by manipulation of a variety of tools, such as formulation, route, and dose, but another tool in the kit is regimen, that is to say, over what duration and frequency of dosing and at what levels.

Chronobiology – the circadian and seasonal rhythm of life – has been recognized and modestly accommodated since the 18th century (the Carl Linneaus flower clock using different flowering plants in 1751), but the knowledge is almost never utilized in medical therapeutics. Chapter 7 will introduce the reader to this emerging field.

The fundamentals of nonclinical formulation (largely poorly represented in a unified manner in the existing literature) and means to improve it are the focus of Chapter 8.

A great deal is known about vehicles and formulations for nonclinical use, but such information, even as valuable as it is, has been poorly published on and not widely accessed and utilized. This information is even held proprietary by many private concerns. Chapter 9 seeks to correct this.

Not addressed in this volume are the differences between animal species (and humans) in terms of drug absorption, metabolism, disposition, and elimination. These are explored in depth in *Animal Models in Toxicology* (Gad, 2015) and are not addressed here. It should be clear, though, that the differences in metabolic patterns and rates and in tolerance to drugs among species must be carefully considered. It is common practice to evaluate the patterns of major metabolites among species (using hepatocytes) before selecting the rodent and nonrodent species to be used in the general toxicology evaluation of a drug – a process that serves to make sure that major human metabolites are identified and assessed approximately 90% of the time (Gad, 2016). Only later are metabolic rates generally assessed in *in vivo* studies.

Chapter 10 will look at some of the wide range of miscellaneous methods for improving regional therapeutic delivery – and its specificity.

There is also an appendix which provides an overview of vehicles used in nonclinical safety assessment and their tolerable levels in major model species by different routes for different durations of treatment (Gad et al., 2016).

References

Brunton L, Chabner B, Knollman B. (2011) *Goodman & Gilman's The Pharmacological Basis of Therapeutics*, 12th Ed. New York, NY: McGraw Hill.

Copeland RA. (2016) The drug target residence time model: A 10-year retrospective. *Nat Rev Drug Discov*, 15:87–95.

Drews J. (2000) Drug discovery: A historical perspective. *Science*, 287:1960–1964.

Gad SC. (2015) *Animal Models in Toxicology*, 3rd Ed. Boca Raton, FL: CRC Press.

Gad SC. (2016) *Drug Safety Evaluation*, 3rd Ed. Hoboken, NJ: John Wiley & Sons.

Gad SC, Spainhour CB, Shoemake C, Pallman DR, Stricker-Krongrad A, Downing PA, Seals RE, Eagle LA, Polhamus K, Daly J. (2016) Tolerable levels of nonclinical vehicles and formulations used in studies by multiple routes in multiple species with notes on methods to improve utility. *Int J Toxicol*, 35(2):95–178.

Keiser MJ, Setola V, Irwin JJ, Laggner C, Abbas AI, Hufeisen SJ, Jensen NH, Kuijer MB, Matos RC, Tran TB, Whaley R, Glennon RA, Hert J, Thomas KLH, Edwards DD, Shoichet BK, Roth BL. (2009) Predicting new molecular targets for known drugs. *Nature*, 462(12):175–182.

Moslehi J.J. (2016) Cardiovascular toxic effects of targeted cancer therapies. *NEJM*, 375:1457–1467.

2

Pharmacotherapy, Toxicodynamics, and Regulatory Science – Divergent Objectives

Introduction

Pharmacodynamics (PD) and toxicodynamics (TD), while related to pharmacokinetics (PK) and toxicokinetics (TK), are quite different from them. PD and TD are the studies characterizing the dynamic interaction of molecules with one or more biologic targets (proteins acting as receptors) and the resulting biologic effects, be they desirable (PD) or undesirable (TD).

Upon entry into a patient or an animal model, a molecule must travel to its therapeutic target in order to hopefully achieve therapeutically effective levels of action. The desired therapeutic effect is the result of the test article attaining sufficiently high (effective) levels for a long enough period of time at a meaningful population of the target in order to elicit the desired pharmacotherapeutic effect (Copeland, 2016; Blass, 2015; Boroujerdi, 2015; Ehlert, 2015). These interactions usually involve interactions with target receptors to bring about required functional and/or structural alterations in the target tissue. This desired PD interaction is the sole objective of pharmacotherapy, whether we are considering a small molecule or a large biologic entity, such as a monoclonal antibody (Mab). Everything else which occurs is at best a waste of valuable molecules and, of more concern, a potential hazard to the patient and a potential risk which may prevent the drug from reaching the marketplace and meaningful clinical use or cause unintended harm in patients (Jameson and Collins, 1998).

Indeed, the target receptor occupancy has to be at a sufficient level to achieve the desired therapeutic effect in a large portion of the desired patient population but there must not be too much or excessive receptor occupancy. Consider the case of TGN-412, where too many of the "right" receptors are occupied or activated or, alternatively, target receptors that happen to be in the wrong place (such as the brain, as is the case with progressive multifocal leukoencephalopathy [PML], a viral disease in the brain associated with the use of immunosuppressing monoclonal antibodies) are activated.

To achieve the effect that is desired, the drug entering the body must transit to the therapeutic structure or receptor site and not be transformed into an ineffective structure or reduced to a concentration that is either ineffective or below the optimal therapeutic level, that is to say, a level that is not at the OPTIMALCRO (optimal cumulative receptor occupancy). These receptors may be on cell surfaces, but more frequently are within cells, thus intracellular drug delivery (and concentrations) is the critical concern (Brouwer et al., 2015). Both cell surfaces and intracellular targets may be located within difficult, but possible-to-reach organs such as the brain (Hammarlund-Udenaes et al., 2014).

At the same time, a molecule or a metabolite of it may go places that are not intended, that is, achieve an off-target hit by either interacting with a receptor other than the one we want it to or with our desired receptor in a different location of the body than where we want to elicit a change*, resulting in an "adverse event." This is the regional specificity challenge of PD.

The receptor specificity, whether in a microbe or a mammal, is a subset of the selective toxicity challenge (Albert, 1965), but one much harder to parse out for drugs than for pesticides (Hill and Rang, 2013; Ng, 2015).

We have generally advanced our technology to improve this aspect of limiting the toxicity of our therapeutic molecules, and we can now readily screen molecules for receptor specificity rapidly and at low expense. Table 2.1 presents a summary (Keiser et al., 2015).

Until the mid-1980s, animal model and patient exposure were quantitated in terms of administered dose by whatever route utilized. The relationship between administered dose and achieved systemic exposure was generally taken to be linear. With the advent of the common measurement and evaluation of pharmacokinetic data, we came to understand that these relationships could be subproportional or supraproportional, or something more complex. With our improved abilities to identify and quantitate levels of molecules in biological matrices, especially plasma, which was and is readily available for in vivo sampling, PK/TK measurements became the standard for such measurements, moving us much closer to determining actual exposure of any patient or animal model. But from the beginning, it has been clear that the relationship between PK/TK and the biological effects we wanted and those we wanted to avoid is neither linear, proportional, or occurring in the same time dimension. In classical pharmacology, bioassays or the measurement of effects as seen in very specific animal models or isolated tissue preparations were used as tools to understand these relationships. While some of these survive in the USP or various guidances, such as the hERG assay, for the most part they have been largely abandoned.

---

* Nature being conservative, evolution has come to use the same receptor for a closely related set of tasks in multiple parts of the body. The adverse side effects seen in some individuals treated for psoriasis or rheumatoid arthritis are prime examples of this problem.

TABLE 2.1
Receptor Panel

Assay System	Receptor
Adenosine A1	Human recombinant
α1-Andrenergic (non-selective)	Rat brain
α2-Andrenergic (non-selective)	Rat cerebral cortex
β-Andrenergic (non-selective)	Rat brain
Androgen AR	Rat prostate
Ca channel Type L, Phenylalkylamine	Rat cerebral cortex
Cannabinoid CB1	Human recombinant
Dopamine D1	Human recombinant
Dopamine transporter	Human recombinant
Estrogen	Rat uterus
Endothelin ETA	Human recombinant
GABA GABA A (agonist site)	Rat cerebellum
GABA GABA A (BZ central)	Rat brain
GABA GABA B	Rat cerebellum
Glucocorticoid	Human recombinant
Glutamate non-selective	Rat cerebral cortex
Glutamate AMPA	Rat cerebral cortex
Glutamate kainate	Rat brain
Glutamate NMDA (agonist site)	Rat cerebral cortex
Glutamate NMDA (glycine site)	Rat cerebral cortex
Glycine strychnine sensitive	Rat spinal cord
Histamine H1	Human recombinant
Histamine H2	Human recombinant
Histamine H3	Human recombinant
K channel KATP	Rat brain
Leukotriene CysLT2	Human recombinant
Leukotriene BLT1	Human recombinant
Melatonin MT1	Human recombinant
Muscarinic non-selective	Rat cerebral cortex
Muscarinic M1	Human recombinant
Muscarinic M2	Human recombinant
Na channel	Rat brain
Norepinephrine transporter	Human recombinant
Nicotinic Ni	Human receptor (non-recombinant)
Opiate non-selective	Rat cerebral cortex
Opiate ORL1	Human recombinant
PAF	Rabbit platelet
Prostanoid CRTH2	Human recombinant
Prostanoid DP	Human recombinant
Prostanoid EP2	Human recombinant

(Continued)

TABLE 2.1 (*Continued*)

Receptor Panel

Assay System	Receptor
Prostanoid EP3	Human recombinant
Prostanoid EP4	Human recombinant
Serotonin 5HT1 (non-selective)	Rat striatum
Serotonin 5HT1A	Human recombinant
Serotonin 5HT1B	Human recombinant
Serotonin 5HT2A	Human recombinant
Serotonin transporter	Human recombinant
Sigma non-selective	Guinea pig brain
Thromboxane A2	TBD
Vasopressin V1	Rat liver

But PK/TK are rarely related to PD/TD in an obvious manner. For these we either need to measure levels at the receptor sites in the tissues of "dosed" patients or animal models and relate or correlate this data to patient/animal model "effects" or measure administered dose versus effects in vivo or at a molecular level of interest across a wide range of "dose" levels and with sufficient numbers (statistical power) to understand the variability of this relationship. The first of these is more likely to provide a useful broad understanding, that is, the use of a valid model.

Another complication is that drugs are intended to treat diseases, potentially leading to our "normal, healthy, young" animal models being an additional step away from where we want to be in the extrapolation/modeling chain. The use of relevant disease models has, of course, been the standard for understanding PD and has come back to active discussion for understanding TD (Morgan et al., 2013; Gad, 2015). Disease models are now commonly used to understand and "calibrate" therapeutic effects, and are increasingly used to avoid "discovering" adverse effects and the limits thereof only in patients.

References

Albert A. (1965) *Selective Toxicity*, 3rd Ed. New York, NY: John Wiley and Sons.

Blass BE. (2015) *Basic Principles of Drug Discovery and Development*. San Diego, CA: Academic Press, pp. 307–309.

Boroujerdi M. (2015) *Pharmacokinetics and Toxicokinetics*. Boca Raton, FL: CRC Press, pp. 1–3, 295–297.

Brouwer KR, Hsiao P, Rosania GR, Kim LR. (2015) Intracellular drug concentrations: A critical consideration for *in vitro* assays. *AAPS*, 18(4):16–19.

Copeland RA. (2016) The drug–target residence time model: A 10-year retrospective. *Nat Rev Drug Discovery*, 15:87–95.

Ehlert FJ. (2015) *Affinity and Efficacy: The Components of Drug Receptor Interactions.* London: World Scientific.

Gad SC. (2015) *Animal Models in Toxicology*, 3rd Ed. Boca Raton, FL: CRC Press.

Hammarlund-Udenaes M, deLange ECM, Thorne RG. (Eds.). (2014) *Drug Delivery to the Brain.* New York, NY: Springer.

Hill RG, Rang HP. (2013) *Drug Discovery and Development*, 2nd Ed. New York, NY: Churchill Livingstone, pp. 291–301.

Jameson JL, Collins FS. (Eds.). (1998) *Principles of Molecular Medicine.* Totowa, NJ: Humana Press.

Keiser MJ, Setola V, Irwin JJ, Laggner C, Abbas AI, Hufeisen SJ, Jensen NH, Kuijer MB, Matos RC, Tran TB, Whaley R, Glennon RA, Hert J, Thomas KLH, Edwards DD, Shoichet BK, Roth BL. (2015) Predicting new molecular targets for known drugs. *Nature*, 462:175–182.

Morgan SJ, Elangbam CS, Berens S, Janovitz E, Vitsky A, Zabka T, Conour L. (2013) Use of animal models of human disease for nonclinical safety assessment of novel pharmaceuticals. *Toxicol Pathol*, 41(3):508–518.

Ng R. (2015) *Drugs: From Discovery to Approval*, 3rd Ed. Hoboken, NJ: Wiley, pp. 154–156.

3

Nonclinical Pharmacokinetics and Toxicokinetics – A Primer

Introduction

Among the cardinal principles of both toxicology and pharmacology is that the means by which an agent comes in contact with or enters the body (i.e., the route of exposure or administration) does much to determine the nature and magnitude of its effects (Goldstein et al., 1974; La Du et al., 1972; Pratt and Taylor, 1990). Accordingly, an understanding of route(s) of administration and their implications for absorption is essential. Our tools to measure absorption and systemic exposure are pharmacokinetics and toxicokinetics. The therapeutic index is a calculated parameter based on plasma, and therefore absorbed levels of test article, and is the ratio between what levels cause adverse effects and the levels which have the desired therapeutic effect in humans. This is the clinical analog of the margin of safety in toxicology studies, which used to be based on administered doses but now is based on the plasma levels at different animal doses.

Probably the most important thing to remember is that what we have historically measured and still currently measure and model in pharmacokinetics are the levels of free molecules in plasma, which are a more readily available surrogate for what we are really interested in: the levels of drug molecules at the sites of action. The sites of action are almost always within cells, and intracellular concentrations are much more difficult to measure (Dollery, 2013). But this is not a measure of either actual total systemic exposure or of exposure at specific target tissues of interest, for the purposes of either therapy or toxicity.

Safety assessment studies usually involve a control group of animals (untreated and/or dosed with formulation only) and at least three treated groups receiving "low," "intermediate," and "high" dose levels of the drug entity of interest via a route that is intended to be used in man (at least as closely as possible). Frequently there will also be "recovery" groups to determine if any observed effects are reversible and, if so, to what extent. In most instances the high dose level is expected to elicit some sort of toxic effects in the animals, expressed as anything from decreased food consumption and/or below-normal bodyweight gain to mortality or actual body weight loss and severe clinical signs. The high dose has typically been selected after consideration of earlier data, perhaps from dose range-finding studies, or at least to dose at a level that is as high as possible by the intended route. The other two dose levels are anticipated to not cause toxic effects or only cause moderate effects. Generally, but not always (e.g., non-steroidal anti-inflammatory drugs in rodents), the "low" dose level is a several-fold multiple of the expected human therapeutic or exposure level (generally more so in rodents than non-rodents). However, without knowing the true relationship of these dose levels to each other with respect to the absorption, distribution, and elimination of the new molecular entity as reflected by its pharmacokinetics, it is difficult to see how meaningful extrapolations concerning safety margins can be made from the toxicity data obtained.

Pharmacokinetic studies can provide information on several aspects of the above factors, knowledge of which greatly facilitates assessment of the safety profile of the therapeutic entity. Six such aspects can be mentioned.

1. Relationship between the dose levels used and the relative extent of absorption of the test compound
2. Relationship between the protein binding of the test compound and the dose levels used
3. Relationship between pharmacological or toxicological effects and the kinetics of the test compound
4. Effect of repeated doses on the kinetics of the test compound
5. Relationship between the age of the animal and the kinetics of the test compound
6. Relationship between the dose regimens of the test compound used in the toxicity studies and those employed clinically in humans

ICH guidelines (ICH S3, 2004a, b) dictate a clearly defined set of objectives for toxicokinetic studies:

- **Primary:** To describe the systemic exposure achieved in animals and its relationship to dose level and the time course of toxicity studies
- **Secondary:**
    - To relate the exposure achieved in toxicity studies to toxicological findings and contribute to the assessment of the relevance of these findings to clinical safety
    - To support the choice of species and treatment regimen in non-clinical toxicity studies

- To provide information, which, in conjunction with the toxicity findings, contributes to the design of subsequent nonclinical toxicity studies

These data may be obtained from all animals on a toxicity study, or from representative subgroups, or from satellite groups, or from separate studies.

If toxicology can be described as being the study of the effects of a chemical on an organism, metabolism can be described as the opposite, essentially the effects of the organism on the chemical. Metabolism refers to a process by which a drug or xenobiotic is chemically modified by an organism (Roberts and Renwick, 2014). It is part of the overall process of disposition of a xenobiotic (absorption, distribution, metabolism, and excretion [ADME]), the process by which a chemical gains access to the inner machinery of an organism (absorption), how it moves around inside an organism (distribution), how it is changed by the organism (metabolism), and how it is eventually eliminated from the organism (elimination). The EPA definition of biotransformation or metabolism is "the sum of processes by which a xenobiotic (foreign chemical) is handled by a living organism." The mathematical formulae used to describe and quantify these processes are collectively known as pharmacokinetics. The EPA definition of pharmacokinetics is "quantitation and determination of the time course and dose dependency of the absorption, distribution, biotransformation and excretion of chemicals." The acronym ADME has been used to describe the multifaceted biological processes undergone by a potential therapeutic agent associated with its administration. The term "metabolism" has also come into common jargon to describe the entire process. This science has long played a central role in pharmaceutical development but has played less of a role in the development of other types of products. This chapter seeks to introduce the basic concepts of PK (ADME) and describe the practices in the conduct of studies.

Regulations

The FDA believes that data from studies on the ADME of a chemical provide insight into mechanisms of toxicity and are essential in the design and evaluation of results from other toxicity studies. Such data should be provided for all drugs and significant impurities. Recommendations for obtaining data on the metabolism and pharmacokinetics of these substances are presented in ICH guidelines (ICH 2004a, b) and the FDA Redbook II (2000). In general, it is required that this information be obtained as part of initial and subsequent repeat dose studies with a drug. The European Medicines Agency (EMA) has promulgated separate guidelines (EMEA 2007) for evaluating the pharmacokinetics of protein therapeutics.

Principles

Preliminary Work

Before one ventures into good laboratory practices (GLP) general (systemic) toxicity studies, some initial characterization of comparative species metabolism and understanding of initial routes of metabolism is essential. The studies used for these early characterizations are performed using *in vitro* methods. These can be performed using either enzymatic (microsomal) preparations or isolated cells (e.g., hepatocytes) from the species of interest. These species generally include humans and a panel of potential animal models under consideration for safety testing: mouse, rat, dog, minipig, and nonhuman primate (NHP). Valuable assessments to enable subsequent decisions are metabolic stability (how rapidly the drug is metabolized by the microsomes or hepatocyte of different species), comparative metabolic "fingerprinting" by different species (either microsomes or hepatocytes) (Dow, 2006), identification of potential metabolites, which CYPs (P-450 isozymes) are involved in the drug's metabolism, and if the drug inhibits or induces a higher rate of metabolic activity by a panel of CYPs, therefore allowing a preliminary understanding of potential drug/drug interactions (DDIs). Remembering that it is only the "free" molecules of a drug which have therapeutic or toxicologic potential, one should also evaluate the potential and degree of plasma protein binding *in vitro* (Smith and Kerns, 2010).

An understanding of the design and analysis of pharmacokinetic studies requires a broad understanding of the underlying concepts and principles inherent in the ADME process and in our current and evolving technology for studying such. Each of these four principle areas is overviewed from a practical basis as it relates to toxicology. First, however, one should consider the fundamental terminology used in pharmacokinetic studies (Table 3.1).

Absorption

Absorption describes the process by which a chemical crosses a barrier, typically a biological membrane, to gain access to the inner complexities of an organism. For mammals, this process results in the entry of the chemical into the blood stream, or systemic circulation. In this case the process is also called systemic absorption. Pharmaceutical formulations, procedures, and delivery devices, such as hypodermic needles or catheters, can be used to bypass biological barriers. Other products gain access to the systemic circulation via the oral, dermal, buccal or inhalation routes of administration.

For a material to be toxic, the first requirement is that it be absorbed into the organism. It is important to remember that local tissue effects are largely not true toxicities by this definition, except that being in the cavity of the gastrointestinal GI tract does not generally qualify as a local effect. Most pharmaceuticals are intended to gain such access.

TABLE 3.1

Fundamental Terms Used in Pharmacokinetic/Toxicokinetic Studies

Absolute bioavailability	The bioavailability of a dosage form relative to an intravenous administration.
Absorption	The process by which a xenobiotic and its metabolites are transferred from the site of absorption to the blood circulation.
Accumulation	The progressive increase of chemical and/or metabolites in the body. Accumulation is influenced by the dosing interval and half-life of the chemical. The process can be characterized by an "accumulation factor," which is the ratio of the plasma concentration at steady state to that following the first dose in a multiple dosing regimen.
Analyte	The drug entity assayed in biological samples.
Area under curve (AUC)	The concentration of chemical and/or metabolites in the blood (or plasma/serum) integrated over time. This is typically considered the best indicator of exposure.
Bioavailability	The rate and extent to which a xenobiotic entity enters the systemic circulation intact, following oral or dermal administration. It is sometimes expanded to include therapeutically active metabolites. Also known as the comparative bioavailability.
Biotransformation	The process by which a xenobiotic is structurally and/or chemically changed in the body by either enzymatic or non-enzymatic reactions. The product of the reaction is a different composition of matter or different configuration than the original compound.
Clearance	The volume of biological fluid which is totally cleared of xenobiotic in a unit time.
C_{max}	The maximum mean concentration of the chemical in the plasma. Also known as the peak plasma concentration.
Concomitant toxicokinetics	Toxicokinetic measurements performed in the toxicity study, in all animals, in representative subgroups, or in satellite groups.
Disposition	All processes and factors which are involved from the time a chemical enters the body to the time when it is eliminated from the body, either intact or in metabolite form.
Distribution	The process by which an absorbed xenobiotic and/or its metabolites partition between blood and various tissues/organs in the body.
Dosage form	The formulation (diet, lotion, capsule, solution, etc.) administered to animals or man.
Dose proportionality	The relationship between doses of a chemical and measured parameters, usually including tests for linearity.
Enterohepatic circulation	The process by which xenobiotics are emptied via the bile into the small intestine and then reabsorbed into the hepatic circulation.

(Continued)

TABLE 3.1 (*Continued*)

Fundamental Terms Used in Pharmacokinetic/Toxicokinetic Studies

Enzyme induction	The increase in enzyme content (activity and/or amount) due to xenobiotic challenge, which may result in more rapid metabolism of a chemical.
Enzyme inhibition	The decrease in enzymatic activity due to the effect of xenobiotic challenge.
Excretion	The process by which the administered compound and/or its biotransformation product(s) are eliminated from the body.
Exposure	Exposure is represented by pharmacokinetic parameters demonstrating the local and systemic burden on the test species with the test compound and/or its metabolites. The area under the matrix level concentration-time curve (AUC) and/or the measurements of matrix concentrations at the expected peak-concentration time C_{max}, or at some other selected time $C_{(time)}$, are the most commonly used parameters. Other parameters might be more appropriate in particular cases.
First-order kinetics	Kinetic processes, the rate of which is proportional to the concentration.
First-pass effect	The phenomenon whereby xenobiotics may be extracted or metabolized following enteral absorption before reaching the systemic circulation.
Flux	Term (that takes area into consideration) used to describe the movement of a chemical across a barrier. Most typically used to describe the absorption of a chemical across the skin as ug/cm^2/hr.
Half-life	The time elapsed for a given chemical entity concentration or amount to be reduced by a factor of two.
Hepatic clearance	The rate of total body clearance accounted for by the liver.
Kel	The elimination constant for a chemical in plasma. Typically calculated using the formula $\text{Kel} = -1\text{n}[10] \times b$, where b is the slope of the linear regression line of the log of the mean plasma concentrations vs. time from the t_{max} to 24 hours.
Lag time	The interval between compound administration and when the compound concentration is measurable in blood.
Metabolite characterization	The determination of physiochemical characteristics of the biotransformation product(s).
Metabolite identification	The structural elucidation of the biotransformation product(s).
Metabolite profile	The chromatographic pattern and/or aqueous/nonaqueous partitioning of the biotransformation products of the administered compound.
Monitor	To take a small number of matrix samples (e.g., 1 to 3) during a dosing interval to estimate $C_{(time)}$ and/or C_{max}.
Nonlinear kinetics (saturation kinetics)	Kinetic processes, the rate of which is not directly proportional to the concentration.

(*Continued*)

TABLE 3.1 (*Continued*)

Fundamental Terms Used in Pharmacokinetic/Toxicokinetic Studies

Presystemic elimination	The loss of that portion of the dose that is not bioavailable. This would include, among others, loss through intestinal and gut-wall metabolism, lack of absorption, and first-pass hepatic metabolism.
Profile	To take (e.g., 4 to 8) matrix samples during a dosing interval to make and estimate of C_{max} and/or $C_{(time)}$ and area under matrix concentration time curve (AUC).
Protein binding	The complexation of a xenobiotic and/or its metabolite(s) with plasma or tissue proteins.
Relative bioavailability	The bioavailability relative to a reference or standard formulation or agent.
Renal clearance	The rate of total body clearance accounted for by the kidney. Its magnitude is determined by the net effects of glomerular filtration, tubular secretion and reabsorption, renal blood flow, and protein binding.
Satellite	Groups of animals included in the design and conduct of a toxicity study, treated and housed under conditions identical to those of the main study animals, but used primarily for toxicokinetics.
Steady state	An equilibrium state where the rate of chemical input is equal to the rate of elimination during a given dose interval.
Support	In the context of a toxicity study – to ratify or confirm the design of a toxicity study with respect to pharmacokinetic and metabolic principles. This process may include two separate steps: Confirmation using toxicokinetic principles that the animals on a study were exposed to appropriate systemic levels of the administered compound and/or its metabolite(s). Confirmation that the metabolic profile in the species used was acceptable; data to support this will normally be derived from metabolism studies in animals and in humans.
8 wing	The ratio between the highest and lowest C_{MAX} values in a test group (at any dose level).
T_{max}	The sampling time point at which C_{max} occurs.
Total clearance	The volume of biological fluid totally cleared of xenobiotic per unit time; usually includes hepatic clearance and renal clearance.
Toxicokinetics	The study of the kinetics of absorption, distribution, metabolism, and excretion of toxic or potentially toxic chemicals.
Validate	In the context of an analytical method – to establish the accuracy, precision, reproducibility, response function, and the specificity of the analytical method with reference to the biological matrix to be examined and the analyte to be quantified.
Volume of distribution (V_d)	A hypothetical volume of body fluid into which the chemical distributes. It is not a "real" volume, but is a proportionality constant relating the amount of chemical in the body to the measured concentration in blood or plasma.

There are characteristics which influence absorption by the different routes, and these need to be understood by any person trying to evaluate and/or predict the toxicities of different moieties. Some key characteristics and considerations are summarized below by route.

1. Oral and rectal routes (gastrointestinal tract).
    a. Lipid-soluble or nonionized compounds are more readily absorbed than water-soluble or ionized compounds.
        i. Weak organic bases are in the nonionized, lipid-soluble form in the intestine and tend to be absorbed there.
        ii. Weak organic acids are in the nonionized, lipid-soluble form in the stomach and one would suspect they would be absorbed there, but the intestine is more important because of time and area of exposure.
    b. Specialized transport systems exist for some moieties: sugars, amino acids, pyrimidines, calcium, and sodium.
    c. Almost everything is absorbed – at least to a small extent, if it has a molecular weight below 10,000.
    d. Digestive fluids may modify the structure of a drug.
    e. Dilution generally increases test article toxicity because of more rapid absorption from the intestine, unless stomach contents bind the moiety.
    f. Physical properties are important – for example, dissolution of metallic mercury is essential to permit absorption.
    g. Age – neonates have an immature and poorly developed intestinal barrier.
    h. Effect of fasting on absorption depends on the properties of the chemical of interest.
2. Inhalation (lungs).
    a. Aerosol deposition.
        i. Nasopharyngeal – 5 μm or larger in man, less in common laboratory animals.
        ii. Tracheobronchiolar – 1–5 μm.
        iii. Alveolar – 1 μm.
    b. If a solid, mucociliary transport may serve to clear from lungs to GI tract, resulting in oral absorption.
    c. Lungs are anatomically good for absorption.
        i. Large surface area (50–100 m^2).
        ii. Blood flow is high.
        iii. Close to blood (10 μm between gas media and blood).

d. Absorption of gases is dependent on solubility of the gas in blood.
   i. Chloroform, for example, has high solubility and is all absorbed; respiration rate is the limiting factor.
   ii. Ethylene has low solubility and only a small percentage is absorbed; blood flow limited absorption.
   iii. Parenteral routes.
   iv. Dermal routes.

As a generalization, there is a pattern of relative absorption rates which extends between the different routes that are commonly employed. This order of absorption, by rate from fastest to slowest and, in a less rigorous manner, in degree to absorption from most to least is iv > inhalation > im > ip > sc > oral > id > other dermal. Chapter 4, which speaks of routes, addresses these relationships in more detail.

The total amount and rate of absorption, distribution, metabolism, and species similarity in response are the reasons for selecting particular routes in toxicology. In acute studies, however, these details are rarely known to us. So the cardinal rule for selecting routes of use in acute testing is to use those routes which mirror the intended route for human exposure. If the route of human exposure is uncertain, or if there is the potential for either the number of routes or the human absorption rate and pattern being greater, then the common practice becomes that of the most conservative approach. This approach stresses maximizing potential absorption in the animal species within the limits of practicality, and selecting from among those routes commonly used in the laboratory that which gets the most material into the animal's system as quickly and completely as possible to evaluate the potential toxicity.

In general, drugs cross biological barriers by one of three mechanisms: active transport, facilitated transport, and passive transport. In active transport, the chemical is specifically recognized by the organism, which then expends energy to take the chemical up, even against a concentration gradient. In facilitative transport, the organism produces a carrier molecule which reacts with the target molecule to form a complex which more easily traverses the membrane, but no energy is expended to take up the complex. Such complexes do not flow against a concentration barrier. The simplest mechanism is passive transport or diffusion. Here, a drug flows down a concentration gradient (from high concentration to a lower concentration) and must passively (no energy expended by organism) cross a biological membrane. Passive transfer or diffusion is the most common (but not the only) mechanism involved in the absorption of the vast majority of approved drugs. It should be remembered that for purposes here, "concentration gradient" must be considered in relationship to partition coefficient. That is, a gradient will reflect the relative solubilities of a drug in polar (water) and nonpolar (lipid) matrices or tissues. The other mechanisms involved in absorption will not be further discussed here.

Drugs in solution have a natural tendency (that is more rigorously defined by the laws of thermodynamics) to move down a concentration gradient. That is to say, the individual molecules of a solute tend to move from a region of high concentration toward regions of lower concentration. Also, the movement of a chemical across a permeable barrier, such as a biological membrane, is a process called diffusion, as illustrated by Figure 3.1. For most products, these biological barriers are either the wall of the gastrointestinal tract, the lining of the pulmonary system, and/or the skin.

Absorption from the GI tract is controlled by a variety of factors. These include the acid/base characteristics of the chemical (described as the pKa), the solubility, the nature of the delivery (e.g., diet vs. gavage), the nature of any vehicle (suspensions vs. solution, or aqueous vs. nonaqueous), and the gastrointestinal tract of the species under study. Gad (2015) provides much greater detail on this subject.

Ionized or charged organic moieties do not readily pass through the lipophilic cell membranes of the epithelial cells that line the GI tract. Thus, more acidic molecules tend to be more readily absorbed from the stomach while more alkaline materials tend to be absorbed from the small intestine. This is because at the acidic pH of the stomach, acidic chemicals tend to be nonionized. More alkaline chemicals tend to be more ionized in the stomach and

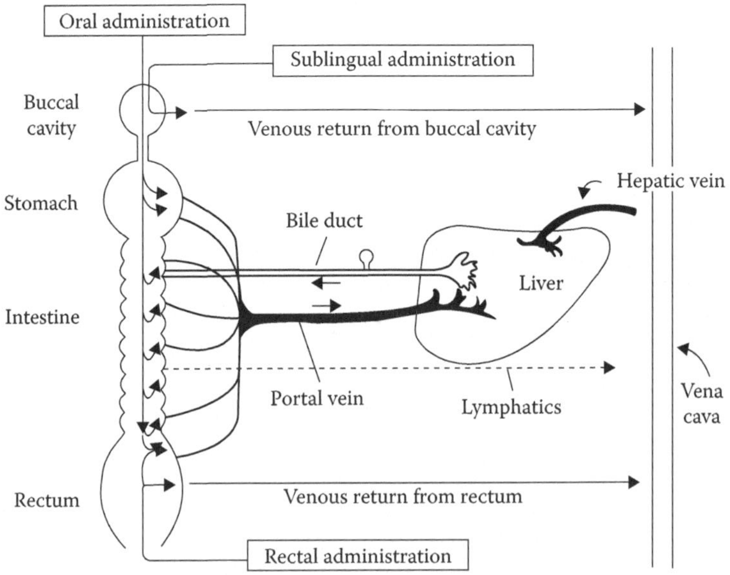

FIGURE 3.1
Stylized paths of distribution of drugs administered by different routes.

less ionized in the gut. The equilibrium reaction for acidic dissociation can be represented by Equation 3.1:

$$
\begin{array}{ccc}
X-C-OH+H_2O & = & X-C-O+H_3O^+ \\
\;\;// & & \;\;// \\
\;\;O & & \;\;O
\end{array}
\tag{3.1}
$$

Like all chemical equations, this one has an equilibrium constant. The discussion of basic chemistry is outside the purview of this book. Readers who may need a refresher are referred to Tse and Jaffe (1991). For every chemical, a pKa can be calculated, based on an equilibrium constant, which represents the proportion of ionized and unionized material in solution. The lower the pKa of a molecule, the more likely it is to be nonionized in the stomach. There is a mathematical relationship that defines this behavior, and it is called the Henderson–Hasselbalch equation.

One can manipulate movement of a drug through the GI tract (and particularly the stomach) by formulation and/or by feeding, concurrent with drug administration. Some of the factors which slow gastric emptying (and therefore some to increase drug absorption) are presented in Table 3.2.

Absorption from the Pulmonary System

Of the three routes discussed here, absorption from the pulmonary system is perhaps the most rapid. Systemic absorption of inhaled materials is highly dependent on the physical properties of the inhaled materials, which dictate how easily the materials reach the alveoli of the deep lung. Proteins may be readily absorbed when instilled in the deep lungs; as an example, note the inhalable insulins. Gases and vapors easily penetrate into the deep lung. For mists and dusts, absorption will be highly dependent on particle size. In general, the larger the particles, the less likely it is that they will penetrate deeply into the pulmonary system. The term "impaction" describes the deposition of particles in the respiratory tract. Particles of less than 0.2 μ are preferentially deposited in the pulmonary portion of the respiratory system and particles over 2 μ do

TABLE 3.2

Receptors Slowing Gastric Emptying

Receptor (Site)	Stimulus	Example	Sensitivity
Osmoreceptor (duodenum)	Osmotic effect (except tryptophan)	Glucose Amino acids Electrolytes	Least
Acid receptor (proximal duodenum & jejunum)	Acids with pKa < 5	Citric acid HCl	Intermediate
Fat receptor (jejunum)	Fatty Acids	Sodium myristate	Most
Tryptophan receptor (duodenum & jejunum)	Tryptophan	Tryptophan	–

not reach the alveolar epithelium in great number. Particles from 1 to 4 μ tend to be distributed over the length of the system and particles over 4 μ tend to be deposited in the nasal region. Aerosolized particles of greater than 20 μ do not commonly occur in nature. Tidal volume will also influence impaction. In general, the larger the tidal volume, and thus the more forceful the inhalation process, the more deeply particles of all sizes tend to be driven into the lung.

Once deposited, materials must be in solution before they can be absorbed. Hence, materials in an aerosolized solution will be more readily absorbed than materials that are delivered as solid (e.g., dust) particles. Solid materials must be able to go into solution *in situ* in order to be absorbed. Particle size influences dissolution rate. Large particles dissolve more slowly, for any given material, than small particles due to the differences in surface area. Once in solution, the same laws of passive diffusion apply to materials in the lung as apply to materials in the GI tract. The large surface area and the rich blood flow at the alveoli make for ideal conditions for rapid absorption into the systemic circulation. Absorption across the mucosal lining the upper airways is less rapid. Materials that do not dissolve are ingested by pulmonary macrophages and either broken down there or moved out of the lungs by the upward movement of the mucociliary escalator.

For gases and vapors, the amount absorbed is highly dependent on the partial pressure of the gas and the solubility of the gas in blood. For example, let's take the simple case of a gas that is not metabolized and is excreted by exhalation (e.g., an anesthetic gas or a Halon-type fire extinguishing agent). At any given concentration or partial pressure in the atmosphere, the concentration in the blood will reach a steady state in the blood. Accordingly, prolonged exposure does not lead to continual buildup.

At equilibrium, the concentration in the blood is depicted by the formula (also known as the Ostwald coefficient) $X_b/X_a = S$, where X_b is the concentration in the blood and X_a is the concentration in the inspired air. Thus, if one knows the S for a given chemical and the target concentration for a given exposure, one can predict what the resulting concentration may be at equilibrium. Additionally, the lower the S value (i.e., the lower the solubility in blood), the more rapidly the chemical will achieve equilibrium.

Absorption across the Skin

An aqueous carrier may be used for a variety of dermal products. In fact, carriers can be designed to limit the transportation of the penetration of the active ingredient (such as an insect repellent), if the desired effect is to keep the active moiety on the surface of the skin. Once again, however, only those materials that are dissolved will be available for penetration across the skin to gain access to the systemic circulation. For almost all drugs in or about to enter clinical trials, dermal penetration is a passive process. The relative thickness of the skin makes absorption into the systemic circulation slower than the absorption across the GI or pulmonary barriers. This is compounded by the fact that the stratum corneum function is impervious to the environment. One of the skin's major functions is protection of the organism from infection. Once a drug penetrates into the dermis, it may partition into the subcutaneous fat. Essentially, absorption across the skin is a two-step process, with the first being penetration and deposition into the skin and the second being release from the skin into the systemic circulation. The pattern of blood levels obtained via dermal penetration is generally one with a delayed absorption and slow buildup to more of a plateau than a peak. Blood levels of chemicals absorbed via the dermal route are generally low.

Given the overwhelming influence of the physical properties of skin in determining bioavailabilities via the dermal route, assessment of dermal penetration is one area in metabolism and toxicology where the use of *in vitro* methods can be effectively used to predict *in vivo* results and to screen chemicals. Apparatus and equipment exist that one can use to maintain sections of skin obtained from euthanized animals or from human cadavers or surgical discard for such experiments (Bronaugh, 1998). These apparatuses are set up to maintain the metabolic integrity of the skin sample between two reservoirs: the one on the stratum corneum side, called the application reservoir, and the other on the subcutaneous side, called the receptor reservoir. One simply places radiolabeled test material in the application reservoir and collects samples at various time points from the receptor fluid.

The rate of penetration can be represented by the traditional kinetic formulae to obtain a penetration rate constant. Given that exposed surface area also plays a role in the amount of material absorbed, the concept of flux is also an important consideration.

The three major considerations in determining the quantity of material that is absorbed into the skin, and eventually released into the systemic circulation, are primarily dependent upon four factors: the surface area exposed, the volume of material applied, the concentration of the material applied, and the nature of the vehicle.

- *Surface area* – all things being equal, it is clear that the greater the surface exposed, the higher the internal dose that is attained.
- *Volume* – the volume of material will obviously play a role in total dose, but it is not as straightforward as the relationship to surface area. Theoretically, the maximum absorption is obtained when the material is spread as thin and uniform as possible. Troweling on material so that it is literally rolling off the animal serves no practical purpose. In fact, it is not a sound practice when dealing with an *in vivo* animal experiment as it makes it more likely for the material to be available for oral ingestion.
- *Concentration* – the higher the concentration in a formulation, the higher the flux achieved of drug molecules across the skin. Of course, the nature of the vehicle that the drug is being carried in may also have a profound influence on absorption.

- *Vehicle* – there are a variety of mechanisms by which vehicles can either aid or hinder the movement of test articles across the skin.

Parameters Controlling Absorption

The absorption of a chemical into the skin is a function of the nature of the molecule, the behavior of the vehicle, and the status of the skin. Three major variables account for differences in the rate of absorption or flux of different topical chemicals or of the same molecule in different vehicles: the concentration of the molecule in the vehicle, the partition coefficient of chemical between the stratum corneum and the vehicle, and the diffusion coefficient of a particular molecule in the stratum corneum.

The rate of diffusion is proportional to the concentration of molecule in the vehicle. The relationship is linear only at low molecule concentrations and applies only to molecules that are insoluble in the vehicle. The latter factor may explain the variable therapeutic effects of different formulations of the same drug molecule. The partition coefficient is a measure of the molecule's ability to escape from the vehicle and is defined as the equilibrium solubility of the molecule in the surface of the stratum corneum relative to its solubility in the vehicle. Increased lipid solubility favors penetration of the molecule through the skin by increasing the solubility in the relatively lipophilic stratum corneum. The diffusion coefficient indicates the extent to which the matrix of the barrier restricts the mobility of the molecule. Increases in molecular size of the molecule will increase frictional resistance and decrease the value of the diffusion coefficient (Bronaugh, 1998); molecules over 1000 daltons usually will not be absorbed easily into normal adult skin.

Finally, intact stratum corneum is an excellent barrier, but in disease states that compromise the skin barrier, the resistance to absorption is rapidly lost and absorption can be facilitated. Such compromised skin can be simulated in humans by using either a dermatome or tape-striping the skin site in question.

Distribution

Once the chemical gains access to the body, it is carried by the bloodstream and lymphatics and distributed to the different organs. The preferential organ of deposition is determined by a variety of factors: the two most important are blood flow to the organ and the affinity of the chemical for that organ. The most efficient means of tracking distribution of a xenobiotic is to radiolabel drug molecules. Different isotopes are selected for inclusion in drug molecules based on their characteristics (Table 3.3). Affinity is governed by two general characteristics. Firstly, the product may be designed to have a specific affinity for a specific molecular entity in a target cell. For example, an anti-cholinesterase insecticide will tend to accumulate in the cells that have the highest concentration of cholinesterase. Secondly, the product may have a nonspecific or

TABLE 3.3

Properties of Primary Radioisotopes Employed in Pharmacokinetics

Property	^{3}H	^{51}Cr	^{14}C	^{125}I
Half-life (yr)	12.3	27.8 (d)	5730	13 (d)
Maximum beta energy (MeV)	0.0186	0.752	0.156	2.150
Average beta energy (MeV)	0.006	0.049	–	–
Range in air (mm)	6	300	–	–
Range in unit density material (mm)	0.0052	0.29	–	–

general chemical attraction for a specific cell type. The more highly lipophilic a chemical, the more likely it is to distribute and remain in adipose tissue. Blood flow will also have a major impact on distribution, as chemicals will be distributed more readily to those organs that are more highly perfused. A highly lipophilic chemical may first be deposited in the brain due to the fact that it is richly perfused, and then be distributed to body fat with time.

Once a material is absorbed, distribution of a compound in most early toxicology studies is usually of limited interest. This is unfortunate, as it is the preferential distribution of the drug to the therapeutic target that is desired. Some factors which can serve to alter distribution are listed in Table 3.4.

For most drugs, the rate of disposition or loss from the biological system is independent of rate and input, once the agent is absorbed. Disposition is defined as what happens to the active molecule after it reaches a site in the blood circulation where concentration measurements can be made (the systemic circulations, generally). It must be noted that different species have characteristic volumes (and half-lives) of body matter (Table 3.5) and bodily fluids (Table 3.6). Disposition processes may be independent of input, the inverse is not necessarily true because disposition can markedly affect the extent of availability. Agents absorbed from the stomach and the intestine must first pass through the liver before reaching the general circulation (Figure 3.1). Thus, if a compound is metabolized in the liver or excreted in bile, some of the active molecule absorbed from the gastrointestinal tract will be inactivated by hepatic processes before it can reach the systemic circulation and be distributed to its sites of action. If the metabolizing or biliary excreting capacity of the liver is great, the effect on the extent of availability will be substantial. Thus, if the hepatic blood clearance for the chemical is large, relative to hepatic blood flow, the extent of availability for this chemical will be low when it is given by a route that yields first-pass metabolic effects.

Likewise, metabolism is generally of only limited concern in most acute studies. There are some special cases, however, in which metabolic considerations must be factored into the process, such as when seeking to understand differences between routes and the effects that may be seen.

The first special case is parenteral routes, where the systemic circulation presents a peak level of the moiety of interest to the body at one time, tempered only by the results of a single pass through the liver.

TABLE 3.4

Selected Factors That May Affect Chemical Distribution to Various Tissues

Factors relating to the chemical and its administration
Degree of binding of chemical to plasma proteins (i.e., agent affinity for proteins) and tissues
Chelation to calcium, which is deposited in growing bones and teeth (e.g., tetracyclines in young children)
Whether the chemical distributes evenly throughout the body (one compartment model) or differentially between different compartments (two- or more compartment model)
Ability of chemical to cross the blood–brain barrier
Diffusion of chemical into the tissues or organs and degree of binding to receptors that are and are not responsible for the drug's beneficial effects
Quantity of chemical given
Route of administration/exposure
Partition coefficients (nonpolar chemicals are distributed more readily to fat tissues than are polar chemicals)
Interactions with other chemicals that may occupy receptors and prevent the drug from attaching to the receptor, inhibit active transport, or otherwise interfere with a drug's activity
Molecular weight of the chemical
Factors relating to the test subject
Body size
Fat content (e.g., obesity affects the distribution of drugs that are highly soluble in fats)
Permeability of membranes
Active transport for chemicals carried across cell membranes by active processes
Amount of proteins in blood, especially albumin
Pathology or altered homeostasis that affects any of the other factors (e.g., cardiac failure and renal failure)
The presence of competitive binding substances (e.g., specific receptor sites in tissues bind drugs)
pH of blood and body tissues
pH of urine[a]
Blood flow to various tissues or organs (e.g., well-perfused organs usually tend to accumulate more chemical than less well-perfused organs)

[a] The pH of urine is usually more important than the pH of blood.

The second special case arises from inhalation exposures. Because of the arrangements of the circulatory system, inhaled compounds enter the full range of systemic circulation without any "first pass" metabolism by the liver. Keberle et al. (1971) and O'Reilly (1972) have published reviews of absorption, distribution, and metabolism that are relevant.

Protein Binding

The degree to which a drug binds to plasma proteins will highly influence its distribution. Albumin, the most prominent of the many proteins found in

TABLE 3.5

Volume and Half-Life of Body Water in Selected Species

Species	Sex	Exchangeable Body Water (% of Body Weight)	Half-Life (Days)
Mouse	F	58.5	1.13
Rat	M	59.6	2.53
Rabbit	F	58.4	3.87
Dog	M	66.0	5.14
Cynomolgus monkey	M	61.6	7.80
Rhesus monkey	M	61.6	7.80
Man	M,F	55.3	9.46

TABLE 3.6

Approximate Volumes of Pertinent Biological Fluids in Adult Laboratory Animals

Fluid	Rat	Mouse	Dog	Rabbit	Monkey
Blood (mL/kg)	75	75	70	60	75
Plasma (mL/kg)	40	45	40	30	45
Urine (mL/kg/day)	60	50	30	60	75
Bile (mL/kg/day)	90	100	12	120	25

mammalian plasma, carries both positive and negative charges with which a polar compound can associate by electrostatic attraction. As with all such reactions, it can be described by Equations 3.2 and 3.3. The more avidly bound the material, the less will be distributed to surrounding fluids as part of a solution, and only that portion that is free in solution will be available for diffusion into the tissues. Another reason that the extent of protein binding of a drug is important is that a test article may displace other drugs that are protein bound, creating drug/drug interactions that could in turn cause adverse events.

Water Solubility

The solubility of a chemical has a direct bearing on its distribution. Recall that only molecules that are in solution will be available for absorption.

As mentioned above, only that portion that is free in solution will be available for diffusion into the tissues. Hence, the more material that is in solution, the more that will be available for diffusion.

Volume of Distribution

If one takes the dose administered (mgs) and divides it by the plasma concentration of the test material (mg/mL), the result is a volume number:

$$\frac{\text{Dose}}{\text{Concentration}} = \text{Volume} \qquad (3.2)$$

TABLE 3.7

Typical Organ Weights in Adult Laboratory Animals

	Percent of Body Weight				
Organ	**Rat**	**Mouse**	**Dog**	**Rabbit**	**Monkey**
Liver	3.5	6	3.5	3	2.5
Kidney	0.8	1.6	0.5	0.8	0.5
Heart	0.4	0.4	0.8	0.3	0.4
Spleen	0.3	0.5	0.3	0.04	0.1
Brain	0.5	0.6	0.8	0.4	3
Adrenals	0.02	0.01	0.01	0.02	0.03
Lung	0.6	0.6	1	0.6	0.7

One can take this process a step further and extrapolate back from a plasma time curve to the y axis. This is theoretically the plasma concentration (C_0) that would occur if, upon being administered, the material is instantly distributed throughout the body. The volume number obtained with the above equation becomes

$$\frac{\text{Dose}}{C_0} = V_D \tag{3.3}$$

In this formula V_D represents the apparent volume of distribution, a proportionality constant that reflects the relation of the concentration of a xenobiotic in plasma to the total amount of the entity in the body. Materials that are avidly bound to plasma proteins will have a high volume of distribution, while materials that are avidly taken by the tissues (deposit fat, for example) will have a low volume of distribution. V_D is a parameter that is simple to calculate yet gives very important information about the distribution of the chemical under investigation.

The available volumes and masses for distribution vary from species to species, as summarized in Table 3.7.

Metabolism/Biotransformation

Metabolism describes the process by which chemicals are changed by the body. In fact, very few drugs that come to enter the body are excreted unchanged. Most are chemically modified. In general, metabolism results in chemicals that are more polar and water soluble, and more easily excreted. Examples of more common metabolic conversions are shown in Table 3.8. In general, the vast majority of lipophilic chemicals are first oxidized via the cytochrome P-450-dependent mixed-function oxidase system of the liver. This is the process classically called phase I metabolism. Cytochrome P-450 exists as a collection or family of isozymes that are generally referred to as the CYP gene superfamily. These isozymes demonstrate varying but overlapping substrate affinities and varying responses to different induction

TABLE 3.8

Summary of Prominent Phase I Biotransformation Reactions

Reaction	Enzyme	Location	Example/Comments
Hydrolysis	Carboxylesterase	Ubiquitous	Vinyl acetate to acetate and acetaldehyde
	Peptidase	Blood, lysomes	Amino-, carboxy- and endo peptidase which cleave peptides at specific amino acid linkages
	Epoxide hydroplase	Microsomes, cytosol	Conversion of styrene 7,8 epoxide to styrene 7,8 glycol
Reductions	Azo and nitro reduction	Gut microflora	Sequential conversion of nitrobenzene to aniline
	Carbonyl reductase	Cytosol	Conversion of haloperidol to reduced haloperidol (a secondary alcohol)
	Disulfide reduction	Cytosol	Glutathine-dependent reduction of disulfiram to deithyldithiocarbamate
	Sulfoxide reduction	Cytosol	Thioredoxin dependent of sulindac to sulindac sulfide
	Quinone reduction	Cytosol, microsomes	DT diaphorase reduction of menadione to hydroquinone
	Reductive dehalogenation	Microsomes	Conversion of pentabromoethane to tetra bromoethane (releasing free bromide ion)
Oxidation	Alcohol dehydrogenase	Cytosol	Conversion of ethaol to acetaldehyde (DAD/DADH-dependent reversible reaction)
	Aldehyde dehydrogenase	Mitochondria/ cytosol	Conversion of acetaldehyde to acetate
	Aldehyde oxidase	Liver cytosol	FAD-dependent metalloenzyme, oxidation of benzaldehyde to benzoic acid
	Xanthene oxidase	Cytosol	Oxidation of purine derivative, conversion of allopurinol to alloxanthene
	Monamine oxidase	Mitochondria	FAD-dependent oxidative deamination of monoamines, e.g., primaquine
	Diamine oxidase	Cytosol	Pyridoxal-dependent, copper-containing enzyme. Conversion of allylamine to acrolein
	Prostaglandin oxidase	Microsomes	Cooxidation reaction, can activate chemical in tissues low in cytochrome P-450, e.g., nephrotoxicity of acetaminophen, oxidation of phenylbutazone
	Flavin-mono-oxygenase	Microsomes	FAD dependent oxidation of nucleophilic nitrogen, sulfur, and phosphorus heteroatoms, e.g., conversion of nicotine to nicotine 1′-N-oxide, cimetidine to cimetidine S-oxide
	Cytochrome P-450	Microsomes	

agents. For a review of the molecular biology of the CYP gene superfamily, the reader is referred to Meyer (1994). Induction is the process whereby exposure to a chemical leads to increased activity of the MFO system due to an increase in the amount of the cytochrome P-450 isozyme. The isoenzymes induced by a variety of different chemicals are given in Table 3.9, as well as example compounds which inhibit the activity of specific CYPs. In a practical sense, a drug can induce its own metabolism. Hence, repeated dosing with a chemical may ultimately lead to lower blood levels at the end of the dosing period, for example, of a thirteen-week study than at the beginning. There could also be alterations in the spectrum of metabolites produced, such that an agent could become more or less toxic with repeated dosing depending on the nature of the metabolites. It is not unusual during a subchronic or chronic toxicity test for tolerance to develop. There may be signs of toxicity early in the study but even with continued daily dosing, the signs abate. This phenomenon, particularly in rodents, is frequently due to microsomal induction, whereby the chemical has induced its own metabolism, and more rapid clearance of the parent chemical occurs. It should be noted that the P450 system is not the only drug metabolizing system. As Table 3.10 summarizes, there are at least five major metabolic systems in mammals.

After the drug has been metabolically oxidized, it can be further metabolized. In fact, it is possible for the metabolites to also be substrates of the MFO or the other metabolic systems shown in Table 3.10 and thus be further metabolized themselves.

The route of metabolic activation of the classic carcinogen benzo[a]pyrene is due to such a mechanism. The biology and handling of these reactive intermediates has been extensively studied. Glutathione is a common antioxidant in the body and among the most common organic intracellular chemicals in all mammalian species, being present at a concentration of up to 10 mM, and its associated enzyme, glutathione S-transferase, is very active. Glutathione is a tripeptide (glutamine-cysteine glycine). The sulfhydryl group of cysteine is the business end of the molecule where the reaction with the nucleophilic reactive intermediate takes place. After that, the glutathione conjugate is further metabolized to a cysteinyl-acetyl moiety. These moieties are called mercapturic acids, and are generally found in the urine. The relative predominance of mercapturic acid over other metabolites may be considered a rough indication of how "reactive" the intermediates may have been. Teleologically, it is tempting to speculate that it is a very well-designed protective mechanism. So long as intracellular glutathione concentrations remain above a critical level, the destructive actions of active metabolites can be held in check. Thus, a small dose of a chemical (e.g., bromo-benzene) may cause no liver damage while a large dose may. This is also a good example of one of the aspects of toxicokinetics versus pharmacokinetics where a high dose of a chemical can become toxic due to saturation of a detoxification pathway.

The glutathione S-transferase pathway is sometimes in biochemical competition with the epoxide hydratase pathway, in that both deactivate

TABLE 3.9

Examples of Xenobiotics Metabolized by Human P450

CYP1A1
Benzo[a]pyrene and other polycyclic aromatic hydrocarbons
CYP1A2
Acetaminophen
2-Acetylaminofluorene
4-Aminobiphenyl
2-Aminofluorene
2-Naphthylamine
Amino acid pyrolysis products
(DiMeQx, MelQ, MelQx, Glu P-1, Glu P-2, IQ, PhlP, Trp P-1, Trp P-2)
CYP2A6
6-Aminochursene
Cyclephosphemide
Isosphamine
N-Nitrosodiethylamine
CYP2B6
6-Aminochrysene
Cyclophosphamide
Ifosphamine
CYP2C8, 9, 18
(Note: 2C9 is absent in 15–30% of Asians)
Talbutamide
Taxol
CYC2C19
Diazepam
Diphenzlhydantion
Hexaburbitol
Propanolol
CYP2D6 (Note: absent in 7% of Caucasians)
Bufuralol
CYP2E1
Acetaminophen
Acrylonitrile
Benzene
Carbon tetrachloride
Chloroform
Chlorzoxzone
Dichloromethane
1,2-Dichloropropane
Ethylene dibromide

(Continued)

TABLE 3.9 (*Continued*)

Examples of Xenobiotics Metabolized by Human P450

Ethylene dichloride
Ethyl carbamate
N-Nitrosodimethylamine
Styrene
Trichloroethylene
Vinyl chloride
CYP3A4 (50% of all marketed drugs are marketed here)
Acetaminophen
Aflatoxin B_1 and G_1
6-Aminochrysene
Benzo[a]pyrene 7,8-dihydrodiol
Cyclophosphamide
Ifosphamide
Nidazolam
Nifedipine
Testosterone
1-Nitropyrene
Sterigmatocystin
Senecionine
Tris(2,3-dibromopropyl) phosphate
CYP4A9/11
None known

TABLE 3.10

A Comparison of the Key *In Vitro* Drug Metabolizing Experimental Systems (Liver Microsomes), Liver Postmitochondrial Supernatant (S9), Liver Cytosol (Cytosol) and Hepatocytes in Their Contents of the Major Drug-Metabolizing Enzymes

In Vitro System	P450	MAO	UGT	ST	GST
Microsomes	+	–	+[a]	–	+[b]
S9	+	–	+[a]	+[a]	+
Cytosol	–	–	–[a]	+[a]	+[c]
Hepatocytes	+	+	+	+	+

Cytochrome P450 isoforms (P450); monoamine oxidase (MAO); UDP-glucuronsyl transferase (UGT); sulfotransferase (ST); and glutathione-S-transferase (GST)

[a] Activity of this drug-metabolizing enzyme requires the addition of specific cofactors, for instance, UDP-glucuronic acid (UDPGA) for UGT activity, and 3′-phosphoadenosine 5′-phosphosulfate (PAPS) for ST activity.

[b] Membrane-bound GST but not soluble GST is found in the microsomes.

[c] Soluble GST but not membrane-bound GST is found in the cytosol.

intermediates of the MFO system. Epoxide hydratase is a microsomal enzyme that acts specifically to deactivate epoxide intermediates, by the addition of water across the C-O bond to form a diol. As a very broad generality, the glutathione S-transferase pathway tends to be more prominent in rodents, while the epoxide hydratase pathway tends to be more dominant in non-rodents.

The hydroxyl or diol containing metabolites of the MFO can be further metabolized by so-called phase II (synthetic) metabolism whereby they are conjugated to/from glucuronides and/or sulfates (so-called ethereal sulfates). Amines can also be substrates. The net effect of phase II reactions is to create a more polar molecule that is more water soluble and therefore readily excretable. While there are species differences, glucuronides are actively transported and excreted into the bile and subsequently into the GI tract. Sulfates are excreted more predominantly in the urine. Both glucuronides and sulfates, however, can be found in both the urine and the feces. Like the MFO pathway, glutathione 3-transferase, UDP-glucuronyl transferase, and epoxide hydralase are inducible, i.e., treatment with exogenous chemicals will increase the amount of enzyme protein present.

Outside of the MFO-mediated (phase I) reactions there are a few other major reactions that are worthy of note. The two major ones involve ester hydrolysis and alcohol and aldehyde dehydrogenases. All mammalian species have an extensive ability to hydrolyze the ester bond. The products of the reactions then can go on to be further metabolized. In the pharmaceutical industry, this property has been utilized to synthesize pro-drugs, i.e., chemicals that have desirable pharmaceutical properties (generally increased water solubility) that are not converted to their active moiety until hydrolyzed in the body.

The activity of alcohol dehydrogenase is one with which we should all be familiar. It oxidizes alcohols to aldehydes. The aldehydes produced by this reaction can go on to be further metabolized to a carbocylic acid, if they are not sterically hindered. Sidechain constituents of aromatic compounds can also be a substrate for this reaction sequence, producing side chain carboxylates. The oxidation of alcohols to aldehydes can also be a form of metabolic activation as aldehydes can have potent physiological actions. Fortunately, aldehyde dehydrogenase has a very high activity when compared to alcohol dehydrogenase, so that the aldehydes do not accumulate. Inhibition of aldehyde dehydrogenase by disulfiram (Antabuse®) leads to the accumulation of acetaldehyde, causing nausea, dizziness, and flushing. Like disulfiram, some pesticides contain dithiocarbamates and have the potential of causing this type of reaction.

Hopefully, this brief description of the major metabolic pathways has given some appreciation of the richness of the processes. The different sites of oxidation, the possibility of additional oxidative metabolism of metabolites, and differences in phase II reactions all lead to a multiplicity of possible metabolites. Over 100 different metabolites of the human pharmaceutical chlorpromazine have been isolated and identified. When analyzed by HPLC, for example, the parent chemical and the different (detectable) metabolites will form a pattern of different peaks. This is referred to as the metabolic fingerprint or profile of a chemical. Different species may have different profiles. Ideally, in doing a risk assessment, one would like to know the similarity in this metabolic pattern between the animals used in the toxicology studies and that produced by human beings. This is only infrequently available for most non-pharmaceutical products. Pesticides (for example) are rarely given intentionally to human subjects for the purposes of such a metabolic study. The technology now exists, however, to address this potential problem. Cell lines with human cytochrome P-450 have been developed that can provide some indication of the similarities of human metabolism of a chemical to that of experimental animals. They may be able to at least assist in identifying the major oxidative metabolite(s). For non-pharmaceutical products, it may be an unusual circumstance that would require one to identify potential human metabolites as part of a marketing application; however, it may be useful for one to know that the technology exists to do so.

The processes of metabolic conversion are frequently involved in the mechanisms of toxicity and carcinogenicity.

Metabolic Activation

As mentioned, most non-nutritive chemicals while moving through the GI tract then enter the mesenteric circulation by the process of passive absorption. The venous circulation from the mesentery flows through the portal vein into the liver. The metabolic action of the liver literally sits between the GI tract and the general systemic circulation. Thus, even chemicals that may be highly absorbed from the GI tract could appear only sparingly in systemic circulation if they are highly metabolized by the liver. The combination of absorption from the GI tract and intestinal and hepatic metabolism leads to what is called the first-pass effect. An extension of this is the fact that the flora in the gastrointestinal trace contain glucuronidases that can cleave glucuronides of chemicals and/or metabolites that are then available to be reabsorbed. This process is called entero-hepatic circulation.

Induction of P-450 Metabolism and Isoenzymes

When organisms are exposed to certain xenobiotics their ability to metabolize a variety of chemicals is increased. This phenomenon can produce either a transitory reduction in the toxicity of a drug or an increase in toxicity if the metabolite is the more toxic species. However, this may not be the case with compounds that require metabolic activation. The exact toxicological outcome of such increased metabolism is dependent on the specific xenobiotic and its specific metabolic pathway. Since the outcome of a xenobiotic exposure can depend on the balance between those reactions that represent detoxification and those that represent activation, increases in metabolic capacity may at times produce unpredictable results.

The ability of different drugs to differentially inhibit and/or induce individual cytochrome P-450 isoenzymes has become critical in assessing the potential safety of drug molecules. Table 3.11 presents a summary overview of some of what we have come to know about differential metabolism by P-450 isoenzymes. Draper et al. (1998) have published on the use of human liver microsomes for determining the levels of activity, induction, or inhibition a drug has on the formation of 6β-hydroxy testosterone as a model for CYP3A activity (1) and chlorzoxazone for CYP2E1 activity (2). If, for example, a chemical under study competitively inhibits the metabolism of these model substrates in these systems, then it is a substrate for that human isozyme. Using these more recently available *in vitro* systems, it is much easier to perform cross-species comparisons with regard to biotransformation. It is now easier to determine how similar the routes of metabolism are in the experimental animals with comparison to that in humans without having to

TABLE 3.11

Examples of Xenobiotics Activated by Human Cytochrome P-450 Isoenzymes

CYP1A1	CYP2D6
Benzo[a]pyrene and other polycyclic	Buforolol
aromatic hydrocarbons	Codeine
CYP1A2	Timololmetoprolol
Acetaminophen	CYP2E1
2-Acetylaminofluorine	Acetaminophen
4-Aminobiphenyl	Acrylonitrile
2-Aminofluorene	Benzene
2-Naphthylamine	Carbon tetrachloride
CYP2A6	Chloroform
N-nitrosodiethylamine	Dichloromethane
Butadiene	1,2-dichloropropane
Coumarin	Ethylene dibromide
CYP2B6	Ethylene dichloride
6-Aminochrysene	Ethyl carbamate
Cyclophosphamide	*N*-Nitrosodimethylamine
Ifosphamine	Styrene
CYP2C8	Trichloroethylene
Taxol	Vinyl chloride
CYP2C9	CYP3A4
Diclofenac	Acetaminophen
Phenytoin	Aflatoxin B_1 and G_1
Piroxicam	6-Aminochrysene
Tolbutamide	Benzo[a]pyrene 7,8-dihydrodiol
CYP2C19	Cyclophosphamide
Diazepam	Ifosphamide
Diphenylhydantoin	1-Nitropyrene
Hexabarbitol	Sterigmatocystin
Propanolol	Senecionine
	Tris(2,3-dibromopropyl) phosphate
	CYP4A9/11
	None known

Source: Adapted from Parkinson A., *Cassarett & Doull's Toxicology: The Basic Science of Poisons*, McGraw-Hill, New York, 1996.

administer the chemical to human subjects. Human and animal microsome preparations may be used as models to identify patterns of metabolites *in vitro*, allowing for better selection of model species for safety studies and competition for or inhibition activation of specific isoenzymes that can be evaluated to identify potential problems of drug/drug interaction in patients. (Levy et al., 2000)

Species Differences

Species differences in metabolism are among the principal reasons that there are species differences in toxicity. Differences between species in cytochrome P450 isozyme content and amounts are one of the most common reasons for differences in metabolism (Table 3.12). For example, Monostory et al. (1997) published a paper comparing the metabolism of Panomifene, a tamoxifen analog, in four different species. These data serve to address that the rates of metabolism in the nonhuman species were most rapid in the dog and slowest in the mouse. Thus, one should not *a priori* make any assumptions about which species will have the more rapid metabolism. Of the seven metabolites, only one was produced in all four species. Both the rat and the dog produced the two metabolites (M5 and M6) produced by human microsomes. So how does one decide which species best represents humans? One needs to consider the chemical structure of the metabolites, the rates at which they are produced, and the amounts and ratios in which they are produced. In this particular case, M5 and M6 were found to be relatively minor metabolites in the dog, which produced three other metabolites in larger proportion. The rat produced the same metabolites at a higher proportion, with fewer other metabolites than the dog. Thus, in this particular instance the rat, rather than the dog, was a better model. Table 3.13 presents a summary of interspecies differences between species in drug metabolism and pharmacokinetics.

TABLE 3.12

Total Liver Content of CYP Enzymes in Humans, Monkeys, Pigs, and Dogs

Species	Total Liver Content of CYP (nmol/mg Protein)	References
Humans	0.29 ± 0.06 (n = 12) 0.307 ± 0.16 (n = 18)	Stevens et al. (1993), Shimada et al. (1997)
Rhesus monkeys	0.95 ± 0.08 (n = 6)	Stevens et al. (1993)
Cynomolgus monkeys	1.03 ± 0.11 (n = 5)	Shimada et al. (1997)
Minipigs	0.81 ± 0.15 (n = 9)	Nebbia et al. (2003)
Pigs	0.22 ± 0.12 (n = 3) 0.46 ± 0.07 (n = 12)	Shimada et al. (1997), Myers et al. (2001), Skaanild and Friis (1999)
Dogs	0.39 ± 0.04 (n = 6)	Shimada et al. (1997)

Source: Extracted from Dalgaard L, *J Pharmacol Toxicol Methods*, 74, 80–92, 2015.

TABLE 3.13

"General Rules" on Interspecies Differences in DMPK

Species	DMPK characteristics
Human	Polymorphisms (e.g., CYP 2C9, CYP 2C19, CYP 2D6, NAT1, NAT2)
Dog	Low acetylation, high capacity for deacetylation Different absorption due to higher pH in gastrointestinal tract than in humans (consider use of synthetic gastric fluid to mimic human situation)
Rat	Often gender differences which are not observed in other species Abundant tetrahydrofolate (protects, e.g., against methanol ocular damage)
Rabbit	Low sulfation
(Mini)Pig	Low sulfation Gastrointestinal conditions similar to humans
Cat	Low glucuronidation High sulfation

A more thorough review on species differences in pharmacokinetics has been presented by Smith (1991) and in Gad and Chengelis (1998) and Gad (2015).

Sex-Related Differences in Rodents

Not only are there differences in absorption, distribution, and metabolism between species, there may also be differences between sexes within a species. Griffin et al. (1997), for example, has demonstrated sex-related differences in the metabolism of 2, 4-dichlorophenoxyacetic acid. They noted that while there were differences between sexes, they tended to be quantitative (rates), not qualitative (metabolites). Differences between species were greater than sex-related differences. With regard to sex-related differences, it is noteworthy that males do not always have the higher rates, as Griffin et al. (1997) have shown; in hamsters, the female metabolizes 2, 4-D more rapidly than males. In general, male rats tend to have higher activity than female rats, especially with regard to CYP-dependent activity.

In the case of 2, 4-D, the only urinary metabolite is 2, 4-D glucuronide, but the half-life of 2, 4-D was 138 min in males and 382 in females.

Excretion

Excretion encompasses the process by which chemicals or their metabolites are transported out of the body. There are three possible major routes of excretion, and a handful of minor ones. The major routes of excretion for chemicals, and in particular their metabolites, are urine, feces, and expired air.

Urine

The kidneys filter the entire cardiac output multiple times each day, and thus provide a large opportunity for the removal of chemicals from the bloodstream. How much of a xenobiotic is actually excreted is dependent on three factors or processes:

1. The glomerular membrane has pores of 70–80 Å, and under the positive hydrostatic conditions in the glomerulus, all molecules smaller than about 20,000 D are filtered. Proteins and protein-bound compounds thus remain in the plasma, and about 20% of the nonbound entity is carried with 20% of the plasma water into the glomular filtrate.
2. Because the glomerular filtrate contains many important body constituents (e.g., glucose), there are specific active uptake processes for them. Also, lipid-soluble chemicals diffuse back from the tubule into the blood, especially as the urine becomes more concentrated because of water reabsorption. The pH of the urine is generally lower than that of the plasma, and therefore pH partitioning tends to increase the reabsorption of weak acids. The pH of the urine can be altered appreciably by treatment with ammonium chloride (decreases pH) or sodium carbonate (increases pH); the buffered plasma shows little change.
3. Xenobiotics may be secreted actively into the renal tubule against a concentration gradient by anion and cation carrier processes. These processes are generally capable of saturation and of relatively low specificity. Many basic or acidic compounds along with metabolites of these compounds, including conjugation products, are removed by these carrier processes. Because the dissociation rate for the chemical-albumin complex is rapid, it is possible for highly protein-bound compounds to be almost completely cleared with a single passage through the kidney.

Feces

The most important mechanism allowing circulating foreign compounds to enter the gut is in the bile. The biological aspects of this mechanism have been reviewed (80), and certain pertinent points have emerged. The bile may be regarded as a complementary pathway to the urine, with small molecules being eliminated by the kidney and large molecules in the bile. Thus the bile becomes the principal excretory route for many drug conjugates. Species differences exist in the molecular weight requirement for significant biliary excretion, which has been estimated as 325 ± 50 in the rat, 440 ± 50 in the guinea pig, and 475 ± 50 in the rabbit. In the rat, small molecules (less than 350 D) are not eliminated in the bile or large molecules (more than 450 D) in the urine, even if the principal excretory mechanism is blocked by ligation of the renal pedicles or bile duct, respectively. Compounds of intermediate molecular weight (350–450 D) are excreted by both routes, and ligation of one pathway results in increased use of the other.

Foreign compounds may also enter the gut by direct diffusion or secretion across the gut wall, elimination in the saliva, pH partitioning of bases into the low pH of the stomach, and elimination in the pancreatic juice.

Expired Air

Volatile compounds or metabolites can be extensively excreted by passage across pulmonary membranes into the airspace of the lungs, then expulsion from the lungs in expired air.

Minor routes for excretion can include tears, saliva, sweat, exfoliated keratinocytes, hair, and nasal discharge. These are of concern or significance only in rare cases. Accordingly, quantitation of excretion typically requires collection of urine and feces (and occasionally expired air) over a period of time.

Pharmacokinetics

The interplay of the processes of absorption, distribution, metabolism, and excretion results in changes in concentration of the test chemical in different organs with time. With regard to the practical concerns of monitoring human exposure, the organ of interest is the blood. Blood is generally considered the central compartment. Determining the concentration of the chemical in plasma gives one an assessment of exposure. Mathematical formulas are used to quantitatively describe this exposure. (Bauer, 2001)

Sampling Methods and Intervals

Blood

Since blood (plasma and serum) is the most easily accessible body compartment, the blood concentration profile is most commonly used to describe the time course of drug disposition in the animal. With the development of sensitive analytical methods that require small volumes (100–200 µl) of blood, ADME data from individual rats can be obtained by serial sample collection. Numerous cannulation techniques have been utilized to facilitate repeated blood collection, but the animal preparation procedures are elaborate and tedious and are incompatible with prolonged sampling periods in studies involving a large number of animals. In contrast, noncannulation methods such as collection from the tail vein, orbital sinus, or jugular vein are most practical. Significant volumes of blood can be obtained from the intact rat by cardiac puncture, although this method can cause shock to the animal system and subsequent death; therefore cardiac puncture is usually only a terminal collection method.

Three new approaches for blood sampling for PK/TK analysis have, however, come into consideration for use in rodents to facilitate use of fewer animals.

The first of these is microsampling, by means of a microdialysis or capillary sampling needle. There is minimal trauma to the rodents and, in taking minimal amounts of blood each time, this method does not compromise the physiological status of the rat or mouse (or hamster). Accordingly, "main study" animals can be used and satellite groups of animals are not required, significantly reducing the number of animals required for a study (particularly for longer duration studies where sampling at multiple time points is required).

A variation on this is the use of dried blood spot sampling (DBSS); since with liquid plasma samples a significant amount of each sample collected is lost in adherence to the walls of a series of glass or plastic collection and handling vessels, DBSS collects directly onto a chemically inert medium from a capillary sampling tube or the culex collecting device. As long as the molecule(s) to be analyzed are chemically stable under these conditions, again smaller samples (and fewer animals) are required.

The third approach is sparse sampling, where only a small number (usually three) of animals are sampled at each time point, with sampling being rotated between time points (such as the traditional 0.5, 1, 2, 4, 8, and 12 hours), so that levels of blood collected from any one animal are sufficient to avoid notable traumatization and physiologic change. Again, either no satellite group of animals is required or only a smaller such group is required.

Excreta

Excretion samples commonly collected from the rat include urine, feces, bile, and expired air. By using properly designed cages and techniques, the samples can be completely collected so that the mass balance is readily determined. These samples also serve to elucidate the biotransformation characteristics of the compound.

These samples can be easily collected through the use of suitable metabolism cages. Since rodents are coprophagic, the cage must be designed to prevent the animal from ingesting the feces as it is passed. Other main features of the cage should include the ability to effectively separate urine from feces with minimal cross-contamination, a feed and water system that prevents spillage and subsequent contamination of collected samples, and collection containers that can be easily removed without disturbing the animal. Also, the cage should be designed so that it can be easily disassembled for cleaning or autoclaving.

Following dose administration, rats are placed in individual cages. The urine and feces that collect in containers are removed at predetermined intervals. The volume or weight of urine and the weight of feces are measured. After the final collection, the cage is rinsed, normally with ethanol or water, to assure complete recovery of excreta and radioactivity. If the rats are also used for serial blood sampling, it is important that bleeding be performed inside the cage to avoid possible loss of urine or feces.

Bile

The bile is the pathway through which an absorbed compound is excreted in the feces. In order to collect this sample, surgical manipulation of the animal is necessary.

Expired Air

For ^{14}C-labeled chemicals, the tracer carbon may be incorporated *in vivo* into carbon dioxide, a possible metabolic product. Therefore, when the position of the radiolable indicates the potential for biological instability, a pilot study to collect expired air and monitor its radioactivity content should be conducted prior to initiating a full-scale study. Expired air studies should also be performed in situations where the radiolabel has been postulated to be stable and analyses of urine and feces from the toxicokinetic study fail to yield complete recovery (mass balance) of the dose.

Following drug administration, the rat is placed in a special metabolism cage. Using a vacuum pump, a constant flow of room air (approx. 500 mL/min) is drawn through a drying column containing anhydrous calcium sulfate impregnated with a moisture indicator (cobalt chloride) and passed into a second column containing Ascarite® II, where it is rendered carbon dioxide free. The air is then drawn in through the top of the metabolism cage. Exhaled breath exiting the metabolism cage is passed through a carbon dioxide adsorption tower, where the expired $^{14}CO_2$ is trapped in a solution such as a mixture of 2-ethoxyethanol and 2-aminoethanol (2:1). The trapping solution is collected, replaced with fresh solution, and assayed at designated times post dose so that the total amount of radioactivity expired as labeled carbon dioxide can be determined.

Milk

The study of passage of a xenobiotic into milk serves to assess the potential risk to breast-fed infants in the absence of human data. The passage into milk can be estimated as the milk:plasma ratio of drug concentrations at each sampling time or that of the AUC (area under curve) values. Approximately 30 rats in their first lactation are used. The litter size is adjusted to about 10 within 1 to 2 days following parturition. The test compound is administered to the mothers 8 to 10 days after parturition. The rats are then divided into groups for milk and blood collection at designated times post dose. All sucklings are removed from the mother rats several hours before milking. Oxytocin, 1 IU per rat, is given intramuscularly 10 to 15 min before each collection of milk to stimulate milk ejection. The usual yield of milk is about 1 mL from each rat. Blood is obtained immediately after milking. In order to minimize the number of

TABLE 3.14

Blood Samples Required So That Certain Toxicokinetic Parameters Can Be Obtained and Calculated

Parameter	Blood Sample Required	Information Obtained
Cmin (C_{24})	24-h	Accumulation
CT	T-h	Proof of absorption
Cmax (C peak)	Several[a]	Rate of absorption
Tmax (T peak)	Several[a]	Rate of absorption
AUC	Several[a]	Extent of absorption
t½	Several[a]	Various
Accumulation ratio	Several after first and repeated doses	Extent of accumulation

[a] Several samples to define concentration-time profile

animals used, the sucklings can be returned to the mother rat, which can then be milked again 8 to 12 hours later.

In all the fluid sampling techniques above, the limitations of availability should be kept in mind. Table 3.14 presents a summary of such availability for the principle model species.

Analysis of Data

Data from all metabolism and pharmacokinetic studies should be analyzed with the same pharmacokinetic model and results should be expressed in the same units (see Table 3.15). Concentration units are acceptable if the organ or sample size is reported, but percent of dose/organ is usually a more meaningful unit. In general, all samples should be analyzed for metabolites that cumulatively represent more than 1% of the dose.

A variety of rate constants and other parameters can be obtained from IV and oral dosing data sets, provided that good coverage of the distribution, elimination, and absorption (oral dose) phase is achieved. Typical parameters calculated to characterize the disposition of a test substance are half-lives of elimination and absorption; area under the concentration-versus-time curve (AUC) for blood (or plasma); total body, renal, and metabolic clearance (CL); volume of distribution (V_d); bioavailability (F); and mean residence and absorption times (MAT, MRT). Some of these parameters, such as half-lives and elimination rates, are easily computed from one another; the half-life is more easily visualized than the rate constant.

Computation of oral absorption (k_a) and elimination (E) rates is often complicated by the "flip-flop" of the absorption and elimination phases when they differ by less than a factor of 3. Because of these analysis problems, computation of absorption and elimination rates should not be attempted on the basis of oral dosing results alone, and therefore it is preferable to have

TABLE 3.15

Definitions and Units of Symbols

Symbol	(Typical) Units	Definition
A	*mg*	*amount of drug in body*
AUC	*mg*hour/L*	*total area under the plasma drug concentration-time curve*
$AUC_{0-\infty}$	*mg*hour/L*	*area under the plasma concentration-time curve from zero to infinite time*
AUC_{iv}	*mg*hour/L*	*area under the plasma drug concentration-time curve after an IV dose*
AUC_{oral}	*mg*hour/L*	*area under the plasma drug concentration-time curve after an oral dose*
C	*mg/L*	*concentration of drug in plasma*
C_0	*mg/L*	*concentration of drug in plasma after a single dose extrapolated back to zero time*
C_b	*mg/L*	*concentration of drug in blood*
CL	*L/hour*	*total clearance of drug from plasma*
CL_{CR}	*L/hour*	*creatinine clearance*
CL_{GF}	*L/hour*	*renal drug clearance by glomerular filtration*
CL_H	*L/hour*	*hepatic clearance of drug from plasma*
CL_{int}	*L/hour*	*intrinsic clearance of drug in an organ of elimination*
CL_R	*L/hour*	*renal clearance of drug*
CL_S	*L/hour*	*renal drug clearance by tubular secretion*
C_t	*mg/L*	*concentration of drug in plasma at a specific time t*
C_{max}	*mg/L*	*maximum plasma drug concentration during a dosing interval*
C_{min}	*mg/L*	*minimum plasma drug concentration during a dosing interval*
C_{ss}	*mg/L*	*concentration of drug in plasma at steady state during a constant-rate intravenous infusion*
C_{uss}	*mg/L*	*unbound concentration of drug in plasma at steady state during a constant-rate intravenous infusion*
C_u	*mg/L*	*unbound drug concentration in plasma*
DR	*mg/hour*	*dosing rate*
E	*ratio*	*extraction ratio for an organ*
E_H	*ratio*	*hepatic extraction ratio*
EC_{50}	*mg/L*	*concentration giving one-half the maximum effect*
E_{max}	*varies*	*maximum effect*
F	*ratio*	*bioavailability of drug*
f_e	*ratio*	*fraction of drug systemically available that is excreted unchanged in urine*
f_g	*ratio*	*fraction of an oral dose that is absorbed intact into the portal circulation*
f_H	*ratio*	*fraction of drug entering the liver that escapes extraction*
fm	*ratio*	*fraction of drug systemically available that is converted to a metabolite*
FR	*ratio*	*fraction of drug reaching the renal tubular fluid that is reabsorbed*
fu	*ratio*	*fraction of drug unbound in plasma*

(*Continued*)

TABLE 3.15 (*Continued*)

Definitions and Units of Symbols

Symbol	(Typical) Units	Definition
fu_T	*ratio*	*fraction of drug unbound in tissues*
GFR	*L/hour*	*glomerular filtration rate*
k	*/hour*	*elimination rate constant*
K_a	*mM*	*association constant for the binding of drug to protein*
K_m	*mg/L*	*Michaelis-Menten constant*
λ (lambda)	*ratio*	*ratio of concentration of drug in whole blood to that in plasma*
P_u	*mM*	*concentration of protein that does not bind drug*
Q_H	*L/hour*	*hepatic blood flow (portal vein plus hepatic artery)*
τ (tau)	*hour*	*dosing interval*
V	*L*	*volume of distribution (apparent) based on drug concentration in plasma*
v	*mg/hour*	*velocity of enzyme reaction (conversion of drug to metabolite)*
V_{max}	*mg/hour*	*maximal rate of enzyme reaction at saturating substrate concentration*
V_p	*L*	*plasma volume*
V_T	*L*	*physiologic volume outside plasma into which drug distributes*

both oral and IV PK data in hand when determining oral absorption and elimination rates.

Blood-tissue uptake rates (k_{jl}) can often be approximated from data at early ($t < 10$ minutes) time points in IV studies, provided that the blood has been washed from the organ (e.g., liver) or the contribution from blood to the tissue residue is subtracted (fat). High accuracy is not usually required since these parameters can be optimized to fit the data when they are used in more complex models. Tissue/blood recycling rates (k_{lj}) and residence times can be computed from partition coefficients if estimates of uptake rates are available.

Tissue/blood partition coefficients (R_{jl}) should be determined when steady state has been achieved. Estimates based on samples obtained during the elimination phase following a single dose of the test substance may lead to underestimates of this ratio in both eliminating and non-eliminating tissues unless its half-life is very long. Correction of these values for elimination has been described by several authors.

It may be important to determine the degree of plasma protein and red blood cell binding of the test substance; calculation of blood clearance rates using plasma or serum concentrations of the substance that have not been adjusted for the degree of binding may under- or overestimate the true rate of clearance of the test substance from the blood. This is usually done through experiments *in vitro*.

Two classical methods used in the analysis of pharmacokinetic data are the fitting of sums of exponential functions (two- and three-compartment mammillary models) to plasma and/or tissue data and, less frequently, the fitting of arbitrary polynomial functions to the data (non-compartmental analysis).

Non-compartmental analysis is limited in that it is not descriptive or predictive; concentrations must be interpolated from data. The appeal of non-compartmental analysis is that the shape of the blood concentration-versus-time curve is not assumed to be represented by an exponential function and, therefore, estimates of metabolic and pharmacokinetic parameters are not biased by this assumption. In order to minimize errors in parameter estimates that are introduced by interpolation, a large number of data points that adequately define the concentration-versus-time curve are needed.

Analysis of data using simple mammillary, compartmental models allows the estimation of all the basic parameters mentioned above, if data for individual tissues are analyzed with one-or two-compartment models, and combined with results from two- to three-compartment analyses of blood data. Optimization of the coefficients and exponents estimated may require the use of more sophisticated software: a number of scientific data analysis packages such as RS/1 and SigmaPlot have the necessary capabilities. Specialized programs such as NONLIN, CONSAM, or SIMUSOLV will be needed when more complex models must be analyzed. Coefficients and exponents from mammillary models can be used to calculate other parameters; however, they should not be taken too literally, since mammillary models assume that all inputs are to a central pool (blood), which equilibrates without limitation into other compartments. This approach does not include details such as blood flow limitations, anatomical volumes, or other physiological limits in the animal.

Physiologically based pharmacokinetic (PBPK) models were developed to overcome the limitations of simple mammillary models. Physiologically based models describe the disposition of test substances via compartmental models which incorporate anatomical, biochemical, and physiological features of specific tissues in the whole animal. The types of information added include organ-specific blood flows, volumes, growth models, and metabolism rates. Metabolic parameters often are obtained from *in vitro* studies (i.e., enzyme reaction rates in cultured hepatocytes, plasma protein binding, etc.), while other parameters are becoming available as standard parameters in the literature. Parameters from mammillary models can be used to compute the value of parameters used in physiological pharmacokinetic models, using tissue-specific blood flows, anatomical volumes, and other information (literature values). Estimation of parameters for a simple mammillary model is often the first data reduction step in creating a physiological model.

Because PBPK models are based on physiological and anatomical measurements and all mammals are inherently similar, they provide a rational basis for relating data obtained from animals to humans. Estimates of predicted disposition patterns for test substances in humans may be obtained by adjusting biochemical parameters in models validated for animals;

adjustments are based on experimental results of animal and human *in vitro* tests and by substituting appropriate human tissue sizes and blood flows. Development of these models requires special software capable of simultaneously solving multiple (often very complex) differential equations, some of which were mentioned earlier. Several detailed descriptions of data analysis have been reported.

Use of Data from Metabolism and Pharmacokinetic Studies

Information from metabolism and pharmacokinetic studies can be used in the design and analysis of data from other toxicity studies. Some examples are described below.

Design of toxicity studies. The concentration-versus-time curve, peak, and steady-state concentrations of the test substance in blood or plasma provide information on the distribution and persistence of the substance in the animal, which may suggest essential elements in the design of the toxicity studies. For example, when metabolic and pharmacokinetic studies indicate that the test compound accumulates in the bone marrow, long-term toxicity tests should include evaluation of the test compound's effect on hematopoietic function and morphology. If a test compound is found to accumulate in milk, an investigator may need to plan to perform reproductive toxicity studies with *in utero* exposure and a nursing phase (cross-fostering study). In addition, information from metabolic and pharmacokinetic studies can be used to predict the amount of test compound that enters biological compartments (tissues, organs, etc.) that may not suffer a toxic insult but may serve as depots for indirect or secondary exposure.

Physiologically Based Pharmacokinetic (PBPK) Modeling

Pharmacokinetic parameters are descriptive in nature. They quantitatively describe the manner in which a test material is absorbed and excreted, such that a specific blood or tissue level is achieved or maintained. In the past, experiments had to be done by every route of administration to gather the data appropriate for describing the pharmacokinetic behavior of a chemical administered by different routes. The development of more sophisticated and readily accessible computers has led to the development of a different approach, that of pharmacokinetic modeling. In this computerized model, different compartments are represented as shown in boxes and the movement of the material in and out of the compartments is defined by the rate constants. These can be determined by methods that are either *in vivo* or *in vitro*. Other physiological parameters are brought into play as well, such as octanol/water partition coefficient, blood flow through an organ, respiration rate (for the inhalation route of exposure), and rate of microsomal metabolism.

Pharmacokinetic modeling is the process of developing mathematical explanations of absorption, distribution, metabolism, and excretion of chemicals in organisms. Two commonly used types of compartmental pharmacokinetic models are (a) data-based and (b) physiologically based. The data-based pharmacokinetic models correspond to mathematical descriptions of the temporal change in the blood/tissue level of a xenobiotic in the animal species of interest. This procedure considers the organism as a single homogeneous compartment or as a multicompartmental system with elimination occurring in specific compartments of the model. The number, behavior, and volume of these hypothetical compartments are estimated by the type of equation chosen to describe the data, and not necessarily by the physiological characteristics of the model species in which the blood/tissue concentration data were acquired. It can provide a powerful tool for both the discovery of new drugs and the optimization of their development (Lavé et al., 2007).

While these data-based pharmacokinetic models can be used for interpolation, they should not be used for extrapolation outside the range of doses, dose routes, and species used in the study on which they were based. In order to use the data-based models to describe the pharmacokinetic behavior of a chemical administered at various doses by different routes, extensive animal experimentation would be required to generate similar blood-time course data under respective conditions. Even within the same species of animal, the time-dependent nature of critical biological determinants of the disposition (e.g., tissue glutathione depletion and resynthesis) cannot easily the included or evaluated with the data-based pharmacokinetic modeling approach. Furthermore, due to the lack of actual anatomical, physiological, and biochemical realism, these data-based compartmental models cannot easily be used in interspecies extrapolation, particularly to predict pharmacokinetic behavior of chemicals in humans. These various extrapolations, which are essential for the conduct of dose-response assessment of chemicals, can be performed more confidently with a physiologically based pharmacokinetic modeling approach.

PBPK modeling is the development of mathematical descriptions of the uptake and disposition of chemicals based on quantitative interrelationships among the critical biological determinants of these processes. These determinants include partition coefficients, rates of biochemical reactions, and physiological characteristics of the animal species. The biological and mechanistic basis of the PBPK models enable them to be used, with limited animal experimentation, for extrapolation of the kinetic behavior of chemicals from high dose to low dose, from one exposure route to another, and from test animal species to people.

The development of PBPK models is performed in four interconnected steps: model representation, model parameterization, model stimulation, and model validation. Model representation involves the development of conceptual, functional, and computational descriptions of the relevant compartments of the animal as well as the exposure and metabolic pathways of the chemical. Model parameterization involves obtaining independent measures of the mechanistic determinants, such as physiological, physicochemical, and biochemical parameters, which are included in one or more of the PBPK model equations. Model simulation involves the prediction of the uptake and disposition of a chemical for defined exposure scenarios, using a numerical integration algorithm, simulation software, and a computer. Finally, the model validation step involves the comparison of the *a priori* predictions of the PBPK model with experimental data to refute, validate, or refine the model description, and the characterization of the sensitivity of tissue dose to changes in model parameter values. PBPK models, after appropriate testing and validation, can be used to conduct extrapolations of the pharmacokinetic behavior of chemicals from one exposure route/scenario to another, from high dose to low dose, and from one species to another.

The PBPK model development for a chemical is preceded by the definition of the problem, which in toxicology may often be related to the apparent complex nature of toxicity. Examples of such apparent complex toxic responses include nonlinearity in dose-response, sex/species differences in tissue response, differential response of tissues to chemical exposure, qualitatively and/or quantitatively different responses for the same cumulative dose administered by different routes/scenarios, etc. In these instances, PBPK modeling studies can be utilized to evaluate the pharmacokinetic basis of the apparent complex nature of toxicity induced by the chemical. One of the values of PBPK modeling, in fact, is that accurate description of target tissue dose often resolves behavior that appears complex at the administered dose level.

The principal application of PBPK models is in the prediction of the target tissue dose of the toxic parent chemical or its reactive metabolite. Use of the target tissue dose of the toxic moiety of a chemical in risk assessment calculations provides a better basis of relating to the observed toxic effects than the external or exposure concentration of the parent chemical. Because PBPK models facilitate the prediction of target tissue dose for various exposure scenarios, routes, doses, and species, they can help reduce the uncertainty associated with the conventional extrapolation approaches. Direct application of modeling includes the following:

- High-dose/low-dose extrapolation
- Route-route extrapolation
- Exposure scenario extrapolation
- Interspecies extrapolation

Points to Consider

Stereoisomerism will influence metabolism and toxicity. For example, Lu et al. (1998) reported a comparison of (S)-(–)Ifosfamide and (R)-(+)-Ifsosfamide. They demonstrated that there were significant differences between the two stereoisomers with regard to pharmacokinetic behavior and major metabolite formation, as shown in Table 3.16.

When considering the safe and efficacious use of drugs in children (pediatrics), it is important to realize that neonates and juveniles are not just scaled-down versions of adults – both their physiologies and their metabolisms/pharmacokinetics are frequently different. Subsequently the pharmacokinetics, pharmacodynamics, and toxicodynamics for neonates and juveniles require their own evaluations (Kearns, 2015).

In addition, treatment of animals with phenobarbital not only increased overall rates of metabolism and clearance, but also shifted the metabolite patterns. One of the more common methods used for determining an exposure to or the amount of a metabolite produced is to determine an AUC for the metabolite. Further, one of the more common methods for representing a racemically preferred metabolite is to calculate the ratio of the R to the S. For example, the 3-decholoro metabolite of ifosfamide was produced in higher amounts from the R enantiomer while the 2-decholorometabolite was the major metabolite produced from the S enantiomer in naive animals. Treatment with phenobarbital shifted the metabolism so that the 3-dechloro metabolite was no longer the major metabolite for the S enantiomer.

TABLE 3.16

Example of Stereoselective Differences in Metabolism (R) versus (S) Ifosfamide

Parameter	Phenobarb	R	S	R/S
Term half-life	–	34.3	41.8	.820
(min)	+	19.8	19.41	1.02
AUC	–	4853	6259	.820
(uM*min)	+	1479	1356	1.03
2-dehloro metabolite	–	799	2794	.287
AUC	+	229	1205	.186
3-dehloro metabolite	–	1380	996	1.41
AUC	+	192	1175	.159

Source: Adapted from Lu H et al., *Drug Metab Disp*, 26, 476–482, 1998.
Animals were pretreated with phenobarbital (80 mg.kg) for four days.

Biologically Derived Materials

The progress and products of biotechnology have brought some new challenges to the assessment of pharmacokinetics and toxicokinetics. While the reasons for why this data is needed (demonstrating exposure, displaying dose dependency, correlating any findings of toxicity to exposure, and determining steady state for systemic agent levels) are certainly as compelling as with traditional drugs, there are a whole set of special problems involved (Baumann, 2006; Dennis et al., 2002).

These special concerns for biologically derived products are as follows:

Assay Sensitivity/Specificity

- Needs to be at 1 ng/mL or lower.
- Cross reactivity to native protein may confound results.
- If test article is the same as native protein, how do you tell the difference?
- Western blot can be used to demonstrate specificity.
- Antibody interference may occur with assay.

Low Systemic Levels

- Rapid metabolism.
    - Metabolites may be endogenous proteins or amino acids.
- Extensive metabolism.
    - Metabolites may be incorporated into cell structures rapidly.
- Rapid distribution.
- Rapid hepatic clearance.
    - Route of administration may bypass first-pass metabolism.
        - SC
        - ICV; IT
        - Buccal

Endogenous Protein

- May cross react and lead to false positive blood levels.
- Can radiolable tell the difference between administered molecule and endogenous molecule?
    - However, the label may lead to different distribution pattern.
    - What is the specific activity if diluted with unlabeled endogenous material?

Sample Volume

- May need to be large to increase sensitivity.
- But also may need to be small because of competing assays:
  - Immune factors (antibodies, globulins)
  - Hormones
  - Disease state modifiers
  - In humans, concomitant medications

Distribution

- Rapidly cleared from blood.
- Frequently distributed via lymphatics.
- Target and off-target receptor-bound portion of drug molecule population in the body rapidly predominates.
- Pharmacodynamics very different from pharmacokinetics.
- Delivery rarely by oral route.
- Available test material supply will be very limited in early development.

The upshot of these points is that it may not be practical to follow established guidelines for ADME evaluation. Binding proteins, immunoreactive metabolites, and antibodies could interfere with the immunoassays used to measure the activity of biotechnologically derived pharmaceuticals. The link between immunoreactivity and pharmacological activity may be difficult to establish, making the data difficult to interpret. In radiolabeled distribution studies, if the label alters the physicochemical and biological properties of the test material, its pharmacokinetic behavior may change. These analytical difficulties make accurate characterization of the distribution, metabolism, and excretion of a protein more difficult. However, as immune assay–based methods of measuring levels of large molecules have improved, this situation has improved, and as more information on species-specific pharmacokinetics of large molecules (such as monoclonal antibodies) has become available, cross-species scaling of the pharmacokinetics of such molecules has also become possible (Zhao et al., 2015).

AUC and C_{max} are commonly measured to identify safety ratios for new chemical entities. Since the analytical methods used for biotechnologically derived pharmaceuticals may lack specificity, a clinical marker of biological activity or efficacy may sometimes be more appropriate than exposure data.

It is therefore essential that before pivotal (repeat-dose) preclinical studies are initiated, bioanalytical assay development must be completed. This has to cover potential test species as well as normal and diseased humans. The assays must be validated in the sampling matrix of the toxicity test species, and one should also develop suitable assays for antibodies to the test article.

Immunoassay Methods

Competitive inhibition of radiolabeled hormone antibody binding by unlabeled hormone either as a standard or an unknown mixture is the principle of most RIAs. A standard curve for measuring antigen (hormone) binding to antibody is constructed by placing known amounts of radiolabeled antigen and the antibody into a set of test tubes. Varying amounts of unlabeled antigen are added to the test tubes. Antigen-antibody complexes are separated from the antigen and the amount of radioactivity from each sample is measured to detect how much unlabeled antigen is bound to the antibody. Smaller amounts of radiolabeled antigen-antibody complexes are present in the fractions containing higher amounts of unlabeled antigen. A standard curve must be constructed to correlate the percentage of radiolabeled antigen bound with the concentration of unlabeled antigen present.

Two methods are commonly employed in RIA methodology to separate antigen-antibody complexes. The first, the double-antibody technique, precipitates antigen-antibody complexes out of solution by utilizing a second antibody, which binds to the first antibody. The second most commonly used method is the dextran-coated activated charcoal technique. Addition of dextran-coated activated charcoal to the sample followed immediately by centrifugation absorbs free antigen and leaves antigen-antibody complexes in the supernatant fraction. This technique works best when the molecular weight of the antigen is 30 kilodaltons (kDa) or less. Also, sufficient carrier protein must be present to prevent adsorption of unbound antibody.

Once a standard curve has been constructed, the RIA can determine the concentration of hormone in a sample (usually plasma or urine). The values of hormone levels are usually accurate using the RIA, but certain factors (e.g., pH or ionic strength) can affect antigen binding to the antibody. Thus, similar conditions must be used for the standard and the sample.

Problems of RIAs include lack of specificity. This problem is usually due to nonspecific cross-reactivity of the antibody. RIA represents an analytical approach of great sensitivity. Unlike assays that often require large amounts of tissue (or blood), greater sensitivity of the RIAs or monoclonal antibody techniques can be achieved using small samples of biological fluids. Some of these RIA methodologies are more useful than others and to some extent depend on the degree of hormonal cross-reactions or, in the case of monoclonal antibody methods, their degree of sensitivity.

Enzyme-linked immunosorbent assay (ELISA) is comparable to the immunoradiometric assay except that an enzyme tag is attached to the antibody instead of a radioactive label. ELISAs have the advantage of no radioactive materials and produce an end product that can be assessed with a spectrophotometer. The molecule of interest is bound to the enzyme-labeled antibody, and the excess antibody is removed for immune-radiometric assays. After excess antibody has been removed or the second antibody containing the enzyme has been added (two-site assay), the substrate and cofactors necessary are added in order to visualize and record enzyme activity. The level of molecule of interest present is directly related to the level of enzymatic activity. The sensitivity of the ELISAs can be enhanced by increasing the incubation time for producing substrate.

Immunoradiometric assays (IRMAs) are like RIAs in that a radiolabeled substance is used in an antibody-antigen reaction, except that the radioactive label is attached to the antibody instead of the hormone. Furthermore, excess of antibody, rather than limited quantity, is present in the assay. All the unknown antigen becomes bound in an IRMA rather than just a portion, as in an RIA; IRMAs are more sensitive. In the one-site assay, the excess antibody that is not bound to the sample is removed by addition of a precipitating binder. In a two-site assay, a molecule with at least two antibody-binding sites is adsorbed onto a solid phase, to which one of the antibodies is attached. After binding to this antibody is completed, a second antibody labeled with ^{125}I is added to the assay. This antibody reacts with the second antibody-binding site to form a "sandwich," composed as antibody-hormone-labeled antibody. The amount of hormone present is proportional to the amount of radioactivity measured in the assay.

With enzyme-multiplied immunoassay technique (EMIT) assays, enzyme tags are used instead of radiolabels. The antibody binding alters the enzyme characteristics, allowing for measurement of target molecules without separating the bound and free components (i.e., homogeneous assay). The enzyme is attached to the molecule being tested. This enzyme-labeled antigen is incubated with the sample and with antibody to the molecule. Binding of the antibody to the enzyme-linked molecule either physically blocks the active site of the enzyme or changes the protein conformation so that the enzyme is no longer active. After antibody binding occurs, the enzyme substrate and cofactor are added, and enzyme activity is measured. If the sample contains subject molecules, it will compete with enzyme-linked molecules for antibody binding, enzyme will not be blocked by antibody, and more enzyme activity will be measurable.

Most protein drug entities can now be assessed using monoclonal antibody (MAb) techniques. It is possible to produce antisera containing a variety of polyclonal antibodies that recognize and bind many parts of a given molecule. Polyclonal antisera can create some nonspecificity problems such as cross-reactivity and variation in binding affinity. Therefore, it is oftentimes desirable to produce a group of antibodies that selectively bind to a specific region of the molecule (i.e., antigenic determinant). In the past, investigators produced antisera to antigenic determinants of the molecule by cleaving the molecule and immunizing an animal with the fragment of the hormone containing the antigenic determinant of interest. This approach solved some

TABLE 3.17

Advantages and Disadvantages of Monoclonal Antibodies Compared to Polyclonal Antisera

Advantages	Disadvantages
Sensitivity	Overly specific
Quantities available	Decreased affinity
Immunologically defined	Diminished complement fixation
Detection of neoantigens on cell membrane	Labor-intensive; high cost

problems with cross-reactivity of antisera with other similar antigenic determinants, but problems were still associated with the heterogeneous collection of antibodies found in polyclonal antisera.

The production of MAbs offers investigators a homogenous collection of antibodies that could bind selectively to a specific antigenic determinant with the same affinity. In addition to protein isolation and diagnostic techniques, MAbs have contributed greatly to RIAs.

While MAbs offer a highly sensitive, specific method for detecting antigen, sometimes increasing MAb specificity compromises affinity of the antibody for the antigen. In addition, there is usually decreased complement fixation, and costs are usually high for preparing and maintaining hybridomas that produce MAbs (Table 3.17).

The monoclonal antibody techniques provide a means of producing a specific antibody for binding antigen. This technique is useful for studying protein structure relations (or alterations) and has been used for devising specific RIAs.

Metabolism and Elimination

Biologics are usually not excreted unchanged in urine. They are degraded to small peptides and individual amino acids with pathways equally and generally understood for endogenous compounds. Their metabolites (amino acids) are reutilized in the endogenous amino acid pool for the *de novo* biosynthesis of structural or functional body proteins.

The metabolism of biologics is highly dependent on structure, including sugars, charge density and distribution, size, and hydrophilicity/lipophilicity. Sites of metabolism of biologics are the liver, the kidneys, the blood, and the extravascular sites of administration. In the liver, hepatocytes are mainly responsible for the catabolism of biologics using carrier-mediated membrane transport as well as endocytosis/pinocytosis for transport process. Kidneys play a major role in the catabolism of many small polypeptides. After being filtered by the glomeruli, some proteins are reabsorbed by the proximal tubule by endocytosis, while small amino acid chains are hydrolyzed at the brush border. Controversy exists surrounding glomerular filtration selectivity regarding size, molecular conformation, and charge of the protein (Tang et al., 2004). The often-observed incomplete bioavailability of biologics after extravascular injection can be attributed to local metabolism. Catabolism at extravascular sites has been observed, e.g., for insulin, calcitonin, and interferon-β (for review see Mohler et al., 1992).

Several biologics, especially antibodies, show high inter-individual variation of pharmacokinetics parameters, which is mainly clearance driven. It is now clear that MAbs which target cellular antigens have far more complex, nonlinear pharmacokinetics, such that the half-life of these drugs can be both dose and time dependent (Lobo et al., 2004). When antigen concentration is high, half-life is short because the MAb is rapidly cleared from the blood through antigen–MAb interaction. As the antigen is depleted, clearance decreases and half-life is consequently prolonged. As the MAb accumulates, a new steady state is reached. Eventually, the target is totally depleted, at which time the clearance of MAb will be at its slowest. At this point, half-life will be at its longest, approaching the half-life of endogenous IgG (~21 days). More probable than total target depletion is saturation of the target–MAb binding with similar consequences.

Due to catabolism of proteins to (mostly) endogenous amino acids, classical biotransformation studies performed for small molecules are not needed. Additionally, limitations of current analytical methods to detect and distinguish metabolites, and the putative lack of pharmacological or toxicological activity of the metabolites, remain obstacles. Similarly, mass balance studies usually used to determine the excretion pathways of small molecules and their metabolites are not used for biologics.

The majority of therapeutic biologics, especially after chronic administration, elicit an immune response in test animal species and often also in humans. This is an inherent property when administering nonhuman sequence proteins of sufficient size. Initial success in reducing immunogenicity has been achieved by replacing biologics obtained from nonhuman sources with human sequences. Antibody formation can also occur in immunocompetent recipients after treatment with products derived from human sera and tissues, and also with recombinant human proteins that are identical or nearly identical in sequence to native human proteins. The mechanism for generation of antibodies of recombinant human proteins is not well understood. In most cases, the underlying mechanism is the breaking of immune tolerance that typically exists to self-antigens (Baumann, 2006). Other reasons for immunogenicity relate to manufacturing, formulation, and storage (e.g., aggregates). These are especially addressed when modifications of these processes are performed which might influence the physicochemical properties of the product. An immune response to a product does not mean it cannot be developed. However, the development and use of a product may be complicated and, in rare cases, also impossible. In certain cases, some patients develop antibodies which neutralize the biological activity of the therapeutic product and become unresponsive to treatment. Alterations in the pharmacokinetic profile due to immune-mediated clearance mechanisms may affect the PK profiles and the interpretation of the pre-clinical toxicity data. Last but not least, safety issues like immune-mediated toxicity may be raised. Detection and characterization of the immune response in patients is expected by the authorities (FDA, 2002).

The development and usefulness of appropriate animal models for testing immunogenicity is still unclear. Conventional animal models have predicted immunogenicity problems in humans poorly. One limitation of traditional animal models is that tolerance, a key aspect of the immune response, is highly species-specific. However, characterization of the immune response in research and pre-clinical development is necessary to get a valid interpretation of the pre-clinical efficacy and safety data.

Not only the compound structure itself, but also the administration route may affect immunogenicity. Extravascular injection is known to stimulate antibody formation more than intravenous application. This is most likely due to increased immunogenicity of protein aggregates and precipitates formed at the injection site.

To lower the systemic clearance and increase elimination half-lives, several strategies have been developed, including polyethylene glycol (PEG) attachment (PEGylation), glycosylation or fusion to proteins with decreasing clearance, and prolonged serum half-lives. PEGylation improves the PK behaviors by increasing the effective size of the protein, with the most significant effects for proteins smaller than 70 kDa. However, conjugation of the protein may also alter receptor affinity and biodistribution, changing the concentration-response profile for the protein independent of effects on pharmacokinetics. PEGylation can also reduce immunogenicity and aggregation.

Points to Consider

In the last decade, our understanding of the function and significance of transporters in both systemic and organ-specific absorption of drug molecules and their retention in potential therapeutic or toxicologically significant organs has become clear. Table 3.18 presents a summary comparison of these functions in humans and the three major nonrodent species.

As has been discussed already, distribution of therapeutic molecules to the desired specificity of target tissues is the actual goal of their administration. Traditionally, quantitative autoradiography has been the best tool for quantifying such exposures (Waddell, 1972). With advances in imaging technology, much more accurate noninvasive methods have become available.

TABLE 3.18

Selected Human Transporters Compared to Monkeys, Pigs, and Dogs

Selected Human Transporters Name (Synonym) (Super Family) (*Gene*)	Accession Numbers (GenBank) for Human, Monkey, Pig, and Dog	% Homology (Zhang, Schwartz, Wagner & Miller, 2000) of Amino Acids Number of Amino Acids: Humans/Animal		
Pgp (MDR1) (ABC efflux-transporter) (*ABCB1*)	NP_000918.2 NP_001028059.1 XP_003130253.2 NP_001028059.1	96 1280/1283	89 1280/1286	90 1280/1280
BCRP (ABC efflux-transporter) (*ABCG2*)	NP_004818.2 NP_001028091.1 NP_999175.1 NP_001041486	97 655/654	84 655/656	83 655/655
OATP1B1 (SLC influx-transporter) (*SLCO1B1*)	NP_006437.3 XP_001097704.1 - -	92 691/691	No homologue	No homologue
OATP1B3 (SLC influx-transporter) (*SLCO1B3*)	NP_062818.1 NP_001028113.1 XP_003355596.1 NP_001159519.1	93 702/702	72 702/684	72 702/692
OAT1 (SLC influx-transporter) (*SLC22A6*)	NP_004781.2 XP_001115901 NP_001001261.1 XP_533258.1	95(97)[a] 563/550	90(89)[b] 563/547	90 563/550
OAT3 (SLC influx-transporter) (*SLC22A8*)	NP_004245.2 BAD99108 NP_999620.1 XP_533257.3	96(96)[a] 542/542	82(81)[c] 542/543	78 542/613 Insignificantly expressed in liver[d]
OCT1 (SLC influx-transporter) (*SLC22A1*)	NP_003048.1 XP_002747211.1 NP_999154.1 XP_850971.2	91 554/506	75 554/554	74 554/521
OCT2 (SLC influx-transporter) (*SLC22A2*)	NP_003049.2 XP_002804005.1 NP_999067.1 XP_533466.2	94 555/708?	86 555/554	84 555/533

Source: Extracted from Dalgaard L, *J Pharmacol Toxicol Methods*, 74, 80–92, 2015.
Values were derived from (NCBI (2012)) using BLAST (http://blast.ncbi.nlm.nih.gov/Blast.cgi) and from [a]Tahara et al. (2005), [b]Hagos et al. (2002), [c]Hagos et al. (2005), and [d]Bleasby et al. (2006).

References

Bauer LA. (2001) *Applied Clinical Pharmacokinetics*. New York, NY: McGraw-Hill.

Baumann A. (2006) Early development of therapeutic biologics-pharmacokinetics. *Curr Drug Metab*, 7:15–21.

Bleasby K, Castle JC, Roberts CJ, Cheng C, Bailey WJ, Sina JF, Kulkarni AV, Hafey MJ, Evers R, Johnson JM, Ulrich RG, Slatter JG. (2006) Expression profiles of 50 xenobiotic transporter genes in humans and pre-clinical species: A resource for investigations into drug disposition. *Xenobiotica*, 36(10–11):963–988.

Bronaugh RL. (1998) Methods for in vitro percutaneous absorption. In: Marzulli FN and Maibach HI (Eds.) *Dermatotoxicology Methods*. Philadelphia, PA: Taylor & Francis.

Dalgaard L. (2015) Comparison of minipig, dog, monkey and human drug metabolism. *J Pharmacol Toxicol Methods*, 74:80–92.

Dennis MS, Zhang M, Meng YG, Kadkhodayan M, Kirchokofer D, Combs D, Damicot LA. (2002) Albumin binding as a general strategy for improving the pharmacokinetics of proteins. *J Bio Chem*, 277:35035–35043.

Dollery CT. (2013) Intracellular drug concentrations. *Clin Pharmacol Ther*, 93:263–266.

Dow P. (2006) In vitro species comparisons and metabolic identification. In: Enna SJ and Williams M (Eds.) *Current Protocols in Pharmacology*. Hoboken, NJ: Wiley, pp. 7.10.1–7.10.14.

Draper A, Madan A, Smith K, Parkinson A. (1998) Development of a non-high pressure liquid chromatography assay to determine testosterone hydroxylase (CYP3A) activity in human liver microsomes. *Drug Metab Dispos*, 26:299–304.

EMEA. (2007) Guideline on the Clinical Investigation of the Pharma-cokinetics of Therapeutic Proteins., *European Medicines Agency (EMEA)*, 24 January 2007, London.

FDA. (2000) *Redbook II*, pp. 138–148. College Park, MD: FDA.

FDA. (October 2002) *US Food and Drug Administration Guidance for Industry-Immunotoxicological Evaluation of Investigation New Drugs*. Washington, DC: Center for Drug Evaluation and Research.

Gad SC. (Ed.) (2015) *Animal Models in Toxicology*, 3rd Ed. New York, NY: Marcel Dekker.

Gad SC, Chengelis CP. (1998) *Acute Toxicology Testing*, 2nd Ed. San Diego, CA: Academic Press.

Goldstein A, Aronow L, Kalman S. (1974) *Principles of Drug Action: The Basis of Pharmacology*. New York, NY: John Wiley & Sons.

Griffin R, Godfrey V, Kim Y, Burka L. (1997) Sex-dependent differences in the disposition of 2, 4-dichlorophenoxyacetic acid in Sprague-Dawley rats, B6C3F1 mice and Syrian hamsters. *Drug Metab Dispos*, 25:1065–1071.

Hagos Y, Bahn A, Asif AR, Krick W, Sendler M, Burkhardt G. (2002) Cloning of the pig renal organic anion transporter 1 (pOAT1). *Biochimie*, 84(12):1221–1224.

Hagos Y, Braun IM, Krick W, Burckhardt G, Bahn A. (2005) FUnctional expression of pig renal organic anion transporter 3 (pOAT3). *BIochimie*, 87:421–424.

ICH. (2004a) *S3A: Toxicokinetics: The Assessment of Systemic Exposure in Toxicity Studies*. International Conference on Harmonisation, Geneva, Switzerland.

ICH. (2004b) *S3B: Pharmacokinetics: Guidance for Repeated Dose Tissue Distribution Studies*. International Conference on Harmolnisation, Geneva, Switzerland.

Kearns GL. (2015) Selecting the proper pediatric dose: It is more than size that matters. *Clin Pharmacol Ther*, 98:238–240.

Keberle G, Brindle S.D, Greengard P. (1971) The route of absorption of intraperitoneally administered compounds. *J Pharmacol Exp Ther*, 178:562–566.

La Du B, Mandel H, Way E. (1972) *Fundamentals of Drug Metabolism and Drug Disposition*. Baltimore, MD: The Williams & Wilkins Co.

Lavé T, Parrott N, Grinm HP, Fleury A, Reddy M. (2007) Challenges and opportunities with modeling and simulation in drug discovery and drug development. *Xenobiotica*, 37:1295–1310.

Levy RH, Thummel KD, Tauger WF, Hansten PD, Meichelbaum E. (2000) *Metabolic Drug Interactions*. Philadelphia, PA: Lippincott, Williams & Wilkins.

Lobo ED, Hansen RJ, Balthasar JP. (2004) Antibody pharmacokinetics and pharmacodynamics. *J Pharmaceut Sci*, 93:2645–2268.

Lu H, Wang J, Chan K, Young D. (1998) Effects of phenobarbital on stereoselective metabolism of ifosfamide in rats. *Drug Metab Disp*, 26:476–482.

Meyer UA. (1994) The molecular basis of genetic polymorphisms of drug metabolism. *J Pharm Pharmacol*, 46(suppl 1):409–415.

Mohler AM, Cook JE, Baumann G. (1992) In: BL Ferraiolo, MA Mohler, and CA Gloff (Eds.) *Protein Pharmacokinetics and Metabolism*. New York, NY; London: Plenum Press, pp. 35–71.

Monostory K, Jemnitz K, Vereczkey L, Czira G. (1997) Species differences in metabolism of panomifene, an analogue of tamoxifen. *Drug Metab Dispos*, 25:1370–1378.

Myers MJ, Farrell DE, Howard KD, Kawalek JC. (2001) Identification of multiple consitutive and inducible hepatic cytochrome P450 enzymes in market weight swine. *Drug Metab Dispos*, 29(6):908–915.

Nebbia C, Dacasto M, Rossetto G, Giuliano A, Carletti M. (2003) Comparative expression of liver cytochrome P450-dependent monooxygenases in the horse and in other agricultural and laboratory species. *Vet J*, 165(1):53–64.

O'Reilly WJ. (1972) Pharmacokinetics in drug metabolism and toxicology. *Can J Pharm Sci*, 7:66–77.

Parkinson A. (1996) Biotransformation of xenobiotics. In: C Klaassen (Ed.) *Cassarett & Doull's Toxicology: The Basic Science of Poisons*, 5th ed. New York, NY: McGraw-Hill.

Pratt WB, Taylor P. (1990) *Principles of Drug Action: The Basis of Pharmacology*, 3rd Ed. New York, NY: John Wiley & Sons.

Roberts A, Renwick AG. (2014) Toxicokinetics. In: AW Hayes and CL Kreuger (Eds.) *Principles and Methods of Toxicology*, 6th Ed. Boca Raton, FL: CRC Press, pp. 215–246.

Shimada T, Mimura M, Inoue K, Nakamura S, Oda H, Ohmori S, Yamazaki H. (1997) Cytochrome P450-dependent drug oxixation activities in liver microsomes of various animal species including rats, guinea pigs, dogs, monekys, and humans. *Arch Toxicol*, 71(6):401–408.

Skaanild MT and Friis C. (1999) Cytochrome P450 sex differences in minipigs and conventional pigs. *Pharmacol Toxicol*, 85(4):174–180.

Smith D. (1991) Species differences in metabolism and pharmacokinetics: Are we close to an understanding? *Drug Metab Rev*, 23:355–373.

Smith DA, Kerns EH. (2010) The effect of plasma protein binding on *in vivo* efficacy: Misconceptions in drug discovery, *Nat Rev Drug Discov*, 9:929–939.

Stevens JC, Shipley LA, Cashman JR, Vandenbranden M, Wrighton SA. (1993) Comparison of human and rhesus monkey in vitro phase I and phase II hepatic drug metabolism activities. *Drug Metab Dispos*, 21(5):753–760.

Tahara H, Shono M, Kusuhara H, Kinoshita H, Fuse E, Takadate A, Otagiri M, Sugiyama Y. (2005) Molecular cloning and funcitonal analyses of OAT1 and OAT3 from cynomolgus monkey kidney. *Pharm Res*, 22(4):647–660.

Tang L, Persky AM, Hochhaus G, Meibohm B. (2004) Pharmacokinetic aspects of biotechnology products. *J Pharmaceut Sci*, 93:2184–2204.

Tse FLS, Jaffe JM. (1991) *Preclinical Drug Disposition*. New York, NY: Marcel Dekker.

Waddell WJ. (1972) Autoradiography in drug dispositions studies. In: BN La Du, HG Mandel, and EL Way (Eds.) *Fundamentals of Drug Metabolism and Drug Disposition*. New York, NY: Williams & Wilkens, pp. 505–514.

Zhao J, Cao Y, Jusko WJ. (2015) Across-species scaling of monoclonal antibody pharmacokinetics using a minimal PBPK model. *Pharm Res*, 32:3269–3281.

4

Routes of Administration: With Considerations for Species Specificity

Each of the routes commonly used in safety assessment studies (here we will examine 12) has its own peculiarities, and for each there are practical considerations and techniques ("tricks") that should either be known or readily available to the practicing toxicologist. Tables 4.1 and 4.2 present both relevant route characteristics and considerations. It should also be kept in mind that disease status may alter regional drug absorption characteristics by routes. An example is increased access across the blood–brain barrier (BBB) in Alzheimer's disease (Mehta et al., 2015) and, as will be covered in Chapter 10, means of administration by specific routes may be modified to emphasize systemic or local development or delivery to specific target tissues.

An alternative approach to bringing drugs to market in the United States is the 505(b)(2) NDA. In this, approved drugs, while they are still on patent, can enter the marketplace if they are intended for administration by an alternate route, have a different formulation than that covered by patent for the already approved drug, or are for a different therapeutic indication.

The US FDA has specific guidance for nonclinical safety evaluation of such drugs (US FDA, 2015), which also serves to provide route-specific safety assessment guidance for drugs to be administered by means other than oral and intravenous.

Bioavailability and Thresholds

The difference between the extent of availability (often designated solely as bioavailability) and the rate of availability is illustrated in Figure 4.1, which depicts the concentration-time curve for a hypothetical agent formulated into three different dosage forms. In the figure, dosage forms A and B are designed so that the agent is put into the blood circulation at the same rate, but twice as fast as for dosage form C. The times at which agent concentrations reach a peak are identical for dosage forms A and B and occur earlier than the peak

TABLE 4.1

Routes: Potential Actions to Improve Objectives

Route	Dose Control	A	B	Local Tissue Tolerance Issues
Oral	Moderate	N	N	High excipient loading causes stasis and blockage of the GI tract • Irritation of the GI tract • Laxative effects • Loss of fluids and electrolytes
IV	Good	N	I	• Hemolysis, phlebitis, thrombosis
Dermal	Poor	Y	N	• Volatile hydrocarbons defat skin • Irritation and corrosion • Sensitization
Buccal		Y	N	• Irritation • Laxative effects • Loss of fluids and electrolytes
Inhalation	Moderate to Good	Y	N	• Pulmonary edema • Fibrosis • Pneumonia
Nasal	Moderate	Y	N	• Irritation
Rectal	Moderate	Y	N	• Irritation • Edema • Dehydration
Vaginal/ Uterine	Moderate	Y	N	• Irritation • Edema • Dehydration
Ocular	Topical: Moderate Intraocular: Good	Y Y	N I/N	• Corneal ulceration • Irritation • Edema of surrounding tissue (conjunctiva and sclera) • Glaucoma • Retinal damage
SC	Moderate	N	I	• Inflammation • Edema • Tissue necrosis • More effective at evoking and raising innate and adaptive immune responses
IM	Moderate	N	I	• Inflammation • Hemolysis
Implantation	Good	Y	I	• Foreign body response • Granuloma

Note: In general, any action which reduces systemic and local C_{max} levels improves tolerance. *A* = Bypass first-pass metabolism; B = Invasive (I) or Noninvasive (N)

TABLE 4.2
Route-Specific Considerations

Route	Points of Consideration	Possible Adjustments to Regimen to Improve Tolerance	References
Oral (PO)	1) Stomach of some species (dog, rabbit) easily irritated	1) Split dose over course of day	Strickley, 2004, 2008
	2) Stomach presents acid environment which may compromise stability of active drug	2) Use of capsules in dogs and mini-pigs to limit stomach irritation/ emesis and to avoid acid stability of drug issues	
	3) Most absorption actually occurs in the intestines	3) Dose with food to improve absorption and tolerance in stomach	
	4) Species-specific volume considerations	4) Intubation an option	
Intra-Articular (IA) *Injection into knee joints*	1) Species-specific volume considerations		
	2) Administration in pigs is limited as they only have such joints in their front limbs		
Intra-Dermal (ID)			
Intravenous (IV)	1) Care must be taken to confirm that drug will stay in solution in bloodstream	1) Infusion better tolerated than bolus	Lee et al., 2003
	2) Need to ensure sterility of formulation	2) Decrease rate of infusion to decrease systemic/ local C_{max} of drug	
	3) Vascular irritation, thrombosis and hemolysis are concerns. Need to perform hemolysis and flocculation studies in advance	3) Need to ensure sterility of formulation	
	4) Species-specific volume considerations	4) 50 Distal to site of injection vascular damage is now a field of significant concern. May not always be due to IV-administered drug.	
		5) For repeated injections, need to move site around	
Intramuscular (IM)	1) Injection site irritation and infection is a concern	1) Adjust injection volume to multiple sites	
	2) Species-specific volume considerations	2) For repeated injections, need to move site around	

(Continued)

TABLE 4.2 ***(Continued)***

Route-Specific Considerations

Route	Points of Consideration	Possible Adjustments to Regimen to Improve Tolerance	References
Intraperitoneal (IP)	1) Scarring and inflammation responses at injection sites 2) Species-specific volume considerations 3) Need to make sure injection is into the peritoneal space		
Subcutaneous (SC)	1) Nonpolar organic vehicles "defat" tissues if injected 2) Species-specific volume considerations	1) Nonpolar organic vehicles all "defat" tissues if injected 2) May need to do multiple sites 3) Adjust volume	
Dermal (Topical)	1) Irritation and sensitization 2) Selection of correct model species essential to be relevant to humans 3) Need to consider formulation viscosity and changes with temperature 4) Need to consider solvents used in formulation, as choices can affect mechanism of absorption	1) May need to wrap 2) May need to do multiple sites 3) May need to collar animals	Zesch, 1988
Periocular	1) Ocular irritation – tissue very sensitive to acid and base 2) Intraocular avoids local immune responses 3) Always keep in mind that anything injected in and around the eye is not going away quickly, so one is stuck with side effects. Be very concerned with pharmacology and half-life before injection 4) Length of needle changes with species 5) Angle of injection changes with species		Ubels et al., 2004
Inhalation	1) Delivery mode/particle size dictate where in respiration tree delivery of drug occurs 2) Consider insufflation as alternative		Warheit et al., 2003

Source: Gad SC et al., *Int J Toxicol.*, 25, 499–521, 2006; *Int J Toxicol.*, 35, 95–178, 2016.

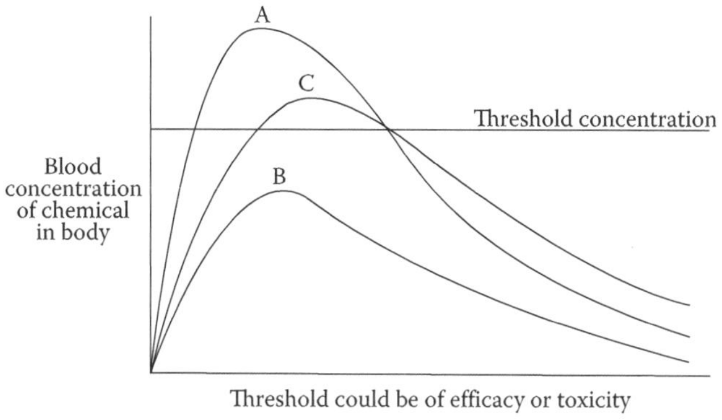

FIGURE 4.1
Examples of three different systemic absorption curves (as determined by plasma levels).

time for dosage form C. In general, the relative order of peak times following the administration of different dosage forms of the drug corresponds to the rates of availability of the chemical moiety from the various dosage forms. The extent of availability can be measured by using either chemical concentrations in the plasma or blood or amounts of unchanged chemical in the urine. The area under the blood concentration-time curve for an agent can serve as a measure of the extent of its availability. In Figure 4.1, the areas under curves A and C are identical and twice as great as the area under curve B. In most cases, where clearance is constant, the relative areas under the curves or the amount of unchanged chemical excreted in the urine will quantitatively describe the relative availability of the agent from the different dosage forms. However, even in nonlinear cases, where clearance is dose dependent, the relative areas under the curves will yield a measurement of the rank order of availability from different dosage forms or from different routes of administration.

Because there is usually a critical concentration of a chemical in the blood that is necessary to elicit either a pharmacological or toxic effect, both the rate and extent of input or availability can alter the toxicity of a compound. In the majority of cases, the duration of effects will be a function of the length of time the blood-concentration curve is above the threshold concentration; the intensity of the effect for many agents will be a function of the elevation of the blood-concentration curve above the threshold concentration.

Thus, the three different dosage forms depicted in Figure 4.1 will exhibit significant differences in their levels of "toxicity." Dosage form B requires that twice the dose be administered to attain blood levels equivalent to those for dosage form A. Differences in the rate of availability are particularly important for agents given acutely. The dosage for A reaches the target concentration earlier than the chemical from dosage form C; concentrations from A reach a higher level and remain above the minimum effect concentration for a longer period of time. In a multiple dosing regimen, dosage forms A and

C will yield the same average blood concentrations, although dosage form A will show somewhat greater maximum and lower minimum concentrations.

For most chemicals, the rate of disposition or loss from the biological system is independent of rate of input, once the agent is absorbed. Disposition is defined as what happens to the active molecule after it reaches a site in the blood circulation where concentration measurements can be made (the systemic circulations, generally). Although disposition processes may be independent of input, the inverse is not necessarily true, because disposition can markedly affect the extent of availability. Table 4.1 summarizes some key characteristics of the major routes, including whether they are subject to the first-pass effect or not. Agents absorbed from the stomach and the intestine must first pass through the liver before reaching the general circulation (Figure 4.2). Thus, if a compound is metabolized in the liver or excreted in bile, some of the active molecule absorbed from the GI tract will be inactivated by hepatic processes before it can reach the systemic circulation and be distributed to its sites of action. If the metabolizing or biliary excreting capacity of the liver is great, the effect on the extent of availability will be substantial. Thus, if the hepatic blood clearance for the chemical is large, relative to hepatic blood flow, the extent of availability for this chemical will be low when it is given by a route that yields first-pass metabolic effects. This decrease in availability is a function of the physiological site from which absorption takes place, and no amount of modification to dosage form can improve the availability under linear

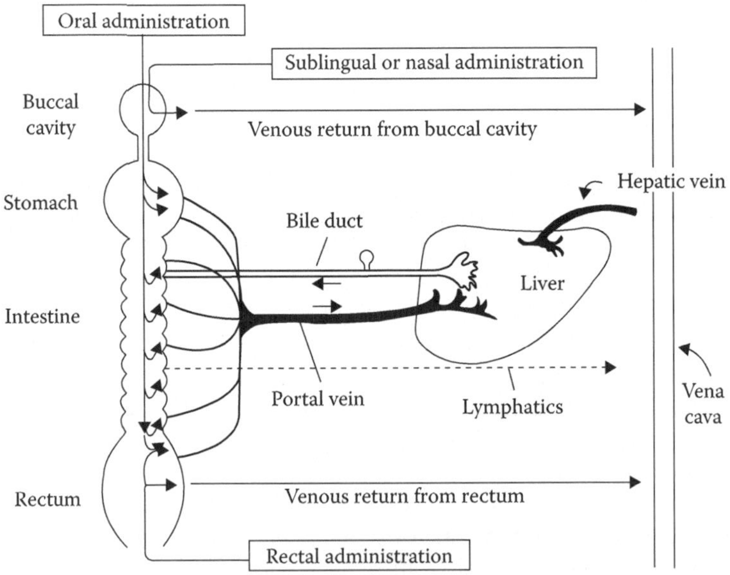

FIGURE 4.2
Path of drugs through the body after absorption by one of three routes of administration.

conditions. Of course, toxic blood levels can be reached by this route of administration if larger doses are given.

It is important to realize that chemicals with high extraction ratios (i.e., greater extents of removal by the liver during first-pass metabolism or "first-pass effects") will exhibit marked intersubject variability in bioavailability because of variations in hepatic function or blood flow or both. For a chemical with an extraction ratio of 0.90 that increases to 0.95, the bioavailability of the agent will be halved, from 0.10 to 0.05. These relationships can explain the marked variability in plasma or blood drug concentrations that occurs among individual animals given similar doses of a chemical that is markedly extracted. Small variations in hepatic extraction between individual animals will result in large differences in availability and plasma drug concentrations.

The first-pass effect can be avoided, to a great extent, by use of the sublingual route and by topical preparations (e.g., nitroglycerine ointment), and it can be partially avoided by using rectal suppositories. The capillaries in the lower and middle sections of the rectum drain into the interior and middle hemorrhoidal veins, which in turn drain into the inferior vena cava, thus bypassing the liver. However, suppositories tend to move upward in the rectum into a region where veins that lead to the liver predominate, such as the superior hemorrhoidal vein. In addition, there are extensive connections between the superior and middle hemorrhoidal veins, and thus probably only about 50% of a rectal dose can be assumed to bypass the liver. The lungs represent a good temporary clearing site for a number of chemicals (especially basic compounds) by partition into lipid tissues, as well as serve a filtering function for particulate matter that may be given by intravenous injection. In essence, the lung may cause first-pass loss by excretion and possible metabolism for chemicals input into the body by the nongastrointestinal routes of administration.

Biological (test subject) factors that can influence absorption of a chemical from the gastrointestinal tract are summarized in Table 4.3. There are also a number of chemical factors that may influence absorption from the GI tract. These are summarized in Table 4.4.

Dermal Route

For all agents of concern in occupational toxicology (except therapeutic agents), the major route by which the general population is most frequently exposed is via the percutaneous (dermal) route. Brown (1980) has previously reviewed background incidence data on pesticides that shows such a route of exposure to be common. Dermal (or topical) drugs, although not as common, are certainly numerous in occurrence.

TABLE 4.3

Test Subject Characteristics That Can Influence GI Tract Absorption[a]

A. General and inherent characteristics
   1. General condition of the subject (e.g., starved versus well-fed, ambulatory versus supine).
   2. Presence of concurrent diseases (i.e., diseases may either speed or slow gastric emptying).
   3. Age.
   4. Weight and degree of obesity.

B. Physiological function
   1. Status of the subject's renal function.
   2. Status of the subject's hepatic function.
   3. Status of the subject's cardiovascular system.
   4. Status of the subject's gastrointestinal motility and function (e.g., ability to swallow).
   5. pH of the gastric fluid (e.g., affected by fasting, disease, food intake, drugs).
   6. Gastrointestinal blood flow to the area of absorption.
   7. Blood flow to areas of absorption for dose forms other than those absorbed through gastrointestinal routes.

C. Acquired characteristics
   1. Status of the subject's anatomy (e.g., previous surgery).
   2. Status of the subject's gastrointestinal flora.
   3. Timing of drug administration relative to meals (i.e., presence of food in the gastrointestinal tract).
   4. Body position of subject (e.g., lying on the side slows gastric emptying).
   5. Psychological state of subject (e.g., stress increases gastric emptying rate, and depression decreases rate).
   6. Physical exercise of subject may reduce gastric emptying rate.

D. Physiological principles
   1. Food enhances gastric blood flow, which should theoretically increase the rate of absorption.
   2. Food slows the rate of gastric emptying, which should theoretically slow the rate of passage to the intestines where the largest amounts of most agents are absorbed. This should decrease the rate of absorption for most agents. Agents absorbed to a larger extent in the stomach will have increased time for absorption in the presence of food and should be absorbed more completely than in fasted patients.
   3. Bile flow and secretion are stimulated by fats and certain other foods. Bile salts may enhance or delay absorption depending on whether they form insoluble complexes with drugs or enhance the solubility of agents.
   4. Changes in splanchnic blood flow as a result of food depend on direction and magnitude of the type of food ingested.
   5. Presence of active (saturable) transport mechanisms places a limit on the amount of a chemical that may be absorbed.

[a] The minimization of variability due to these factors rests on the selection of an appropriate animal model, careful selection of healthy animals, and use of proper techniques.

Percutaneous entry into the body is really by separate means (Marzulli, 1962; Bland and Scheuplein, 1964; Scheuplein, 1965, 1967; Lien et al., 1971):

- Between the cells of the stratum corneum
- Through the cells of the stratum corneum
- Via the hair follicles
- Via the sweat glands
- Via the sebaceous glands

TABLE 4.4

Chemical Characteristics of a Drug That May Influence Absorption

A. Administration of chemical and its passage through the body
   1. Dissolution characteristics of solid dosage forms, which depend on formulation in addition to the properties of the chemical itself (e.g., vehicle may decrease permeability of suspension or capsule to water and retard dissolution and diffusion).
   2. Rate of dissolution in gastrointestinal fluids. Chemicals that are inadequately dissolved in gastric contents may be inadequately absorbed.
   3. Chemicals that are absorbed into food may have a delayed absorption.
   4. Carrier-transported chemicals are more likely to be absorbed in the small intestine.
   5. Route of administration.
   6. Chemicals undergo metabolism in the gastrointestinal tract.

B. Physiochemical properties of chemicals
   1. Chemicals that chelate metal ions in food may form insoluble complexes and will not be adequately absorbed.
   2. pH of dosing solutions – weakly basic solutions are absorbed to a greater degree in the small intestine.
   3. Salts used.
   4. Hydrates or solvates.
   5. Crystal form of chemical (e.g., insulin).
   6. "Pharmaceutical" form (e.g., fluid, solid, suspension).
   7. Enteric coating.
   8. Absorption of quaternary compounds (e.g., hexamethonium, amiloride) are decreased by food.
   9. Molecular weight of chemical (e.g., when the molecular weight of a drug is above about 1000, absorption is markedly decreased).
   10. pK_a (dissociation constant).
   11. Lipid solubility (i.e., a hydrophobic property relating to penetration through membranes).
   12. Particle size of chemical in solid dosage form – smaller particle sizes will increase the rate and/or degree of absorption if dissolution of the chemical is the rate-limiting factor in absorption. Chemicals that have a low dissolution rate may be made in a micronized form to increase their rate of dissolution.
   13. Particle size of the dispersed phase in an emulsion.
   14. Type of disintegrating agent in the formulation.
   15. Hardness of a solid (granule, pellet, or tablet) (i.e., related to amount of compression used to make tablet) or capsule if they do not disintegrate appropriately.

Specific important aspects of the dosage forms of interest, as well as those of the test animals, involving application and absorption (Bland and Scheuplein, 1964) are as follows:

1. Small molecules penetrate skin better than large molecules.
2. Undissociated molecules penetrate skin better than do ions.
3. Preferential solubility of the toxicant in organic solvents indicates better penetration characteristics than preferential solubility in water.
4. The less viscous or more volatile the toxicant, the greater its penetrability.
5. The nature of the vehicle and the concentration of the toxicant in the vehicle both affect absorption (vehicles are discussed later in this chapter).
6. Hydration (water content) of the stratum corneum affects penetrability.
7. Ambient temperature can influence the uptake of toxicant through the skin. The warmer it is, the greater the blood flow through the skin and, therefore, the greater the degree of percutaneous absorption.
8. Molecular shape (particularly symmetry) influences absorption (Medved and Kundiev, 1964).

There are at least two excellent texts on the subject of percutaneous absorption (Brandau and Lippold, 1982; Bronaugh and Maibach, 1985) that go into great detail.

Parenteral Route

The parenteral routes include four major ones – IV (intravenous), IM (intramuscular), and SC (subcutaneous), and IP (intraperitoneal) – and a number of minor routes (such as intra-arterial) than are not considered here. Administration by the parenteral routes raises a number of special safety concerns in addition to the usual systemic safety questions. These include irritation (vascular, muscular, or subcutaneous), pyrogenicity, blood compatibility, and sterility (Avis, 1985). The background of each of these, along with the underlying mechanisms and factors that influence the level of occurrence of such an effect, are discussed in Chapter 16 of Gad (2016).

The need for a rapid onset of action (and/or clearance) usually requires that an IV route be used, although at a certain stage of cardio-pulmonary resuscitation (for example), the need for an even more rapid effect may require the use of an intracardiac injection. The required site of action may influence the choice of route of administration (e.g., certain radiopaque dyes are given intra-arterially near the site being evaluated; streptokinase is sometimes injected experimentally into the coronary arteries close to coronary vessel occlusion during a myocardial infarction to cause lysis of the thrombus and therefore re-establish coronary blood flow).

The characteristics of the fluid to be injected will also influence the choice of parenteral routes. The drug must be compatible with other fluids (e.g., saline, dextrose, Ringer's lactate) with which it may be combined for administration to the patient, as well as with the components of the blood itself.

There are certain clinical situations in which a parenteral route of administration is preferred to other possible routes. These include the following:

When the amount of drug given to a subject must be precisely controlled (e.g., in many pharmacokinetic studies), it is preferable to use a parenteral (usually IV) route of administration.

When the "first-pass effect" of a drug going through the liver must be avoided, a parenteral route of administration is usually chosen, although a sublingual route or dermal patch will also avoid the first-pass effect.

When one requires complete assurance that an uncooperative subject has actually received the drug and has not rejected it (e.g., via forced emesis).

When subjects are in a stupor, coma, or otherwise unable to take a drug orally.

When large volumes (i.e., more than a liter) of fluid are injected (such as in peritoneal dialysis, hyperalimentation, fluid replacement, and other conditions). Special consideration of fluid balance must be given to patients receiving large volumes, as well as careful consideration of the systemic effects of injection fluid components (e.g., amino acids and their nephrotoxicity).

Each of the three significant parenteral routes considered here has a specific set of either advantages and disadvantages or specific considerations that must be kept in mind.

Intravenous Route (IV)

The IV route is the most common method of introducing a drug directly into the systemic circulation. It offers the following advantages:

1. Rapid onset of effect.
2. Usefulness in situations of poor gastrointestinal absorption.
3. Avoidance of tissue irritation that may be present in IM or other routes (e.g., nitrogen mustard).
4. More precise control of levels of drug than with other routes, especially of toxic drugs, where the levels must be kept within narrow limits.

5. Ability to administer large volumes over time by a slow infusion.
6. Ability to administer drugs at a constant rate over a long period of time.

And it suffers from the following disadvantages:

1. Higher incidence of anaphylactic reactions than with many other routes.
2. Possibility of infection or phlebitis at site of injection.
3. Greater pain to patients than with many other routes.
4. Possibility that embolic phenomena may occur – either air embolism or vascular clot – as a result of damage to the vascular wall.
5. Impossibility of removing or lavaging a drug after it is given, except by dialysis.
6. Inconvenience in many situations.
7. Possibility that rapid injection rates may cause severe adverse reactions.
8. Patient dislike of, and psychological discomfort with, the injection procedure.

For IV fluids, it must be determined how the dose will be given (i.e., by bolus or slow injection, intermittent or constant infusion, or by constant drip) and whether special equipment will be used to control and monitor the flow. Drugs with short half-lives are usually given by a constant drip or infusion technique. All IV fluids given immediately subsequent to an IV drug must be evaluated for their compatibility with the study drug. Suspensions are generally not given intravenously because of the possibility of blocking the capillaries.

In the IV route, anaphylactic reactions (caused by administration of an agent to an animal previously sensitized to it or to a particularly sensitive species such as a guinea pig) may be especially severe – probably because of sudden, massive antigen-antibody reactions. When the drug is given by other routes, its access to antibody molecules is necessarily slower; moreover, its further absorption can be retarded or prevented at the first sign of a serious allergic reaction. Subcutaneous administration actually tends to be more effective in initiating or achieving a heightened immune response (a reason why it is preferred for vaccine administration).

Embolism is another possible complication of the IV route. Particulate matter may be introduced if a drug intended for intravenous use precipitates for some reason, or if a particular suspension intended for IM or SC use is inadvertently given into a vein. Hemolysis or agglutination of erythrocytes may be caused by injection of hypotonic hypertonic solutions, or by more specific mechanisms (Gray, 1978).

Bolus versus Infusion

Technically, for all the parenteral routes (but in practice only for the IV route), there are two options for injecting a material into the body. The bolus and infusion methods are differentiated on the single basis of rate of injection, but they actually differ on a wide range of characteristics.

The most commonly exercised option is the bolus, "push," injection, in which the injection device (syringe or catheter) is appropriately entered into the vein and a defined volume of material is introduced through the device. The device is then removed. In this operation, it is relatively easy to restrain an experimental animal and the stress on the animal is limited. Though the person doing the injection must be skilled, it takes only a short amount of time to become so. And the one variable to be controlled in determining dosage is the total volume of material injected (assuming dosing solutions have been properly prepared). For infusion, a port (particularly over a period of weeks or more) (Garramone, 1986) is typically installed to minimize cumulative injection site trauma.

Cell therapy delivered by a parenteral route (usually IC) can also be used to delivery therapeutics directly into cellular tissue targets by carrying connexosomes loaded with drug (Arnauld, 2016).

Intramuscular Route (IM)

The IM route is frequently used for drugs dissolved in oily vehicles or for those in a microcrystalline formulation that are poorly soluble in water (e.g., procaine or penicillin G). Advantages include rapid absorption (often in under 30 min), the opportunity to inject a relatively large amount of solution, and a reduction in pain and local irritation compared with SC injections. Potential complications include infections and nerve damage. The latter usually results from the choice of an incorrect site for injection.

Although the time to peak drug concentration is often on the order of 1–2 hours, depot preparations given by IM injection are absorbed extremely slowly. Numerous physiochemical properties of a material given intramuscularly will affect the rate of absorption from the site within the muscle (e.g., ionization of the drug, lipid solubility, osmolality of the solution, volume given). The primary sites used for IM injections in people are the gluteal (buttock), deltoid (upper arm), epaxial and lateral vastus (lateral thigh) muscles, with the corresponding sites in test animals being species specific. The rate of drug absorption and the peak drug levels obtained will often differ between sites used for IM injections because of differences in blood flow between muscle groups. The site chosen for an IM injection in humans and some animals may be a critical factor in whether or not the drug exhibits an effect (Schwartz et al., 1974). Agents injected into the larger muscle masses are generally absorbed rapidly.

Blood flow through muscles in a resting animal is about 0.02–0.07 mL min^{-1} g^{-1} of tissue, and this flow rate may increase many times during exercise, when additional vascular channels open. Large amounts of solution can be introduced intramuscularly, and there is usually less pain and local irritation than is encountered by the SC route. Ordinary aqueous solutions of chemicals are usually absorbed from an intramuscular site within 10–30 min, but faster or slower absorption is possible, depending on the vascularity of the site, the ionization and lipid solubility of the drug, the volume of the injection, the osmolality of the solution, animal temperature, and other variables. Small molecules are absorbed directly into the capillaries from an intramuscular site, whereas large molecules (e.g., proteins) gain access to the circulation by way of the lymphatic channels. Radiolabeled compounds of widely differing molecular weights (maximum 585) and physical properties have been shown to be absorbed from rat muscle at virtually the same rate, about 16% per minute (i.e., the absorption process is limited by the rate of blood flow.)

Drugs that are insoluble at tissue pH, or that are in an oily vehicle, form a depot in the muscle tissue, from which absorption proceeds very slowly.

Subcutaneous Route (SC)

Drugs given by the SC route are forced into spaces between connective tissues, as with IM injections. Vasoconstrictors and drugs that cause local irritation should not be given subcutaneously under usual circumstances, since inflammation, abscess formation, or even tissue necrosis may result. When daily or even more frequent SC injections are made, the site of injection should be continually changed to prevent local complications. Fluids given subcutaneously must have an appropriate tonicity/osmolarity to prevent pain. Care must be taken to prevent injection of the drug directly into veins.

The vascular absorption of drugs from an SC route is influenced by blood flow to the area, as with IM injections. The rate of absorption may be retarded by cooling the local area to cause vasoconstriction, adding epinephrine to the solution for the same purpose (e.g., with local anesthetics), decreasing blood flow with a tourniquet, or immobilizing the area. The opposite effect may be achieved by warming the injection region or by using the enzyme hyaluronidase, which breaks down mucopolysaccharides of the connective tissue matrix to allow the injected solution to spread over a larger area and thus increase its rate of absorption (Ballard, 1968). Absorption/transport into the lymphatic system is also well served by this route.

Absorption from SC injection sites is affected by the same factors that determine the rate of absorption from IM sites (Schou, 1971). Blood flow through these regions is generally poorer than in muscles, so the absorption rate is generally slower.

The rate of absorption from an SC injection site may be retarded by immobilization of the limb, local cooling to cause vasoconstriction, or application of a tourniquet proximal to the injection site to block the superficial venous drainage and lymphatic flow. In small amounts, adrenergic stimulants, such as epinephrine, will constrict the local blood vessels and, therefore, slow systemic absorption. Conversely, cholinergic stimulants (such as methacholine) will induce very rapid systemic absorption subcutaneously. Other agents may also alter their own rate of absorption by affecting local blood supply or capillary permeability.

A prime determinant of the absorption rate from an SC injection is the total surface area over which the absorption can occur. Although the subcutaneous tissues are somewhat loose and moderate amounts of fluid can be administered, the normal connective tissue matrix prevents indefinite lateral spread of the injected solution. These barriers may be overcome by agents that break down mucopolysaccharides of the connective tissue matrix; the resulting spread of injected solution leads to a much faster absorption rate.

In addition to fluids, solid forms of drugs may be given by SC injection. For example, this has been done with compressed pellets of testosterone placed under the skin, which are absorbed at a relatively constant rate over a long period.

Intraperitoneal Route (IP)

Kruger et al. (1962) demonstrated the efficiency of absorption of some chemicals injected IP, while Lukas et al. (1971) showed that compounds administered IP are absorbed primarily through the portal circulation.

A prime practical consideration in the use of the IP route for acute testing should be the utilization of aseptic techniques to preclude bacterial or viral contamination. If these are not exercised, the resulting infected and compromised animals cannot be expected to produce either valid or reproducible indications or actual chemical toxicity.

Compounds that are very lipophilic will be quickly absorbed systemically by the IP route, but not by the IM or SC route.

Oral Routes (PO)

Strictly speaking, oral (PO) includes buccal and sublingual, but these will be considered separately. The oral route is the most commonly used route for the administration of drugs both because of ease of administration and because it is the most readily accepted route of administration.

Enteral routes technically include any that will put a material directly into the GI tract, but the use of enteral routes other than oral (such as rectal) are uncommon in clinical medicine and safety assessment. Though there are a number of variations of technique and peculiarities of animal response that are specific to different animal species, there is also a great deal of commonality across species in methods, considerations, and mechanisms.

Mechanisms of Absorption

Ingestion is generally referred to as oral or peroral (PO) exposure and includes direct intragastric exposure in experimental toxicology. The regions for possible agent action and absorption from PO absorption should receive separate consideration.

Absorption of chemicals with widely differing characteristics can occur at different levels in the GI tract (Hogben et al., 1959; Schranker, 1960; Bates and Gibaldi, 1970; Gad, 2007). The two factors primarily influencing this regional absorption are (1) the lipid-water partition characteristics of the undissociated toxicant and (2) the dissociation constant (pK_a) that determines the amount of toxicant in the dissociated form. There are also species differences (Schranker et al., 1957).

Therefore, weak organic acids and bases are readily absorbed as uncharged lipid-soluble molecules, whereas ionized compounds are absorbed only with difficulty, and nonionized toxicants with poor lipid-solubility characteristics are absorbed slowly. Lipid-soluble acid molecules can be absorbed efficiently through the gastric mucosa, but bases are not absorbed in the stomach.

In the intestines the nonionized form of the drug is preferentially absorbed and the rate of absorption is related to the lipid-water partition coefficient of the toxicant. The highest pK_a value for a base compatible with efficient gastric absorption is about 7.8 and the lowest pK_a for an acid is about 3.0, although a limited amount of absorption can occur outside these ranges. The gastric absorption and the intestinal absorption of a series of compounds with different carbon chain lengths follow two different patterns. Absorption from the stomach increases as the chain lengthens from methyl to *n*-hexyl, whereas intestinal absorption increases over the range methyl to *n*-butyl and then diminishes as the chain length further increases. Houston et al. (1974) concluded that to explain the logic of optimal partition coefficients for intestinal absorption it was necessary to postulate a two-compartment model with a hydrophilic barrier and a lipoidal membrane, and that if there is an acceptable optimal partition coefficient for gastric absorption it must be at least ten times greater than the corresponding intestinal value.

Because they are crucial to the course of an organism's response, the rate and extent of absorption of biologically active agents from the GI tract also have major implications for the formulation of test material dosages and also for how production (commercial) materials may be formulated to minimize potential accidental intoxications while maximizing the therapeutic profile.

There are a number of separate mechanisms involved in absorption from the gastrointestinal tract.

Passive absorption. The membrane lining of the tract has a passive role in absorption. As toxicant molecules move from the bulk water phase of the intestinal contents into the epithelial cells, they must pass through two membranes in series, one layer of water and the other the lipid membrane of the microvillar surface (Wilson and Dietschy, 1974). The water layer may be the rate-limiting factor for passive absorption into the intestinal mucosa, but it is not rate-limiting for active absorption. The concentration gradient in addition to the physiochemical properties of the drug along with the lining membrane are the major controlling factors. Chemicals that are highly lipid-soluble are capable of passive diffusion and they pass readily from the aqueous fluids of the gut lumen through the lipid barrier of the intestinal wall and into the bloodstream. The interference in the absorption process by the water layer increases with increasing absorbability of the substances in the intestine (Winne, 1978).

Aliphatic carbamates are rapidly absorbed from the colon by passive uptake, and it is found that there is a linear relationship between log k_a and log P for absorption of these carbamates in the colon and the stomach, whereas there is a parabolic relationship between these two values for absorption in the small intestine. The factors to be considered are as follows:

P = octanol-buffer partition coefficient

k_a = absorption rate constant

t = time

$t^{1/2}$ = half-life = $\underline{\ln 2}\ k_a$

Organic acids that are extensively ionized at intestinal pHs are absorbed primarily by simple diffusion.

Facilitated diffusion. A temporary combination of the chemical with some form of "carrier" occurs in the gut wall, facilitating the transfer of the toxicant across the membranes. This process is also dependent on the concentration gradient across the membrane, and there is no energy utilization in making the translocation. In some intoxications, the carrier may become saturated, making this the rate-limiting step in the absorption process.

Active transport. As above, the process depends on a carrier, but differs in that the carrier provides energy for translocation from regions of lower concentration to regions of higher concentration.

Pinocytosis. This process by which particles are absorbed can be an important factor in the ingestion of particulate formulations of chemicals (e.g., dust formulations, suspensions of wettable powders); however, it must not be confused with absorption by one of the above processes, where the agent has been released from particles.

Absorption via lymphatic channels. Some lipophilic chemicals dissolved in lipids may be absorbed through the lymphatics.

Convective absorption. Compounds with molecular radii of less than 4 nm can pass through pores in the gut membrane. The membrane exhibits a molecular sieving effect.

Characteristically, within certain concentration limits, if a chemical is absorbed by passive diffusion, then the concentration of toxicant in the gut and the rate of absorption are linearly related. However, if absorption is mediated by active transport, the relationship between concentration and rate of absorption conforms to Michaelis-Menten kinetics and a Lineweaver-Burk plot (i.e., reciprocal of rate of absorption plotted against reciprocal of concentration), which graphs as a straight line.

Differences in the physiological chemistry of gastrointestinal fluids can have a significant effect on toxicity. Both physical and chemical differences in the GI tract can lead to species differences in susceptibility to acute intoxication. The antihelminthic pyrvinium chloride has an identical LD_{50} value when administered intraperitoneally to rats and mice (approximately 4 mg kg^{-1}); when administered orally, however, the LD_{50} value in mice was found to be 15 mg kg^{-1}, while for the rat, the LD_{50} values were 430 mg kg^{-1} for females and 1550 mg kg^{-1} for males. It is thought that this is an absorption difference rather than a metabolic difference (Ritschel et al., 1974).

Most of any exogenous chemical absorbed from the GI tract must pass through the liver via the hepatic-portal system (leading to the so-called first-pass effect) and, as mixing of the venous blood with arterial blood from the liver occurs, consideration and caution are called for in estimating the amounts of chemical in both the systemic circulation and the liver itself.

Despite the gastrointestinal absorption characteristics discussed above, it is common for absorption from the alimentary tract to be facilitated by dilution of the toxicant. Borowitz et al. (1971) have suggested that the concentration effects they observed in atropine sulfate, aminopyrine, sodium salicylate, and sodium pentoparbital were due to a combination of rapid stomach emptying and the large surface area for absorption of the drugs.

Major structural or physiological differences in the alimentary tract (e.g., species differences or surgical effects) can give rise to modifications of toxicity. For example, ruminant animals may metabolize toxicants in the GI tract in a way that is unlikely to occur in nonruminants.

The presence of bile salts in the alimentary tract can affect absorption of potential toxicants in a variety of ways, depending on their solubility characteristics.

Factors Affecting Oral Absorption

Test chemicals are given most commonly by either mouth or direct instillation into the stomach (gavage). This is certainly the most convenient route, and it is the only one of practical importance for self-administration. Absorption, in general, takes place along the whole length of the gastrointestinal tract, but the chemical properties of each molecule determine whether it will be absorbed in the strongly acidic stomach or in the nearly neutral intestine. Gastric absorption is favored by an empty stomach, in which the chemical, in undiluted gastric juice, will have good access to the mucosal wall. Only when a chemical would be irritating to the gastric mucosa is it rational to administer it with or after a meal. However, the antibiotic griseofulvin is an example of a substance with poor water solubility, the absorption of which is aided by a fatty meal. The large surface area of the intestinal villi, the presence of bile, and the rich blood supply all favor intestinal absorption of griseofulvin and physiochemically similar compounds.

The presence of food can impair the absorption of chemicals given by mouth. Suggested mechanisms include reduced mixing, complexing with substances in the food, and retarded gastric emptying. In experiments with rats, prolonged fasting has been shown to diminish the absorption of several chemicals, possibly by deleterious effects upon the epithelium of intestinal villi.

Chemicals that are metabolized rapidly by the liver cannot be given for systemic effect by the enteral route because the portal circulation carries them directly to the liver. For example, lidocaine, a drug of value in controlling cardiac arrhythmias, is absorbed well from the gut, but is completely inactivated in a single passage through the liver.

The principles governing the absorption of drugs from the gastrointestinal lumen are the same as for the passage of drugs across biological membranes elsewhere. A lower degree of ionization, high lipid-water partition coefficient of the nonionized form, and a small atomic or molecular radius of water-soluble substances all favor rapid absorption. Water passes readily in both directions across the wall of the gastrointestinal lumen. Sodium ion is probably transported actively from lumen into blood. Magnesium ion is very poorly absorbed and therefore acts as a cathartic, retaining an osmotic equivalent of water as it passes down the intestinal tract. Ionic iron is absorbed as an amino acid complex, at a rate usually determined by the body's need for it. Glucose and amino acids are transported across the intestinal wall by specific carrier systems. Some compounds of high molecular weight (polysaccharides and large proteins) cannot be absorbed until they are degraded enzymatically. Other substances cannot be absorbed because they are destroyed by gastrointestinal enzymes – insulin, epinephrine, and histamine are examples. Substances that form insoluble precipitates in the gastrointestinal lumen or that are insoluble either in water or in lipid clearly cannot be absorbed.

Absorption of weak acids and bases. Human gastric juice is very acid (about pH 1), whereas the intestinal contents are nearly neutral (actually very slightly acid). The pH difference between plasma (pH 7.4) and the lumen of the GI tract plays a major role in determining whether a drug that is a weak electrolyte will be absorbed into plasma, or excreted from plasma into the stomach or intestine. For practical purposes, the mucosal lining of the GI tract is impermeable to the ionized form of a weak acid or base, but the nonionized form equilibrates freely. The rate of equilibration of the nonionized molecule is directly related to its lipid solubility. If there is a pH difference across the membrane, then the fraction ionized may be considerably greater on one side than on the other. At equilibrium, the concentration of the nonionized moiety will be the same on both sides, but there will be more total drug on the side where the degree of ionization is greater. This mechanism is known as *ion-trapping*. The energy for sustaining the unequal chemical potential of the acid or base in question is derived from whatever mechanism maintains the pH difference. In the stomach, this mechanism is the energy-dependent secretion of hydrogen ions.

Consider how a weak electrolyte is distributed across the gastric mucosa between plasma (pH 7.4) and gastric fluid (pH 1.0). In each compartment, the Henderson–Hasselbalch equation gives the ratio of acid-base concentrations. The negative logarithm of the acid dissociation constant is designated here by the symbol pK_a rather than the more precisely correct pK^1.

$$pH = pK_a + \log\frac{(\text{base})}{(\text{acid})}$$

$$\log\frac{(\text{base})}{(\text{acid})} = pH - pK_a$$

$$\frac{(\text{base})}{(\text{acid})} = \text{anti}\log\left(pH - pK_a\right)$$

The implications of the above equations are clear. Weak acids are readily absorbed from the stomach. Weak bases are not absorbed well; indeed, they would tend to accumulate within the stomach at the expense of agent in the bloodstream. Naturally, in the more alkaline intestine, bases would be absorbed better, acids more poorly.

It should be realized that although the principles outlined here are correct, the system is dynamic, not static. Molecules that are absorbed across the gastric or intestinal mucosa are removed constantly by blood flow; thus, simple reversible equilibrium across the membrane does not occur until the agent is distributed throughout the body.

Absorption from the stomach, as determined by direct measurements, conforms, in general, to the principles outlined above. Organic acids are absorbed well since they are all almost completely nonionized at the gastric pH; indeed, many of these substances are absorbed well since they are all almost completely nonionized at the gastric pH. Many of these substances are absorbed faster than ethyl alcohol, which had long been considered one of the few compounds that were absorbed well from the stomach. Strong acids whose pK_a values lie below 1, which are ionized even in the acid contents of the stomach, are not absorbed well. Weak bases are absorbed only negligibly, but their absorption can be increased by raising the pH of the gastric fluid.

As for bases, only the weakest are absorbed to any appreciable extent at normal gastric pH, but their absorption can be increased substantially by neutralizing the stomach contents. The quaternary cations, however, which are charged at all pH values, are not absorbed at either pH.

The accumulation of weak bases in the stomach by ion-trapping mimics a secretory process; if the drug is administered systemically, it accumulates in the stomach. Dogs given various drugs intravenously by continuous infusion to maintain a constant drug level in the plasma had the gastric contents sampled by means of an indwelling catheter. The results showed that stronger bases ($pK_a > 5$) accumulated in stomach contents to many times their plasma concentrations; the weak bases appeared in about equal concentrations in gastric juice and in plasma. Among the acids, only the weakest appeared in detectable amounts in the stomach. One might wonder why the strong bases, which are completely ionized in gastric juice, and whose theoretical concentration ratios (gastric juice/plasma) are very large, should nevertheless attain only about a 40-fold excess over plasma. Direct measurements of arterial and venous blood show that essentially all the blood flowing through the gastric mucosa is cleared of these agents; obviously, no more chemical can enter the gastric juice in a given time period than is brought there by circulation. Another limitation comes into play when the base pK_a exceeds 7.4; now a major fraction of the circulating base is cationic and a decreasing fraction is nonionized, so the effective concentration gradient for diffusion across the stomach wall is reduced.

The ion-trapping mechanism provides a method of some forensic value for detecting the presence of alkaloids (e.g., narcotics, cocaine, amphetamines) in cases of death suspected to be due to overdosage of self-administered drugs. Drug concentrations in gastric contents may be very high even after parenteral injection.

Absorption from the intestine has been studied by perfusing drug solutions slowly through rat intestine *in situ* and by varying the pH as desired. The relationships that emerge from such studies are the same as those for the stomach, the difference being that the intestinal pH is normally very near neutrality. As the pH is increased, the bases are absorbed better, that acids more poorly. Detailed studies with a great many drugs in unbuffered solutions revealed that in the normal intestine, acids with $pK_a > 3.0$ and bases with $pK_a < 7.8$ are very well absorbed; outside these limits the absorption of acids and bases falls off rapidly. This behavior leads to the conclusion that the "virtual pH" in the microenvironment of the absorbing surface in the gut is about 5.3; this is somewhat more acidic than the pH in the intestinal lumen is usually considered to be.

Absorption from the buccal cavity has been shown to follow exactly the same principles as those described for absorption from the stomach and intestine. The pH of human and canine saliva is usually about 6. Bases in people are absorbed only on the alkaline side of their pK_a, that is, only in the nonionized form. At normal saliva pH, only weak bases are absorbed to a significant extent.

Techniques of Oral Administration

There are three major techniques for oral delivery of drugs to test animals. The most common way is by gavage, which requires that the material be in a solution or suspension for delivery by tube to the stomach. Less common materials may be given as capsules (particularly to dogs) or in diet (for longer-term studies). Rarely, oral studies may also be done by inclusion of materials in drinking water.

Test materials may be administered as solutions or suspensions as long as they are homogeneous and delivery is accurate. For traditional oral administration (gavage), the solution or suspension can be administered with a suitable stomach tube or feeding needle ("Popper" tube) attached to a syringe. If the dose is too large to be administered at one time, it can be divided into equal subparts with 2–4 h between each administration; however, this subdivided dosing approach should generally be avoided.

Test chemicals placed into any natural orifice exert local effects and, in many instances, systemic effects as well. The possibility of systemic effects occurring when local effects are to be evaluated should be considered.

For routes of administration in which the chemical is given orally or placed into an orifice other than the mouth, clear instructions about the correct administration of the chemical must be provided. Many cases are known of oral pediatric drops for ear infections being placed into the ear, and vice versa (ear drops being swallowed) in humans. Errors in test-article administration are especially prevalent when a chemical form is being used in a nontraditional manner (e.g., suppositories that are given by the buccal route).

Administration of drug in a capsule is a common means of dosing larger test animals (particularly dogs). It is labor intensive (each capsule must be individually prepared, though robotic systems are now available for this), but capsules offer the advantages that neat drug may be used (no special formulation need be prepared, and the questions of formulation or solution stability are avoided), the dogs are less likely to vomit, and the actual act of dosing requires less labor than using a gavage tube. Capsules may also be used with primates, though they are not administered as easily.

Incorporation of a drug in the diet is commonly used for longer-term studies (particularly carcinogenicity studies, though the method is not limited to these). Dosing by diet is much less labor intensive than any other oral dosing methodology, which is particularly attractive over the course of a long (13-week, 1-year, 18-month, or 2-year) study.

The most critical factor to dietary studies is the proper preparation of the test chemical-diet admixtures. The range of physical and chemical characteristics of test materials requires that appropriate mixing techniques be determined on an individual basis. Standard practices generally dictate the preparation of a premix, to which is added appropriate amounts of feed to achieve the proper concentrations.

Dietary preparation involving liquid materials frequently results in either wet feed in which the test article does not disperse or formation of "gumballs" – feed and test material that form discernible lumps and chemical "hotspots." Drying and grinding of the premix to a free-flowing form prior to mixing the final diets may be required; however, these actions can affect the chemical nature of the test article.

Solid materials require special techniques prior to or during addition to diets. Materials that are soluble in water may be dissolved and added as described above for liquids. Non-water-soluble materials may require several preparatory steps. The test chemical may be dissolved in corn oil, acetone, or other appropriate vehicle prior to addition to the weighed diet. When an organic solvent such as acetone is used, the mixing time for the premix should be sufficient for the solvent to evaporate. Some solids may require grinding in a mortar and pestle with feed added during the grinding process.

Prior to study initiation, stability of the test chemical in the diet must be determined over a test period at least equivalent to the time period during which animals are to be exposed to a specific diet mix. Stability of test samples under the conditions of the proposed study is preferable. Labor and expense can be saved when long-term stability data permit mixing of several weeks (or a month) of test diet in a single mixing interval.

Homogeneity and concentration analysis of the test-article-diet admixture are performed by sampling at three or four regions within the freshly mixed diet (e.g., samples from the top, middle, and bottom of the mixing bowl or blender).

A variety of feeders are commercially available for rats and mice. These include various-sized glass jars and stainless steel or galvanized feed cups, which can be equipped with restraining lids and food followers to preclude significant losses of feed due to animals digging in the feeders. Slotted metal feeders are designed so that animals cannot climb into the feed, and they also contain mesh food followers to prevent digging.

Another problem sometimes encountered is palatability – the material may taste so strongly that animals will not eat it. As a result, palatability, stability in diet, and homogeneity of mix must all be ensured prior to the initiation of an actual study.

Inclusion in drinking water is rarely used for oral administration of human drugs to test animals, though it sees more frequent use for the study of environmental agents.

Physiochemical properties of the test material should be a major consideration in selecting of drinking water as a dosing matrix. Unlike diet preparation or preparation of gavage dose solutions and suspensions where a variety of solvents and physical processes can be utilized to prepare a dosable form, preparations of drinking water solutions are less flexible. Water solubility of the test chemical is the major governing factor and is dependent on factors such as pH, dissolved salts, and temperature. The animal model itself sets limitations for these factors (acceptability and suitability of pH and salt-adjusted water by the animals as well as animal environmental specifications such as room temperature).

Stability of the test chemical in drinking water under study conditions should be determined prior to study initiation. Consideration should be given to conducting stability tests on test chemical–drinking water admixtures presented to some test animals. Besides difficulties of inherent stability, changes in chemical concentrations may result from other influences. Chemicals with low vapor pressure can volatilize from the water into the air space located above the water of an inverted water bottle; thus, a majority of the chemical may be found in the "dead space," not in the water.

Certain drugs may be degraded by contamination with microorganisms. A primary source of these microorganisms is the oral cavity of rodents. Although rats and mice are not as notorious as the guinea pig in spitting back into water bottles, significant bacteria can pass via the sipper tubes and water flow restraints into the water bottles. Sanitation and sterilization procedures for water bottles and sipper tubes must be carefully attended to.

Many technicians may not be familiar with terms such as sublingual (under the tongue), buccal (between the cheek and gingiva), otic, and so on. A clear description of each of these nontraditional routes (i.e., other than gavage routes) should be discussed with technicians, and instructions may also be written down and given to them. Demonstrations are often useful to illustrate selected techniques of administration (e.g., to use an inhaler or nebulizer). Some chemicals must be placed by technicians into body orifices (e.g., medicated intrauterine devices such as Progesterset).

Buccal and Sublingual

While considered subsets of the oral route, these two routes have substantial differences from the downstream option. Their epithelial tissue surfaces allow for absorption of peptides and proteins (Caon et al., 2015) while avoiding both the acid and digestive enzyme–rich environments of the stomach and first-pass metabolism.

Because of the rich blood supply to the mucous membranes of the mouth (buccal cavity), many compounds can be absorbed through them. Absorption from the buccal cavity is limited to nonionized, lipid-soluble compounds. Buccal absorption of a wide range of aromatic and aliphatic acids and basic drugs in human subjects has been found to be parabolically dependent on log P, where P is the octanol-water partition coefficient. The ideal lipophilic character (log P_0) for maximum buccal absorption has also been shown to be in the range 4.2–5.5 (Lien et al., 1971). Compounds with large molecular weights are poorly absorbed in the buccal cavity, and, since absorption increases linearly with concentration and there is generally no difference between optical enantiomorphs of several compounds known to be absorbed from the mouth, it is believed that uptake is by passive diffusion rather than by active transport chemical moieties.

A knowledge of the buccal absorption characteristics of a chemical can be important in a case of accidental poisoning. Although an agent taken into the mouth will be voided immediately on being found objectionable, it is possible that significant absorption can occur before any material is swallowed. Unless voided, most materials in the buccal cavity are ultimately swallowed.

No significant absorption occurs in the esophagus and the agent passes on to enter the stomach. It is a common practice in safety assessment studies to avoid the possibility of buccal absorption by intubation (gavage) or by the administration of the agent in gelatin capsules designed to disintegrate in the gastric fluid.

Absorption from both is largely determined by area and duration of surface tissue contact, so maximizing these factors serves to promote systemic contact. Both also can be used to achieve delivery of therapeutic moieties over extended periods of time with the assistance of clever dosage forms (gels and films) which adhere to the surfaces and release drug as they dissolve and pass to the lower GI tract. As with the nasal route, a residual material that is not absorbed across the epithelial tissue of the oral cavity proceeds to the stomach and lower GI tract.

Lesser Routes

The lesser routes see less use clinically and therefore their clinical use is increasing in safety assessment. Six of these are briefly presented here.

Perocular and Intraocular Routes

The administration of drugs to the surface of the eyes is not a common route for systemic drug administration due to the small surface area exposed and the efficiency of the protective mechanisms (i.e., blink reflex and tears). Topical administration to the eye is most commonly for irritation, dry eye, and glaucoma. Chapter 8 specifically addresses formulation approaches for these. The limited volumes that can be administered combined with limitations on drug concentration in and physicochemical limitations on formulation are the principal limitations to utilization of the route. Drugs on the surface of the eye are subject to clearance in tears and the lacrimal drainage system. Drugs which gains access to the interior of the eye and the aqueous and vitreous humor are subject to clearance by absorption into the blood stream and can be physically impeded, leading to increased ocular pressure (glaucoma) and damage to the eye. As long as the epithelium of the eyes remains intact, it is impermeable to many molecules, but, if the toxicant has a suitable polar-nonpolar balance, penetration may occur (Kondrizer et al., 1959; Swan and White, 1942).

Holmstedt (1959) and Brown and Muir (1971) have reviewed perocular absorption of pesticides. More recently, Sinow and Wei (1973) have shown that the quarternary herbicide paraquat can be lethal to rabbits if applied directly to the surface of the eyes. Parathion, in particular, is exceedingly toxic when administered via the eye – a concern that must be kept in mind for the protection of pesticide applicators.

Drugs may also be administered directly into the eye (intraocular delivery) by either direct delivery or insertion of a delivery device (such as an intraocular stent – a permeable capillary tube containing drug formulation which slowly diffuses into the eye). This route allows delivery of larger amounts of therapeutics (and larger drug molecules) and a protracted period of administration.

Rectal Administration

Since a number of therapeutic compounds are given in the form of suppositories, an indication of the toxicity after rectal administration is sometimes required. Toxicity studies and initial drug formulations of such compounds are usually performed by the oral route, and the rectal formulation comes late in development and marketing. In view of the difference between laboratory animals and humans in the anatomy and microflora of the colon and rectum, animal toxicity studies late in drug development are of limited value. However, in a case where there is an indication of potential rectal hazard or bioavailability, the compound may be introduced into the rectum of the rat using an oral dosing needle to prevent tissue damage. To avoid the rapid excretion of the unabsorbed dose, anesthetized animals should be used and the dose retained with an inert plug or bung (such as a cork).

Drugs (and, therefore, test chemicals) are occasionally administered by rectum, but most are not as well absorbed here as they are from the upper intestine, though improved formulations are changing this (Prasanna et al., 2012). Aminophylline, used in suppository form for the management of asthma, is one of the few drugs routinely given in this way. Inert vehicles employed for suppository preparations include cocoa butter, glycerinated vehicles, gelatin, and polyethylene glycol. Because the rectal mucosae are irritated by nonisotonic solutions, fluids administered by this route should always be isotonic with plasma (e.g., 0.9% NaCl).

Vaginal Administration

Though not a common one, some materials do have routine exposure by this route (tampons, spermicides, douches, and antibiotics, for example), and, therefore, must be evaluated for irritation and toxicity by this route. The older preferred models used rabbits and monkeys (Eckstein et al., 1969), but, more recently, a model that uses rats has been developed (Staab et al., 1987). McConnell (1974) clearly described the limitations, particularly of volume of test material, involved in such tests.

Nasal Administration

A route that has gained increasing popularity of late for pharmaceutical administration in humans is the intranasal route. The reasons for this popularity are the ease of use (and, therefore, ready patient acceptance and high compliance rate), the high degree and rate of absorption of many substances (reportedly for most substances up to 1000 molecular weight; McMartin et al., 1987), and the avoidance of the highly acid environment in the stomach and first-pass metabolism in the liver (particularly important for some of the newer peptide moieties) (Attman and Dittmer, 1971). The only special safety concerns are the potential for irritation of the mucous membrane and the rapid distribution of administered materials to the CNS.

A number of means may be used to administer materials nasally – nebulizers and aerosol pumps being the most attractive first choices. Accurate dose administration requires careful planning, evaluation of the administration device, and attention to technique.

Implantation

Not generally considered as a route, using implants to deliver drugs has advanced from SC implanted osmatic pumps to pumps with refillable drug delivery reservoirs (both with the advantages of bypassing first-pass metabolism and allowing long-term drug administration at a set rate, reducing C_{max} systemic exposure levels), and now to drug-loaded matrices (cardiovascular stents, degradable matrices releasing bone growth stimulating peptides and proteins, and polymer matrices that are implanted into the target organ or tumor and release oncologic agents directly into the target organ or tissue). Imaging technologies have advanced to allow exact placement of implanted objects in desired locations.

For the surgically implanted delivery objects, the current trend is that they be biodegraded over time to remove the burden of either leaving them in place or future (usually more complicated) surgical removal. Note that while initial drug-coated cardiovascular stents were "bare metal" in composition, the current generation are drug coated or loaded biodegradable polymers.

Additionally, for chronic treatment drugs, repeat daily parenteral administration (with associated discomfort or trauma) is no longer required.

A nonclinical assessment disadvantage is that each of these limits which animal models can be used and requires more careful and experienced administration procedure.

Route Comparisons and Contrasts

The first part of this chapter described, compared, and contrasted th various routes used in toxicology and presented guidelines for their use. There are, however, some exceptions to the general rules that the practicing toxicologist should keep in mind.

The relative ranking of efficacy of routes that was presented earlier in this chapter is not absolute; there can be striking exceptions. For example, though materials are usually much quicker acting and more potent when given by the oral route than by the dermal, this is not always the case. In the literature, Shaffer and West (1960) reported that tetram as an aqueous solution was more toxic when applied dermally than when given orally to rats. LD_{50}s reported were as follows [LD_{50} (mg kg^{-1}) of tetram; 95% confidence limits]:

Rat	Oral (mg k^{-1})	Percutaneous (mg k^{-1})
Male	9 (7–13)	2 (1–3)
Female	8 (6–11)	2 (1–3)

The author has in the past experienced this same phenomenon. Several materials that were found to be relatively nontoxic orally were extremely potent by the dermal route (differences in potency of more than an order of magnitude have been seen at least twice).

A general rule of toxicology, applicable to routes and vehicles, should be presented here:

Vehicles can mask the effects of active ingredients. Particularly for clinical signs, attention should be paid to the fact that a number of vehicles (e.g., propylene glycol) cause transient neurobehavioral effects that may mask similar short-lived (though not necessarily equally transient and reversible) effects of test materials.

Volume Limitations by Route

In the strictest sense, absolute limitations on how much of a dosage form may be administered by any particular route are determined by specific aspects of the test species or dosage form. But there are some general guidelines (determined by issues of humane treatment of animals, accurate delivery of dose and such) that can be put forth. These are summarized in Table 4.5. The Appendix and Chapter 8, "Fundamentals of Nonclinical Formulation," should, of course, be checked to see if there is specific guidance due to the characteristics of a particular vehicle.

TABLE 4.5

Volume Guidelines for Administration of Compounds by Route of Administration to Laboratory Animals[†]

Species	Intranasal µL/nostril	Gavage mL/kg*		IV Bolus** mL/kg*		IP*** mL/kg*		SC**** mL/kg*		IM mL/kg*		ID mL/site*		IV Infusion** mL/kg		Continuous Infusion		Intra Vaginal mL	Intra Vitreal µL/eye	Intra Articular mL/joint		Perivenous/ Periarterial mL/ear
	Vol.	Ideal	Max	Ideal	Max	Ideal	Max	Ideal	Max	Ideal	Max	Ideal	Max	Vol.	Rate mL/min	Max[n] mL/kg/hr	Cath. Maint. Rate mL/hr	Max	Vol.	Joint	Vol.	Max
Mouse	25	10	40[a]	5	25	5–10	50	1–5	20[f]	0.1	1[g,h]	0.1	0.1	50	1	NIA*****		NIA	2	NIA	NIA	NIA
Rat	50	10	20[b]	1	20	5–10	20	1	20[f]	0.1	10[k]	0.1	0.1	50	1	5	0.1–0.5	0.35	5	Stifle	0.1	NIA
																				Tarsal	0.05	
Guinea pig	100	10	30	1	5	1–5	20	1–5	10[f]	0.1	0.5–1	0.1	0.1	10	1	NIA	NIA	NIA	NIA	NIA	NIA	NIA
Rabbit	200	1	20[c]	1[d]	10[d]	3	5[e]	1–2.5	10[f]	0.1–0.5	1	0.1	0.1	20	1	NIA	NIA	2.0	100[m]	Stifle	0.5	0.5
Dog	500	5	20	1	10	3	5	0.5	2[f]	0.1–0.25	1[i,j]	0.1	0.1	20	5	5	1–2	2.0	100	Stifle	1	NIA
Non-human primate	200	5	10	1	10	3	5	0.5	2[f]	0.1–0.5	1[i,j]	0.1	0.1	20	1	5	1–2	NIA	50	NIA	NIA	NIA

(*Continued*)

TABLE 4.5 (*Continued*)

Volume Guidelines for Administration of Compounds by Route of Administration to Laboratory Animals

	Route																				
Species	Intranasal µL/nostril	Gavage mL/kg*		IV Bolus** mL/kg*		IP*** mL/kg*		SC**** mL/kg*		IM mL/kg*		ID mL/site*		IV Infusion** mL/kg		Continuous Infusion		Intra Vaginal mL	Intra Vitreal µL/eye	Intra Articular mL/joint		Perivenous/ Periarterial mL/ear
	Vol.	Ideal	Max	Ideal	Max	Ideal	Max	Ideal	Max	Ideal	Max	Ideal	Max	Vol.	Rate mL/min	Max[n] mL/kg/hr	Cath. Maint. Rate mL/hr	Max	Vol.	Joint	Vol.	Max
Mini-swine	NIA	NIA	NIA	1	10	1	5	1	3[f]	0.25	0.5[j]	0.1	0.1	10	1	5	1–2	NIA	NIA	NIA	NIA	NIA

Source: Adapted with changes from Gad SC et al., *Int J Toxicol.*, 25(6), 499–521, 2016.

[a] To accommodate a larger volume, the dose may be divided over time (e.g., 20 mL/kg administered four times per day to reach a total of 80 mL/kg in a 24 hour period).

[b] To accommodate a larger volume, the dose may be divided over time (e.g., 10 mL/kg administered four times per day to reach a total of 40 mL/kg in a 24-hour period).

[c] Rabbits should not be fed prior to administration. Rabbits should be fed after the completion of dose administration.

[d] These volumes may also be used for intra-arterial injection.

[e] Not often used.

[f] If volumes greater than those cited above are used, the volume must be divided over multiple sites.

[g] May be used if divided over multiple sites and alternating legs, maximum of 5 sites per leg. Final volume not to exceed 0.10 mL.

[h] 0.05 mL total volume limit per site.

[i] 3 mL total volume limit per site.

[j] 5 mL total volume limit per site.

[k] Up to 20 mL/kg if divided over multiple sites.

[l] 3 mL total volume limit.

[m] Repeat dose 67 µL in the rabbit.

[n] Solution properties such as tonicity, pH, etc., need to be taken into account when determining the volume that may be infused IV. Animal health must also be taken into consideration, such as kidney function and cardiovascular function. These systems need to be normal to handle increased fluid volumes.

*Single dose per day except where noted otherwise

**Solution properties such as tonicity, pH, etc. need to be taken into account when approaching the volume limits or determining the volume to be infused IV. The recommended working range for pH is 4.5–8.0. The order of degree of tolerance of pH for different dosing routes is oral>intravenous>intramuscular>subcutaneous>intraperitoneal. Animal health must also be taken into consideration, such as kidney function and cardiovascular function. These systems must be normal to handle increased fluid volumes.

***When administering a solution IP, the viscosity, concentration, tonicity, and pH of the solution need to be taken into account.

****When administering a solution SC, the concentration, tonicity, and pH of the solution must be taken into account.

*****NIA – No information currently available.

References

Arnauld C. (2016) Drug delivery: Depositing drugs straight to the cytoplasm. *C&EN*, October 5, 2016.

Attman PL and Dittmer DS. (Eds.) (1971) *Respiration and Circulation*. Bethesda, MD: FASEB, pp. 56–59.

Avis KE. (1985) Parenteral preparations. In: Gennaro AR (Ed.) *Remington's Pharmaceutical Sciences*. Easton, PA: Mack Publishing Company, pp. 1518–1541.

Ballard BE. (1968) Biopharmaceutical considerations in subcutaneous and intramuscular drug administration. *J Pharm Sci*, 57:357–378.

Bates TR and Gibaldi M. (1970) Gastrointestinal absorption of drugs. In: Swarbrick J (Ed.) *Current Concepts in the Pharmaceutical Sciences: Biopharmaceutics*. Philadelphia, PA: Lea & Febiger, pp. 58–99.

Bland HI and Scheuplein RJ. (1964) The epidermal barrier. In: Rook A and Champion RH (Eds.) *Progress in the Biological Sciences in Relation to Dermatology*, Vol. 2. Cambridge: Cambridge University Press.

Borowitz JL, Moore PF, Yim GKW, Miya TS. (1971) Mechanism of enhanced drug effects produced by dilution of the oral dose. *Toxicol Appl Pharmacol*, 19:164–168.

Brandau R and Lippold BH. (1982) *Dermal and Transdermal Absorption*. Stuttgart: Wissenschaftliche, Verlagsgesellschaft mbH.

Bronaugh RL and Maibach HI. (1985) *Percutaneous Absorption*. New York, NY: Marcel Dekker.

Brown VK and Muir CMC. (1971) Some factors affecting the acute toxicity of pesticides to mammals when absorbed through skin and eyes. *Int Pest Control*, 13:16–21.

Brown VK. (1980) *Acute Toxicity in Theory and Practice*. New York, NY: John Wiley & Sons.

Caon T, Jin L, Simões CM, Norton RS, Nicolazzo JA. (2105) Enhancing the buccal mucosal delivery of peptide and protein therapeutics. *Pharm Res*, 32(1):1–21.

Eckstein P, Jackson MCN, Millman N, Sobrero AJ. (1969) Comparison of vaginal tolerance tests and spermicidal preparations in rabbits and monkeys. *J Reprod Fertil*, 20:85–93.

Gad SC, Cassidy CD, Aubert N, Spainhour B, Robbe H. (2006) Nonclinical vehicle use in studies by multiple routes in multiple species. *Int J Toxicol*, 25(6):499–521.

Gad SC, Spainhour CB, Shoemake C, Stackhouse Pallman DR, Stricker-Krongrad A, Downing PA, Seals RE, Eagle LA, Polhamus K, Daly J. (2016) Tolerable levels of nonclinical vehicles and formulations used in studies by multiple routes in multiple species with notes on methods to improve utility. *Int J Toxicol*¸ 35(2):95–178.

Gad SC. (2007) *Target Organ Toxicity: The GI Tract*. Philadelphia, PA: Taylor & Francis.

Gad SE. (2016) *Drug Safety Evaluation*, 3rd Ed. Hoboken, NJ: Wiley.

Garramone JP. (1986) *Vascular Access Port Model SLA Users Manual*. Skokie, IL: Norfolk Medical Products.

Gray JE. (1978) Pathological evaluation of injection injury. In: Robinson J (Ed.) *Sustained and Controlled Release Drug Delivery System*. New York, NY: Marcel Dekker, p 35105.

Hogben CAM, Tocco DJ, Brodie BB, Schranker LS. (1959) On the mechanism of intestinal absorption of drugs. *J Pharmacol Exp Therap*, 125(4):275–282.

Holmstedt B. (1959) Pharmacology of organophosphorus cholinesterase inhibitors. *Pharmacol Revs,* 11:567–688.

Houston JB, Upshall DG, Bridges JW. (1974) The re-evaluation of the importance of partition coefficients in the gastrointestinal absorption of nutrients. *J Pharmacol Exptl Ther,* 189:244–254.

Kondrizer AA, Mayer WH, Zviblis P. (1959) Removal of sarin from skin and eyes. *Arch Industr Health,* 20:50–52.

Kruger S, Greve DW, Schueler FW. (1962) The absorption of fluid from the peritoneal cavity. *Arch Int Pharmacodyn,* 137:173–178.

Lee Y-C, Zocharski PD, Samas B. (2003) An intravenous decision tree for discovery compound formulation development. *Int J Pharm,* 253(1–2):111–119.

Lien E, Koda RT, Tong GL. (1971) Buccal and percutaneous absorptions. *Drug Intel Clin Pharm,* 5:38–41.

Lukas G, Brindle SD, Greengard P. (1971) The route of absorption of intraperitoneally administered compounds. *J Pharmacol Exptl Ther,* 178:562–566.

Marzulli FN. (1962) Barriers to skin penetration. *J Invest Derm,* 39:387–393.

McConnell RF. (1974) Special requirements for testing spermicides. *Acta Endocrinol Suppl (Copenh),* 185:373–386.

McMartin C, Hutchinson LEF, Hyde R, Peters GE. (1987) Analysis of structural requirements for the absorption of drugs and macromolecules from the nasal cavity. *J Pharm Sci,* 76:535–540.

Medved LL and Kundiev YI. (1964) On the methods of study of penetration of chemical substances through the intact skin (Russian). *Gig Sanit,* 29:71–76.

Mehta DC, Short JL, Hilmer SN, Nicolazzo JA. (2015) Drug access to the central nervous system in Alzheimer's disease: Preclinical and clinical insights. *Pharm Res,* 32: 819–839.

Prasanna L, Deepthi B, Rao RN. (2012) Rectal drug delivery: A promising route for enhancing drug administration. *Asian J Res Pharm Sci,* 2:143–149.

Ritschel WA, Siegel EG, Ring PE. (1974) Biopharmaceutical factors influencing LDP: Pt. 1 Viscosity. *Arzneim.-Forsch,* 24:907–910.

Scheuplein RJ. (1965) Mechanism of percutaneous absorption. (i) Routes of penetration and the influence of solubility. *J Invest Derm,* 45:334–346.

Scheuplein RJ. (1967) Mechanism of percutaneous absorption. (ii) Transient diffusion and the relative importance of various routes of skin penetration. *J Invest Derm,* 48:79–88.

Schou J. (1971) Subcutaneous and intramuscular injection of drugs. In: Brodie BB and Gillette JR (Eds.) *Handbook of Experimental Pharmacology,* Ch. 4. Berlin: Springer-Verlag.

Schranker LS, Shore PA, Brodie BB, Hogben CAM. (1957) Absorption of drugs from the stomach, I. The rat. *J Pharmacol Exp Therap,* 120:528.

Schranker LS. (1960) On the mechanism of absorption of drugs from the gastrointestinal tract. *J Med Pharm Chem,* 2:343–359.

Schwartz ML, Meyer MB, Covino BG, Narang RM. (1974) Antiarrhythmic effectiveness of intramuscular lidocaine: Influence of different injection sites. *J Clin Pharmacol,* 14:77–83.

Shaffer CB, West B. (1960) The acute and subacute toxicity of technical *o,o*-diethyl *s*-2-diethyl-aminoethyl phosphorothioate hydrogen oxalate (TETRAM). *Toxicol Appl Pharmacol,* 2:1–13.

Sinow J and Wei E. (1973) Ocular toxicity of paraquat. *Bull Environ Contam Toxicol,* 9:163–168.

Staab RJ, Palmer MA, Auletta CS, Blaszcak DL, McConnell RF. (1987) A relevant vaginal irritation/subacute toxicity model in the rabbit and ovariectomized rat. *Toxicologist,* 7A:1096.

Strickley RG. (2004) Solubilizing excipients in oral and injectable formulations. *Pharm Res,* 21(2):201–230.

Strickley RG. (2008) Formulation in drug discovery. In: Macor JE (Ed.) *Annual Reports in Medicinal Chemistry,* Vol 43. London: Academic Press, pp. 419–451.

Swan KC and White NG. (1942) Corneal permeability: (1) Factors affecting penetration of drugs into the cornea. *Amer J Ophthalmol,* 25:1043–1058.

Ubels J, Clousing D, Van Haisma T, Hong B-S, Stauffer P, Asgharian B, Meadows D. (2004) Pre-clinical investigation of the efficacy of an artificial tear solution containing hydroxypropyl-guar as a gelling agent. *Curr Eye Res,* 28(6):437–444.

US FDA (2015) *Nonclinical Safety Evaluation of Reformulated Drug Products and Products Intended for Administration by an Alternate Route.* Silver Spring, MD: Center for Drug Evaluation and Research (CDER).

Warheit DB, Reed KL, Webb TR. (2003) Pulmonary toxicity studies in rats with triethoxyoctylsilane (OTES)-coated, pigment-grade titanium dioxide particles: Bridging studies to predict inhalation hazard. *Exp Lung Res,* 29(8):593–606.

Wilson FA and Dietschy JM. (1974) The intestinal unstirred layer: its surface area and effect on active transport kinetics. *Biochem Biophys Acta,* 363:112–126.

Winne D. (1978) Dependence of intestinal absorption *in vivo* in the unstirred layer. *Nauyn-Schmiedelberg's Arch Pharmacol,* 304:175–181.

Zesch A. (1988) Adverse reactions of externally applied drugs and inert substances. *Derm Beruf Umwelt,* 36(4):128–133.

5

Delivery Systems

Drug delivery systems are composed of the nonformulation and novel formulation components used to deliver a drug product that are intended to add specificity to target delivery and improve the efficiency of drug delivery (Vogelson, 2001). These systems have been around since ancient Greece and include such systems as hypodermic syringes, although the first actual recorded medical use was by Christopher Wren in 1656 (Ng, 2015). The last 30 years have seen an explosion in the development of a variety of new approaches. For most of these, what we are speaking of would generally be called devices (Touitou and Barry, 2007; Ng, 2015) (Table 5.1). Until the last 15 years delivery systems have been treated, regulated, and evaluated for safety (biocompatibility) as such; more recently, drugs and devices together have been treated as "combination products" and evaluated as such both for compatibility and biocompatibility of the device component (Gad, 2016).

Here, as the principal focus of this book is applications for nonclinical studies, our discussion will favor these aspects.

Purposes

First, drug delivery systems or devices are primarily, except for syringes and infusion sets, intended to facilitate more specificity in getting to the target organ or delivery of therapeutic agents to a specific system. While this had meant broad specificity in the past (circulating blood for hypodermics and rectal surfaces for suppositories), more recent approaches seek even greater target specificity. Nonclinical scientists tend to overlook these approaches, perhaps because of resistance to change or a lack of sponsor requests to a select CRO.

Second is the desire to be able to deliver a more precisely controlled dose or a dose over a more controlled or extended period of time of administration. Included in this approach is the ability to deliver either much more or much less of a drug.

Third is effecting delivery of a therapeutic molecule while at the same time avoiding or minimizing drug degradation by chemical or metabolic factors.

TABLE 5.1

Delivery Systems and Delivery Devices Summary

Device/System	Routes for Use	Animal Model Species Used In	Maximum Volumes for Delivery	Advantages	Disadvantages
Syringes/infusion pumps (pumps can be implanted, such as the Alzet osmotic)	Any parenteral route – IV, intraperitoneal, intrathecal	All common species	Limited by region of delivery and species	Precise regional and volume control of delivery, though limits as to access of some regions and to duration of administration.	Tissue/trauma along path of placement. Proper placement requires skill, patience and for some routes (such as intrathecal) time.
Suppositories	Rectal	Usually rabbits, dogs, and pigs	3–5 mL for rabbits and dogs	Direct rectal administration by passes. First pass, metabolism, may minimalize systemic exposure	May require extra steps to retain in place.
Tampons	Vaginal	Usually dogs and rabbits; pigs are possible	2 mL for rabbits 5 mL for dogs or pigs	Limited doses, vaginal surface only.	Tampons may be physically abrasive and must be removed physically.
Vascular stents	IV but may also be in other natural channels	Pigs, dogs	Limited by stent size and target tissue	Direct delivery to vascular tissue at controlled rate – minimalized systemic exposure.	Unless retractable, difficult to remove.
Buccal and nasal strips	Buccal and nasal epithelium	Hamsters, dogs	One gram	By pass. First pass, metabolism. Better bioavailability to CNS.	Limited duration of applicability. Some oral exposure also.
Transdermal patches	Dermal	Rats, mice, dogs, rabbits, pigs	5 mL of formulation	Delivery can be metered over time. Limited in extent and by passes. First pass metabolism.	Generally only useful for small molecules. Requires use of formulation.

(Continued)

TABLE 5.1 (*Continued*)

Delivery Systems and Delivery Devices Summary

Device/System	Routes for Use	Animal Model Species Used In	Maximum Volumes for Delivery	Advantages	Disadvantages
Ocular stents	Inside of eye	Rats, rabbits, dogs	Less than 0.1 mL per unit	Direct and extended delivery to the eye.	Limited quantities. Requires insertion into eye.
Liposomes	IV but could be injected parenterally into other tissues	No limit	Route determined	Protects therapeutic from metabolism and clearance. If coupled with a targeting molecule, can be targeted to selected tissues.	May evoke immunologic responses.
Polymer systems	Any parenteral or oral route where delivery formulation/object can be placed	No limit	Route determined	Targets delivery. Placement protects therapeutic from metabolism. Reduced clearance.	Technique required for specific tissue deliveries. Clearance of polymer may be an issue.
Nanoparticles	Any systemic route (or topically)	No limit	As administration route dictates	Increased surface areas presented to body and can have tissue selectivity effects.	Below ~10μM diameter may have cytotoxicity.
Capsules	Oral	Rats, mice, dogs, primates, pigs	1 mL per capsule	Bypass stomach. Protect from acid environment.	Must be either filled or manufactured.
Iontophoresis	Dermal	All species	5 mL of formulation	By passing first pass metabolism while also improving through the skin	Effectiveness hampered by skin thickness.
Plasmonic therapy (using sound oscillation of the skin)	Dermal	All species	Such as to cover up to 10% of BSA (body surface area). In general, 5 mL.	Improved systemic absorption through intact skin. Allow absorption of metals.	Potential disruption of normal structure.

(*Continued*)

TABLE 5.1 (***Continued***)

Delivery Systems and Delivery Devices Summary

Device/System	Routes for Use	Animal Model Species Used In	Maximum Volumes for Delivery	Advantages	Disadvantages
Infusion pumps	Most commonly IV, but can be any other parenteral	All species, but use in mice and hamsters uncommon	As administration route dictates	Allows delivery of larger volumes spread over periods of time.	Catheter or port must be put in place if more than a day.
Microneedle arrays (100–1000 μm in length)	Dermal	All species	Depends on duration of administration	Improved systemic bioavailability; controlled delivery rate.	Limits on flow rates.
Electro perforation	Dermal	Rabbits, dogs, minipigs	NA	Improved passage through the skin.	Some local trauma at site.
Sonophoresis	Dermal	Rats, rabbits, dogs, minipigs	NA	Improved passage through the skin.	Possible effects on test material.

Sources: Gad SC, *Drug Safety Evaluation*, Wiley, Hoboken, 2016; Vogelson CT, *Mod Drug Discovery*, 4, 49–50, 52, 2001; Washington N et al., *Physiological Pharmaceutics – Barriers to Drug Absorption*, Taylor and Francis, New York, 2001.

All three of these considerations are desirable in both nonclinical and clinical applications.

Examples

Syringes: Simple pressure used to push fluid in a rigid tube and then through a delivery needle.

Iontophoresis: A physical process in which ions flow diffusively in a medium driven by an applied electric field across or through the skin.

Suppositories: Lipid-based solid formulation containing the active therapeutic agent.

Nasal strips: Solid gum or polymeric strip that adheres to epithelium inside the nose to release a drug over a period of time.

Transdermal patches: Occlusive patches with a solid reservoir that contains and slowly releases/elutes the drug.

Ocular stents: Permeable polymer or glass rods filled with drug formulation that releases formulation over a period of time (Thassu and Chader, 2013).

Liposomes: Lipoprotein encapsulating therapeutics (particularly proteins).

Polymeric systems: Two new developments are the dendrimers (highly branched, globular, synthetic macromolecules) and modified buckyballs. Together with hydrogels, these agents are tailored to provide targeted delivery. The dendrimers form small micelles, which transport small molecules within their matrices or act as hubs for covalent bonding to drug molecules, extending like dendrites. In this way, they can shepherd high concentrations of drugs to specific targets. Buckyballs are cage-like molecules of fullerenes (carbon form). They are robust and can carry radioactive drugs to specific targets. Research is currently directed at using these buckyballs as delivery systems for the treatment of cancer.

Nanoparticles: Hydrogels are 3D cross-linked polymer networks. They can withstand acidic conditions and release the entrapped drug molecules. Purdue University researchers have used a poly[methacrylic acid-g-poly(ethylene glycol)] hydrogel to encapsulate insulin, which can be released by a pH trigger.

There are a wide range of approaches to clinical drug products and delivery – that is, the actual final form of the drug as administered to patients (Gad, 2014). The key points on the major classes of these are summarized as follows. It should be noted that if a delivery device is used, that device itself must also be reviewed and approved by the Food and Drug Administration.

Liquid Formulation

- These are the most widely used dosage forms in clinical practice.
- These are mainly based on aqueous formulations.
- Humidifying effect is a convenience due to the drying of mucous membranes owing to allergic and chronic diseases.
- Major drawback is the limitation of microbiological stability.
- Reduced chemical stability of the drug and short residence time in the nasal cavity are other disadvantages.
- Deposition site and deposition pattern are dependent on delivery device, mode of administration, and the physicochemical properties of formulation.
- Preparations depend on whether administered for local or systemic application.
- Patient compliance, cost effectiveness, and risk assessment are concerns.

Instillation and Rhinyle Catheter

- Catheters are used for delivery to a defined region.
- The combination of an instillation catheter to a Hamilton threaded plunger syringe has been used in order to compare the deposition of drops, nebulizer, and sprays in rhesus monkeys.
- These are used only for experimental studies and not for commercial clinical products.

Drops

- This is one of the oldest delivery systems for natural channels such as the nose.
- Low-cost devices are utilized.
- Manufacture is facile.

- Disadvantages are related to microbiological and chemical stability.
- The delivered volume cannot be controlled.
- Formulation can be easily contaminated by a pipette, the usual delivery device.

Powder Dosage Forms

- Dry powders are less frequently used in nasal drug delivery than inhalation.
- Major advantages include lack of need for preservatives and improved drug stability.
- This form prolongs retention times in the nasal region when compared to solutions.
- The addition of bioadhesive excipients results in further decreased clearance rates.
- Nasal powders may increase patient compliance, especially for children, if smell and taste of drug are otherwise unacceptable.

Insufflators and Mono-Dose Powder Inhaler

- Many insufflators work with predosed powder doses in capsules.
- The use of a variety of gelatin capsules enables the filling and application of different amounts of powder.
- In a mono-dose powder inhaler, pushing a piston results in a precompression of air in chamber.
- The piston pierces a membrane, and the expanding air expels air into the nostrils.

Pressurized MDIs

- These are manufactured by suspending the drug in liquid propellants with the aid of surfactants.
- Physicochemical compatibility between the drug and propellants must be evaluated.

- A complex approach – phase separation, precipitation, crystal growth, polymorphism, dispersibility, and the absorption of drug influence drug particle size, dose distribution, and deposition pattern.
- The advantages of this approach include the following:
  - Portability, small size, availability over a wide dose range, dose consistency and accuracy, and protection of the contents
- The disadvantages include the following:
  - Nasal irritation by propellants and depletion of ozone layer by CFCs

Nasal Gels

- Nasal administration of gels can be achieved by precompression pumps.
- The deposition of gel in the nasal cavity depends on the mode of administration, due to its viscosity and poor spreading properties.
- Nasal gels containing vitamin B_{12} for systemic administration are available in the marketplace

Patented Nasal Formulations

- West Pharma developed a nasal technology (ChiSys) based on the use of chitosan as an absorption enhancer.
- Chitosan is a natural polysaccharide with bioadhesive properties.
- It prolongs the retention time of the formulation in the nasal cavity.
- It may facilitate absorption through promoting paracellular transport.

Table 5.1 provides a broad overview of current delivery device systems.

The optimal selection of delivery systems or devices requires a complete and thorough understanding of the target tissue/organ and any intermediary tissues involved in the direct administration of the therapeutic agent.

One could also consider techniques for enhancing absorption, such as tape stripping to remove the barrier layer before topical cutaneous application (Tregear and Dirnhuber, 1962). Also, there is the use of ballistic methods (such as powderject) to propel drug through the stratum corneum and into the underlying tissues (Kendall and Wrighton-Smith, 2004).

References

Gad SC. (2014) Safety assessment of therapeutic agents administered by the respiratory routes. In: H Salem and SA Katz (Eds.) *Inhalation Toxicology*, 3rd Ed. Boca Raton, FL: CRC Press, pp. 537–579.

Gad SC. (2016) *Drug Safety Evaluation*, 3rd Ed. Hoboken, NJ: Wiley.

Kendall MT, Wrighton-Smith P. (2004) Intradermal ballistic delivery of microparticles into excised human skin for pharmaceutical applications. *J Biomech*, 37:1733.

Ng R. (2015) *Drugs: From Discovery to Approval*, 3rd Ed. Hoboken, NJ: Wiley Blackwell, p. 182.

Thassu D, Chader GJ. (2013) *Ocular Drug Delivery Systems: Barriers and Application of Nanoparticulate Systems*. Boca Raton, FL: CRC Press.

Touitou E, Barry BW. (2007) *Enhancement in Drug Delivery*. Boca Raton, FL: CRC Press.

Tregear RT, Dirnhuber P. (1962) The mass of keratin removed from the stratum corneum by stripping with adhesive tape. *J Invest Dermatol*, 38:375.

Vogelson CT. (April, 2001) Advances in drug delivery systems. *Mod Drug Discovery*, 4:49–50, 52.

Washington N, Washington C, Wilson CG. (2001) *Physiological Pharmaceutics – Barriers to Drug Absorption*, 2nd Ed, New York, NY: Taylor and Francis.

6

Dosing Regimens as a Means of Improving Drug Delivery and Tolerance

A dosing regimen is a schedule of doses of a therapeutic over a unit of time (Segen, 2012; Putnick and Herzyk, 2014). It may also refer to what times of day a drug is administered, or whether doses are of equal amount (or if there is a loading then subsequent maintenance doses administered). Selection of regimen seeks to achieve several goals – achieving greater systemic exposure, avoiding having to administer too large a volume of formulation (or dosage form), and keeping exposure levels (systemic and adverse tissue) in the optimal range. Additionally, regimens (with multiple/periodic administrations) serve to ensure longer and more consistent systemic and/or target tissue exposure (Kruger-Thiemer, 1966).

The PK/TK analysis of multiple dosing regimens, either nonclinically or clinically, is a bit more complex and requires careful thought as to sampling times and patterns (Williams, 1992), but understanding both these administration/exposure relationships and what the MEC (minimum effective concentration) and MTC (maximum tolerated concentrations) are for the drug and for the species in question is crucial.

While in clinical medicine, regimens may be personalized, this does not service the needs or resources of nonclinical evaluations. It may be the case, however, that regimens may be different for different populations (such as age groups or genders in nonclinical applications).

Regimen is little discussed in the nonclinical or toxicology literature. A major reason is the stated preference of regulatory authorities that the route and regimen or therapeutic administration mirror that of intended clinical practice *as closely as possible*.

But, just as the stricture on route has exceptions (the best example being that drugs intended for dermal administration should also be evaluated by a route that achieves a significant systemic exposure – something dermal routes frequently do not achieve), there are likewise some very practical reasons for varying regimen:

1. Due to limits on concentration of an active in formulation, it may not be possible to achieve the desired or required systemic exposure levels to evaluate safety at supraclinical levels (or even clinical levels).

2. Likewise, clearance of a therapeutic agent from the plasma/body of an animal model may be so rapid, particularly at higher concentration levels, that adequate systemic exposure (particularly in terms of AUC) cannot be achieved in a single daily administration (here, obviously, the oral route is the route of administration).
3. Most short-term toxicity (which may not be relevant to clinical exposure) is due to the C_{MAX} level systemically exceeding some threshold level and overloading a physiologic response (such as cardiovascular or renal function) or triggering an immunological response. For therapeutics intended to treat short-term conditions, this may serve to allow determination of a safety margin (or therapeutic index) and therefore be relevant to assessing potential clinical safety.

   But for many drugs, it is not – the C_{MAX} is a reflection of the need to achieve supratherapeutic exposure levels to meet regulatory expectations. In these cases, dividing the dose (either physically by multiple administrations per day or over a time course) by administering a parenteral (particularly intravenous dose) by infusion over a period of time rather than by a single short-term bolus administration.
4. It may also be the case that an initial higher ("loading") dose is administered to achieve either desired therapeutic levels or to saturate tissue reservoirs (depots) in the body. Eight subsequently lesser doses are administered to maintain desired levels in plasma or reservoir tissues, or desired occupancy levels in target receptors.

Generally the regimen issues for vaccines or immunologically functioning modifying biologics have different rationales – for vaccines, multiple administrations separated by sufficient time intervals to achieve a desired level of indication of an adaptive immune response, for other immune active biologics, to achieve sufficient (but not excessive) levels of increased or suppressed immune response.

References

Kruger-Thiemer E. (1966) Formal theory of drug dosage regimen. *I J Theo Biol*, 13:212–235.

Putnick LM, Herzyk DJ. (2014) *Nonclinical Development of Novel Biologics, Biosimilar, Vaccines and Specialty Biologics*. New York, NY: Elsevier, pp. 232–233.

Segen JC. (2012) *Segens Medical Dictionary*. Boca Raton, FL: CRC Press.

Williams RL. (1992) Dosage regimen design: Pharmacodynamic considerations. *J Clin Pharmacol*, 32:597–602.

7

The Utility of Circadian Oscillations in Formulation and Drug Development

Background and Introduction

Chronobiology is a field of biology that examines periodic (cyclic) phenomena in living organisms and the adaptation of these organisms to solar- and lunar-related rhythms. These cycles are known as biological rhythms. The term "chronobiology" comes from the ancient Greek χρόνος (chrónos, meaning "time"), and biology, which pertains to the study, or science, of life. The related terms "chronomics" and "chronome" have been used in some cases to describe either the molecular mechanisms involved in chronobiological phenomena or the more quantitative aspects of chronobiology, particularly where comparison of cycles between organisms is required. Chronobiological studies include but are not limited to comparative anatomy, physiology, genetics, molecular biology, and behavior of organisms within biological rhythms mechanics. Other aspects include development, reproduction, ecology, and evolution.

The variations of the timing and duration of biological activity in living organisms occur for many essential biological processes. These occur (a) in animals (eating, sleeping, mating, hibernating, migration, cellular regeneration, etc.), and (b) in plants (leaf movements, photosynthetic reactions, etc.), and in microbial organisms such as fungi and protozoa biochemical and metabolic processes). They have even been found in bacteria, especially among the cyanobacteria (e.g., blue-green algae). The most important rhythm in chronobiology is the circadian rhythm, a roughly 24-hour cycle demonstrated in the physiological processes in all these organisms. The term "circadian" comes from the Latin circa, meaning "around" and dies, "day", meaning "approximately a day." It is regulated by circadian clocks.

Knowledge of these patterns (for the specific species in question) facilitates optimal drug delivery in terms of bioavailability and achieving better specificity in delivery.

The circadian rhythm can further be broken down into routine cycles during the 24-hour day:

- Diurnal, which describes organisms active during daytime
- Nocturnal, which describes organisms active in the night
- Crepuscular, which describes animals primarily active during the dawn and dusk hours (e.g., skunks, white-tailed deer, woodchucks)

While circadian rhythms are defined as being endogenously regulated, other biological cycles may be regulated by exogenous signals. In some cases, multi-trophic systems may exhibit rhythms driven by the circadian clock of one of the members (which may also be influenced or reset by external factors such as light). The endogenous plant cycles may regulate the activity of the bacterium by controlling the availability of plant-produced photosynthate.

Many other important cycles are also studied, including the following:

- Infradian rhythms, which are cycles longer than a day, such as the annual migration or reproduction cycles found in certain animals or the human menstrual cycle.
- Ultradian rhythms, which are cycles shorter than 24 hours, such as the 90-minute REM cycle, the 4-hour nasal cycle, or the 3-hour cycle of growth hormone production.
- Tidal rhythms, commonly observed in marine life, which follow the roughly 12.4-hour transition from high to low tide and back.
- Lunar rhythms, which follow the lunar month (29.5 days). They are relevant, e.g., for marine life, as the level of the tides is modulated across the lunar cycle.
- Gene oscillations – some genes are expressed more during certain hours of the day than during other hours.

Within each cycle, the time period during which the process is more active is called the acrophase. When the process is less active, the cycle is in its bathyphase or trough phase. The particular moment of highest activity is the peak or maximum; the lowest point is the nadir. How high (or low) the process gets is measured by the amplitude.

All eukaryotic organisms ranging from insects to mammals have over time developed behavioral, physiological, and/or molecular biological and biochemical processes that have a particular rhythm that is synchronized to time and time periods within each day or even over a period of days (Cambar et al., 1992; Cambar and Pons, 2010). These processes enable an organism to adapt to its own special environment most optimally, quickly and with greater facility. Again, these circadian oscillations or rhythms are endogenous in nature to each organism, specific to each organism and most commonly exhibit a periodicity or single cycle length of 24 hours and are responsive to and can be adjusted by the stimulus of light. Succinctly, the study of the time-related periodic changes in the susceptibility, as well as responses and sensitivity of organisms expressed as the pharmacology, pharmacokinetics, pharmacodynamics, and adverse effects of drugs, xenobiotics, toxins, chemicals, etc. is designated as chronotoxicology, chronopharmacodynamics, chronopharmacokinetics, or chronopharmacology.

It is important to recognize that these temporal rhythms are driven by circadian (24-hour time period) or solar time cycles, which are in turn based upon periodic photo-stimuli received from sunlight and potentially further influenced by other local time signals.

The biological rhythms of living organisms have been well known for several decades as a result of numerous and various reports in the published literature of cell biology, physiology, molecular biology, biochemistry, and pharmacology. Temporal variations can and do simply explain why the same xenobiotics etc., regardless of origin, do not induce the same efficacy or other effects in a living organism, if it is administered at different times of the day or at different seasons of the year (Walker et al., 1983; Bélanger, 1987). Rhythmic or circadian changes in the sensitivity or susceptibility to xenobiotics were actually first reported as far back as 1958 in experiments with mice (Belanger et al., 2010). A few years later it was demonstrated that German cockroaches (*Blatella germanica*) exhibited a rhythmic susceptibility over a 24-hour period of time to a standard dose of potassium cyanide, when organisms were exposed to the toxin at different times within the day (Redfern and Lemmer, 2010).

Circadian Clock

The mammalian circadian system is organized in a prioritized manner with a highly important centrally located pacemaker within the suprachiasmatic nucleus (SCN) of the hypothalamic region of the brain (Hastings, 2010). This structure is located proximal to the optic chiasm and receives transmitted light through the eyes within the wavelength range of 460–480 nm. The reception and processing of this light is through non-visual photosensitive retinal ganglion cells, which contain the substance melanopsin and are located within the retino-hypothalamic tract of the brain (Hastings, 2010; Edmunds, 2010; Moore, 2010; Buhr and Takahashi, 2013). This center controls and synchronizes the circadian receivers or oscillators in millions of cells that are located all over the body. These cells themselves are not responsive to light, but do respond to various neural and neuro-humoral signals from the SCN in the brain. Major timing signals in the synchronization of many of these peripheral clocks include sleep and activity periods and feeding and fasting cycles (Fisher et al., 2013; Albrecht, 2013). This has suggested to researchers in the field that the temporal coordination of metabolism and cell proliferation is a major effort of the mammalian timing, circadian, circannual, or other periodic system. Taken together, the complete chronological biological system directs and controls precisely timed events, which sustain life by regulating physiological function of the cardiovascular, hemodynamic, metabolic, nervous, digestive, immune, reproductive, and endocrine systems (Antoch and Kondratov, 2013; Patel et al., 1982; Rensing, 2010). Rhythmic fluctuations in absorption, distribution, metabolism, and excretion regulate the detoxification of toxic substances, drug toxicity, and drug efficacy (Bélanger, 1987; Marcheva et al., 2013). This watch-like control of drug and xenobiotic catabolism and detoxification provides the molecular and biochemical basis for the dosing time-dependence of both drug toxicities and drug efficacies. The inactivation of noxious food components by intestinal, hepatic, and renal detoxification systems is among the processes regulated in a circadian fashion. Many of these chronologic processes are further regulated at the genetic level by actual clock genes (not discussed here), which establish basic transcriptional and translational feedback oscillation loops under the control of such entities within the body as *Per* 1–3, *Cry* 1,2, Rev-erbα, Rev-erbß, Rorα, Rorß, and cAMP response element-binding protein (CREB) along with the circadian locomotor output cycles involving ARNT-like protein 1 (CLOCK/BMAL 1) heterodimers (Buhr and Takahashi, 2013; Sahar and Sassone-Corsi, 2013; Brown and Azzi, 2013). Knowledge and understanding of these clock-based processes continue to grow and expand, and this information can be used for improving or designing therapeutic treatment regimens (Lévi and Schibler, 2007).

It has not been an uncommon finding that some studies in the area of chronotoxicology are also susceptible to seasonal variability (Cambar and Pons, 2010; Lévi and Schibler, 2007). Reports have shown that some circadian peak times of susceptibility found during one season may be totally different from that observed in a different season.

Reality of the Science of Timed Cyclical Rhythms

If one looks at the subacute toxicity of amikacin, urinary enzyme activity in animals in autumn is enhanced by a factor of about 5 when amikacin is administered at 20:00 hr and by a factor only a little more than 1 when administration occurs at 14:00 hr (Dorian et al., 1988). Alternatively, in the spring of the year this increase in enzymatic activity is very weakly expressed with administration of amikacin at 20:00 hr and significantly elevated by about four to seven times when administration occurs at 14:00 hr. Seasonal variation has been observed in the degree of acute toxicity of X-rays, cyclophosphamide, and arabinosylcytosine in rodents. The exposure of mice via intra-peritoneal injection to mercuric chloride at four different times a day over 6 months of the year revealed that the same dose of mercury (5 mg/kg) induced different mortality rates when administered in January, June, or September (Cal et al., 1985; Dorian et al., 1988). The tolerance of cisplatin in rats is twice as high in the winter as it is in the summer. The toxicity of phenobarbital has been studied in mice (). It has been reported that a dose of 190 mg/kg administered at 16:00 hr killed no animals in the month of July but yet killed 90% of the mice in January. A dose of 270 mg/kg of phenobarbital administered at 04:00 hr killed all animals in October and only 40% of the mice in July. The tolerances to doxorubicin and daunomycin in mice were found to be optimal in autumn and poorest in the spring and summer (Lévi et al., 1988). These studies clearly indicate that not only is the time of the day important, but also the month of the year can also be very important. It has also become obvious from reports in the literature that the wavelike high (acrophase) or low (bathyphase) phases of susceptibility or tolerance in an organism can totally reverse themselves according to the season. It is important to keep in mind that the severity of potential damage to organs exposed to the administration of metals, antibiotics, anticancer agents, etc. is dependent not only on the circadian time but also on the season of the year. These annual cycles or rhythms of tolerance and susceptibility may indeed be controlled at least in part by the levels of hormones (testosterone, estrogen, luteinizing hormone, thyroxin, oxytocin, etc.), which can possibly affect organ or tissue defense mechanisms. Research continues in this area.

These examples that are being provided of alterations in the acute, subchronic, or chronic toxicities or side effects of drugs, chemicals, etc. as a result of the temporality of administration do raise the question as to what exactly the mechanism or mechanisms are that drive or control these differences of tolerance or susceptibility (Batalla et al., 1994; Bruckner et al., 1984; Earnest et al., 1991; Fisch et al., 1984; Motahashi and Miyazaki, 1990; Prat and Bruguerolle, 1988). Original research in the field of circadian oscillations and chronobiology has out of necessity led to investigations of biotransformation and pharmacokinetics. It is known that the liver and kidneys do undergo a circadian-dependent synthesis of their enzymes and changes in membrane structure and function and nuclear components. A complete understanding of the chronophysiology of the liver and kidneys do make it possible to explain the chronopharmacological variations in hepatic or renal elimination and the metabolism of drugs (Batalla et al., 1994; Malmary et al., 1994; Bruckner et al., 1984; Luthra et al., 1984; Earnest et al., 1991; Digiogio, 1991; Fisch et al., 1984; Yonovitz, 1984; Motahashi and Miyazaki, 1990; Prat and Bruguerolle, 1988). Time studies of the activities of glycogen phosphorylase and glycogen synthetase copy exactly the activities of hexobarbital oxidase and barbiturate-induced sleep duration. Similarly, the circadian changes in the hemodynamics of the kidneys are well correlated with cortical accumulation and elimination. This all brings us to the introduction of another new term, "chronesthesy" (Ortiz-Tudela et al., 2013). This is defined as the rhythmic changes of any biological system with regard to its susceptibility or tolerance. The biological system can be any living organism, cell, tissue, or organ.

The immune system has also been found to exhibit a clocklike regulation behavioral pattern. If mice are administered *Bacillus Calmette-Guerrin* (BCG) in the middle of the day, 75% of the exposed animals survived. However, if the mice were injected in the latter portion of the night period, only 50% were found to survive (Arendt, 2010).

Timed Rhythms and the Hepatic System

Over the years a variety of agents have been reported in the literature as being hepatotoxic and the solvent carbon tetrachloride has been a popular agent that has been studied (Bruckner et al., 1984; Luthra et al., 1984). After a single dose exposure of carbon tetrachloride, centrilobular necrosis and lipidosis occurs in the livers of all animals. Mitochondrial damage can be observed by 5 to 6 hours after exposure. Other lesions develop by 12 hours post exposure and severe hepatic necrosis has developed by 24 hours post exposure. Although trichloroethylene is a well known and highly characterized hepatotoxin; it is also a potent neurotoxin (Walker et al., 1983; Motahashi and Miyazaki, 1990). For example, if trichloroethylene is administered to rats via the intra-peritoneal route at four different time points within a day and toxicity assessed by evaluating muscle tone, it was determined that the time of the most significant neurotoxic effect of trichloroethylene was early in the active phase (21:00 hr; nighttime). Interestingly, the plasma concentrations of trichloroethylene and two of its major metabolites are found to be at minimal levels at 21:00 hr (nighttime). Accordingly it appears that neurotoxicity of trichloroethylene is accentuated when hepatic elimination is at its lowest level of activity. Studies have demonstrated that the circadian susceptibility of animals to hepatotoxicity has been found to be at least in part dependent on the timing of toxin uptake and metabolism (Bélanger et al., 2010). Similar correlations with chronotoxicity have been found in studies with the agents chloroform, trichloroethylene, and 1,1-dichloroethylene, and the drug acetoaminophen (Bélanger et al., 2010; Bruguerolle et al., 2010; Labrecque et al., 2010). Rats have been found to be most susceptible to chloroform exposure when the administration is performed within a 2-hour period after the initiation of their active period (nighttime) (21:00 hr) (Bélanger et al., 2010; Bruguerolle et al., 2010). The lowest level of susceptibility was observed when exposure was at 09:00 hr (daytime). Even though rats exposed to chloroform at both 21:00 hr and 09:00 hr revealed depressed levels of glutathione, hepatic levels of glutathione were found to be more significantly depressed at 21:00 hr than at 09:00 hr. If the serum activities of glutamic-pyruvic-transaminase, isocitrate dehydrogenase, ornithine-carbamyl-transferase, and sorbitol dehydrogenase were measured in rats after exposures to carbon tetrachloride at eight different time points within a day, it was revealed that significant increases (in some cases approaching 400%) were observed with the administration of carbon tetrachloride at 18:00 hr or 20:00 hr, but remained unchanged if exposure was at 10:00 hr (Bélanger et al., 2010; Bruguerolle et al., 2010).

Biological Rhythms in Species Other Than Mammals

In all the studies mentioned above, for the most part they have been performed in rodents or humans as part of clinical trials or other evaluations. A big reason for this is the fact that the maintenance of synchronized and strictly defined conditions can be most cost effectively managed with rodents, which also then permit the use of sufficient numbers of animals for statistical power. However, studies in the area of circadian oscillations have also been performed and reported in other species, including both fish and insects (Hayes and Morgan, 1988). The study of chronobiology in insects could lead to the more efficient and efficacious use of pesticides and improved ecological outcomes. For example if one looks at houseflies and yellow fever mosquito (*Aedes aegypti*) larvae and their exposure to malathion and/or dursban, investigations have shown that differing potencies of lethality are observed to the same concentration of pesticide, depending on the timing of exposure (Hayes and Morgan, 1988). Ninety percent of insect larvae can be killed if they are exposed during the latter half of the nighttime and only 20% if exposed at the end of the daytime period. Circadian rhythms have been observed in rainbow trout (Cambar and Pons, 2010). In the case of rainbow trout, a seasonal pattern to the lethality of cyanide exposure was discovered.

Use of the Internal Clock in Pharmacokinetics, Formulation Development, and Drug Delivery

The minimal goals in the development of a formulation for drug delivery is to devise something that solubilizes the test article, is not irritating to tissues via the route of administration, has pH and osmolality values that are biologically compatible, has desirable physical characteristics, is palatable (if oral), provides stability to the test article, does not interfere with the pharmacological activity of the test article, is not toxic, and supports or enhances the bioavailability of the test article. Aside from the pure chemical aspects of formulation development, a big part of the goal in formulation development is to be able to support and enhance the bioavailability of the test article in such a fashion as to achieve pharmacokinetic levels that support the pharmacologic function and therapeutic effectiveness of the test article. While there are a variety of "tricks of the trade" in formulation development, consideration of the normal and natural rhythms of the organism to maximize effectiveness of the test article should be evaluated.

Biological Oscillations in Pharmacokinetics

Pharmacokinetics is a scientific discipline or tool that provides a quantitative description as to what a living organism does to a compound/drug that is administered. Processes such as absorption, distribution, metabolism, and excretion are described by various mathematical functions and equations and a complete pharmacokinetic characterization or profile facilitates the selection of appropriate dosing regimens. Refer to Tables 7.1 and 7.2 (Lehman-McKeeman, 2013). In preclinical drug development, this pharmacokinetic profiling serves as an important tool for advancing drug candidates. Pharmacokinetic parameters are generally considered to be constant in time and do not in and of themselves compensate for variations in plasma volume, metabolic activities, blood flow rates, etc. that may exhibit temporal changes (Bélanger et al., 2010). It is well known from the literature that the efficacy, therapeutic effects, side effects, and toxicities of many drugs can and do vary depending on the hour of the day that administrations of the drugs occur (Reinberg et al., 1990; Bruguerolle and Lemmer, 1993; Bélanger et al., 2010; Lemmer and Bruguerolle, 1994).

This chapter will not include a detailed treatise on the discipline and science of pharmacokinetics. What will be discussed is the impact and potential use of circadian rhythms in the development of dosing regimens to achieve a steady state blood level (Bélanger et al., 2010). Furthermore, we will explore the influence of circadian rhythms on the absorption, distribution, excretion, and elimination of drugs and connections between time and pharmacokinetic parameters (Bélanger, 1993; Bélanger et al., 2010).

Timed Rhythms in Absorption

Absorption can be most simply defined as the transfer of a test article from the site of its administration into the systemic circulation. Generally, the most important pharmacokinetic parameters involved in the process of absorption are the maximal blood/serum/plasma concentration (C_{max}), the time to attain the C_{max} value (t_{max}), the absorption rate constant (k_a), and the area under the curve (AUC) (Bélanger et al., 2010). The AUC value is an estimate of the amount of drug absorbed and serves as an extremely useful tool to compare the bioavailabilities of different formulations, but

TABLE 7.1
Common Symbols Used in Pharmacokinetics

Symbol	(Typical) Units	Definition
A	Mg	amount of drug in body
AUC	mg*hour/L	total area under the plasma drug concentration-time curve
$AUC_{0-\infty}$	mg*hour/L	area under the plasma concentration-time curve from zero to infinite time
AUC_{iv}	mg*hour/L	area under the plasma drug concentration-time curve after an IV dose
AUC_{oral}	mg*hour/L	area under the plasma drug concentration-time curve after an oral dose
C	mg/L	concentration of drug in plasma
C_o	mg/L	concentration of drug in plasma after a single dose extrapolated back to zero time
C_b	mg/L	concentration of drug in blood
CL	L/hour	total clearance of drug from plasma
CL_{CR}	L/hour	creatinine clearance
CL_{GF}	L/hour	renal drug clearance by glomerular filtration
CL_H	L/hour	hepatic clearance of drug from plasma
CL_{int}	L/hour	intrinsic clearance of drug in an organ of elimination
CL_R	L/hour	renal clearance of drug
CL_S	L/hour	renal drug clearance by tubular secretion
C_t	mg/L	concentration of drug in plasma at a specific time t
C_{max}	mg/L	maximum plasma drug concentration during a dosing interval
C_{min}	mg/L	minimum plasma drug concentration during a dosing interval
C_{SS}	mg/L	concentration of drug in plasma at steady state during a constant-rate intravenous infusion
C_{USS}	mg/L	unbound concentration of drug in plasma at steady state during a constant-rate intravenous infusion
C_U	mg/L	unbound drug concentration in plasma
DR	mg/hour	dosing rate
E	ratio	extraction ratio for an organ
E_H	ratio	hepatic extraction ratio
EC_{50}	mg/L	concentration giving one-half the maximum effect
E_{max}	varies	maximum effect
F	ratio	bioavailability of drug
f_e	ratio	fraction of drug systemically available that is excreted unchanged in urine
f_g	ratio	fraction of an oral dose that is absorbed intact into the portal circulation
f_H	ratio	fraction of drug entering the liver that escapes extraction
fm	ratio	fraction of drug systemically available that is converted to a metabolite

(Continued)

TABLE 7.1 (*Continued*)

Common Symbols Used in Pharmacokinetics

Symbol	[Typical] Units	Definition
FR	ratio	fraction of drug reaching the renal tubular fluid that is reabsorbed
fu	ratio	fraction of drug unbound in plasma
fu_T	ratio	fraction of drug unbound in tissues
GFR	L/hour	glomerular filtration rate
k	/hour	elimination rate constant
K_a	mM	association constant for the binding of drug to protein
K_m	mg/L	Michaelis–Menten constant
λ (lambda)	ratio	ratio of concentration of drug in whole blood to that in plasma
P_u	mM	concentration of protein that does not bind drug
Q_H	L/hour	hepatic blood flow (portal vein plus hepatic artery)
τ (taul)	hour	dosing interval
V	L	volume of distribution (apparent) based on drug concentration in plasma
v	mg/hour	velocity of enzyme reaction (conversion of drug to metabolite)
V_{max}	mg/hour	maximal rate of enzyme reaction at saturating substrate concentration
V_p	L	plasma volume
V_T	L	physiologic volume outside plasma into which drug distributes

also to make comparisons from the pharmacokinetic perspective using ratios of AUC values from an intravenous administration compared to other routes of administration (Bélanger et al., 2010). There are a variety of factors that can affect absorption and they include the physicochemical characteristics of the test article itself, the dosage form, formulation, and biological function. Any one of these factors has the potential to involve diurnal oscillations and accordingly subsequent changes in the pharmacokinetic parameters with time.

Membrane transporters can be major determinants of the pharmacokinetic, safety, and efficacy profiles of drugs (The International Transporter Consortium, 2010). This presents several key questions for drug development, including which transporters are clinically important in drug absorption and disposition, and which *in vitro* methods are suitable for studying drug interactions with these transporters. A classification of drug transporters is represented in Table 7.3 (The International Transporter Consortium, 2010). In addition, what criteria should trigger follow-up clinical studies, and which clinical studies should be conducted. Overall, it is advised that the timing of

TABLE 7.2
Common Pharmacokinetic Equations

Elimination rate = $CL * C$

$DR = CL * C_{SS}$

$C_b = C * \lambda$

$$V = \frac{A}{C}$$

$$V = V_p + \frac{fu}{fu_T} * V_T$$

loading dose = $V * C$

$C_t = C_o * e^{-k\tau}$

$$k = \frac{0.693}{t_{1/2}}$$

$$t_{1/2} = \frac{0.693 * V}{CL}$$

$$k = \frac{CL}{V}$$

$CL_H = Q_H * E_H$

$$E_H = \frac{fu * CL_{\text{int}}}{Q_H + fu * CL_{\text{int}}}$$

$$CL_{\text{int}} = \frac{V_{\max}}{K_m}$$

$$CL_H = Q_H * \frac{fu * CL_{\text{int}}}{Q_H + fu * CL_{\text{int}}}$$

$F = f_g * f_H$

$$F = \frac{AUC_{oral}}{AUC_{iv}}$$

$CL = CL_R + CL_H$

$$f_e = \frac{CL_R}{CL}$$

$f_H = 1 - E_H$

$CL_R = fu\,(GFR + CL_S)\,(1 - FR)$

$$CL_{CR} = \frac{(140 - \text{age in years})(\text{weight in } kg)}{814 * \text{serum creatinine}\,(mmol\,/\,L)}$$

(Continued)

TABLE 7.2 (*Continued*)
Common Pharmacokinetic Equations

$$fu = \frac{C_u}{C}$$

$$C_{SS} = \frac{F * DR}{CL}$$

$$CL\,(L/\text{hour}) = \frac{IV\ \text{dose}\,(mg)}{AUC_{0-\infty}\,(mg * hour/L)}$$

$$V(L) = \frac{t_{1/2}\,(\text{hour}) * CL\,(L\,/\,\text{hour})}{0.693} = \frac{CL\,(Lh^{-1})}{k\,(h^{-1})}$$

$$CL = \frac{F * \text{dose}\,(mg)}{AUC_{0-\infty}}$$

$$\frac{C_{\max}\ \text{steady state}}{C_{\max}\ \text{first dose}} = \frac{1}{1 - e^{-k\tau}}$$

$$\frac{C_{\max}\ \text{steady state}}{C_{\min}\ \text{steady state}} = \frac{1}{e^{-k\tau}}$$

transporter investigations should be driven by efficacy, safety, and clinical trial enrollment questions (e.g., exclusion and inclusion criteria), as well as a need for further understanding of the absorption, distribution, metabolism, and excretion properties of the drug molecule, and information required for drug labeling. Membrane transporters can be major determinants of the pharmacokinetic, safety, and efficacy profiles of drugs. This presents several key questions for drug development, including which transporters are clinically important in drug absorption and disposition, and which *in vitro* methods are suitable for studying drug interactions with these transporters. In addition, what criteria should trigger follow-up clinical studies, and which clinical studies should be conducted. The general feeling these days is that the timing of transporter investigations should be driven by efficacy, safety, and clinical trial enrollment questions (e.g. exclusion and inclusion criteria), as well as a need for further understanding of the absorption, distribution, metabolism, and excretion properties of the drug molecule, and information required for drug labeling. Unfortunately, there is no data available on the connection between transporter activity and biological rhythms.

Timed Patterns in the Absorption of Drugs

A large number of drugs, regardless of route of administration (exclusive of intravenous), are absorbed via the simple partition process of passive diffusion (Denvis, 1990). Drugs such as diclofenac, ketoprofen, nifedipine,

TABLE 7.3
Classification of Drug Transporters

Transporter Family	Family Member	Gene Name
Organic cation transporter (OCT)	hOCT1	SLC22A1
	hOCT2	SLC22A2
	hOCT3	SLC22A3
Organic cation/carnitine transporter (OCTN)	OCTN1	SLC22A4
	OCTN2	SLC22A5
	OCTN3	SLC22A21
	CT2	SLC22A16
Organic anion transporter (OAT)	OAT1	SLC22A6
	OAT2	SLC22A7
	OAT3	SLC22A8
	OAT4	SLC22A11
	OAT5	SLC22A10
	OAT6	SLC22A20
	URAT1	SLC22A12
Organic anion transporter polypeptides (OATPs)	OATP1C1	SLCO1C1
	OATP1B1	SLCO1B1
	OATP1A2	SLCO1A2
	OATP1B3	SLCO1B3
	OATP2A1	SLCO2A1
	OATP2B1	SLCO2B1
	OATP3A1	SLCO3A1
	OATP4A1	SLCO4A1
	OATP4C1	SLCO4C1
	OATP5A1	SLCO5A1
	OATP6A1	SLCO6A1
Peptide transporter (PEPT)	PEPT1	SLC15A1
	PEPT2	SLC15A2
	PHT1	SLC15A4
	PHT2	SLC15A3
Monocarboxylate transporters (MCTs, sMCTs)	MCT1	SLC16A1
	MCT2	SLC16A7
	MCT3	SLC16A8
	MCT4	SLC16A3
	SMCT1	SLC5A8
	SMCT2	SLC5A12
Nucleoside transporters (CNTs, ENTs)	CNT1	SLC28A1
	CNT2	SLC28A2
	CNT3	SLC28A3
	ENT1	SLC29A1
	ENT2	SLC29A2
	ENT3	SLC29A3
	ENT4	SLC29A4

TABLE 7.3 (*Continued*)

Classification of Drug Transporters

Transporter Family	Family Member	Gene Name
Bile acid transporters	NTCP	SLC10A1
	ASBT	SLC10A2
	BSEP	ABCB11
	OST-α	–
	OST-β	–
Multidrug resistance protein (MDR)	MDR1	ABCB1
Multidrug resistance – associated protein (MRP)	MRP1	ABCC1
	MRP2	ABCC2
	MRP3	ABCC3
	MRP4	ABCC4
	MRP5	ABCC5
	MRP6	ABCC6
	MRP7	ABCC10
	MRP8	ABCC11
	MRP9	ABCC12
Breast cancer resistance protein (BCRP)	BCRP1	ABCG2

prednisolone, propanolol, salicylic acid, valproic acid, and many others are known to be absorbed to a greater degree and faster if administered in the morning to a human as a single dose in a nonexotic and nonsustained-release formulation (Bélanger, 1993; Bruguerolle and Lemmer, 1993; Lemmer and Bruguerolle, 1994). These reports demonstrate minimally the potential importance of the time of the day that administration is performed and the determination of bioavailability.

To be effectively absorbed, any solid pharmaceutical formulation must first be broken down into very small particles, which then need to be solubilized in an aqueous milieu at the site of absorption to initiate this passive diffusion process and absorption.

Medicines such as gentamicin, midazolam, and acetaminophen have demonstrated no temporal correlation when administered orally (Bélanger, 1993). While it is true that the mechanisms associated with morning administration and differences in bioavailability have never been adequately characterized in detail, it is fair to state that an adequate understanding of the physicochemical properties of a test article coupled with an appreciation of the circadian oscillations of the gastrointestinal tract can explain the results (Bélanger, 1993; Bruguerolle and Lemmer, 1993; Labrecque and Belanger, 1991).

If one looks at the C_{max} and t_{max} values for indomethacin, phenylbutazone, hydrochlorothiazide, and furosemide after single oral administrations at 08.00 hour and 20.00 hour, no time-related change was observed for the C_{max} value of furosemide and hydrochlorothiazide, but the t_{max} values were found to be shorter for the evening administration. The furosemide and hydrochlorothiazide were administered as solutions and the indomethacin and phenylbutazone were administered as suspensions of powder. However, the phenylbutazone and indomethacin demonstrated greater C_{max} values and shorter t_{max} values at the 20.00 hour administration (Shiga et al., 1993; Bélanger et al., 2010). The most significant diurnal variations were obtained with drugs given as suspensions, suggesting that the rate of solubilization of the two drugs is higher in the evening (Shiga et al., 1993; Bélanger et al., 2010).

Biological Rhythms and Heavy Metal Absorption

There is little in the literature with regard to the variations in time with regard to the absorption of inorganic compounds and macromolecules. In general, the mechanisms of absorption of these compounds are poorly understood (Bélanger et al., 2010). However, there are reports in the literature with regard to temporal variations in the absorption of iron, calcium, lead, and vitamin B12 (Tarquini et al., 1979; Wrobel and Nagel, 1979; Aungst and Fung, 1981; Markiewicz et al., 1981).

Diurnal Rhythms and the Gastrointestinal Tract

It is interesting to note that the circadian oscillatory gastrointestinal absorptive behavior in many studies is eliminated if the drugs are administered in a sustained-release form or as a rectal suppository (Bélanger et al., 2010). However, it has been shown for indomethacin and other drugs (e.g. theophylline) that some sustained-release formulations do exhibit variations in time for oral absorption, but this is different from the diurnal pattern of the immediate-release formulation of the drug. Reports also suggest that the rate of drug solubilization from the administered form of the drug is a determining factor in the temporal pattern of absorption for many drugs (Smolensky, 1989; Reinberg et al., 1986; Jespersen et al., 1989; Taggart et al., 1987, Guissou et al., 1983; Bélanger et al., 2010).

In cases of oral administration, whether by powder, liquid, suspension, capsule, tablet, slurry, etc., a number of factors can influence subsequent plasma levels (Moore and Merki, 2010). These factors include but are not limited to gastrointestinal transit times and the rate of gastric emptying. While there are drugs that act on the stomach lining, in general test articles are not absorbed in the stomach, and the absorption process begins only when they enter the intestine, and such biological functions as gastric emptying and gastrointestinal transit times are indeed affected by circadian rhythms, which then in turn affect the rate of absorption of a material (Moore and

Merki, 2010; Lemmer and Bruguerolle, 1994). Not surprisingly, for many drugs the bioavailability does not change with either a daytime or nighttime administration (Moore and Merki, 2010). However, dosage in the evening may be a better approach for drugs exhibiting a lower threshold of toxicity, because side effects are associated with plasma levels and gastric emptying times are slower in the evening.

Studies involving the assessment of gastric diurnal toxicology using gastric mucosal integrity have revealed that there is indeed a time-related damage component caused by the well-characterized aspirin and other nonsteroidal anti-inflammatory (NSAID) agents (Moore and Merki, 2010). Although still somewhat controversial, it appears that the nighttime administration of these types of agents are better tolerated than the daytime administration of the same. In humans, using endoscopic visualization, it was determined that the administration of 1300 mg of aspirin at 08:00 hr versus 20:00 hr resulted in the development of almost 40% fewer gastric mucosal lesions if the medicine was administered in the evening (Labrecque et al., 2010). In a study utilizing osteoarthritic patients, it was discovered that the administration of a sustained-release formulation of indomethacin at 08:00 hr resulted in a significantly greater number of undesirable side effects than if the medication was administered in the evening at 20:00 hr (Labrecque et al., 2010).

Drugs that are completely disintegrated or dissolved in the stomach are completely emptied into the intestine in a liquid phase that does not display a significant circadian variation in absorption if they are compared to drugs that are ingested with solids (Moore and Merki, 2010; Goo et al., 1987). It is well known that the ingestion of a solid meal delays the absorption of drugs because of the inhibitory effects of solids on gastric emptying. Since gastric emptying occurs more slowly in the evening as compared to the morning, this delay is amplified in the evening (Goo et al., 1987). The absorption of enteric or delayed release forms are even more delayed with concurrent meal ingestion. The presence of food in the stomach inhibits the gastric migrating motor complex waves that move solids such as enteric coated capsules out of the stomach and into the small intestine, where their dissolution begins. The speed of migrating motor complex waves is slower at night (Bogentoft et al., 1978). If for some reason there is a decreased frequency of migrating motor complex waves, gastric residence times are increased. As one can see, there are a number of different factors in play here, and it is entirely possible that a sustained-release form administered in the morning on a full stomach may not be absorbed during the day.

The rhythmic motor patterns of the gastrointestinal tract (stomach and intestines) strongly influence the absorption and distribution of orally delivered medicaments. Test articles that are administered via the oral route must exit the stomach before the process of absorption occurs. The transfer of drugs into the small intestine is related to the rate of gastric emptying, and that itself is influenced by a number of factors, including the presence of food, the absence of food, whether the food is in liquid form or solid form, the percent of amount of protein, the percent or amount of carbohydrate, the percent of amount of lipid, body posture, activity level, gender, age, and circadian oscillations (Moore et al., 1990). In the presence of food, gastric motor activity is pretty much dominated by caudally directed peristaltic contractions. These contractions occur as a result of the synergistic activity of muscles in the wall of the stomach, which as a result of their activity grind solids to a very small particle size, which in turn permits facile passage from the stomach into the small intestine (Moore and Merki, 2010; Kumar et al., 1986). Liquids are emptied from the stomach more easily. Liquids exit the stomach generally according to first-order kinetics and digestible solids exit the stomach according to zero-order kinetics. Indigestible solids are not able to exit the stomach as liquids and digestible solids do, and are emptied from the stomach by the migrating motor complex, which is subject to rhythmic oscillations of various durations, frequencies, and periods (Moore and Merki, 2010). In humans, gastric emptying rates were significantly different, when subjects were fed at two different time points. Subjects consuming food at 20.00 hours demonstrated significantly slower gastric emptying rates for the same meal administered to the same subjects as 08.00 hours (Goo et al., 1987).

Gastric acid secretion is stimulated by a number of different routes and pathways, many of which can be independently blocked by various inhibitors (Moore and Merki, 2010). Without stimulation, and in the presence of fasting, gastric acid is secreted in low amounts to maintain a pH in the stomach within the range of 0.8 and 2.5 for a 24-hour period of time (Moore and Halberg, 1986). Gastric acid output (acid and various fluids) is highest in the evening and lowest during the morning hours in the absence of the presence of food (Moore and Halberg, 1986). The activities and kinetic parameters of many enzyme systems of the intestinal lining have also been shown to vary with the time of the day (Vener and Moore, 1987). Both circadian oscillations and the presence of both liquid and solid food play a significant synergistic role in regulation of gastric emptying (Moore and Merki, 2010).

Delayed intestinal absorption rates may result in enhanced first-pass effects (Langner and Lemmer, 1988). However, administration in the morning hours can be a desirable course of action because of a more rapid onset-of-action due to increased gastric emptying activity.

Motility of the gastrointestinal tract and biliary secretion are two physiologic activities that involve circadian rhythms and enhance the rate of drug dissolution (van Berge Henegower and Hoffman, 1978). In humans maximal amounts of bile acids are reported to be secreted between the times of 08.00 hr and 14.00 hr, which correspond to the actual period of the day of maximal absorption. The synthesis of bile acids in humans also follows a circadian oscillation with the maximal synthetic activity being in the morning and the nadir of synthetic activity being late in the afternoon and evening (Pooler and Duane, 1988; Bélanger et al., 2010). Very similar patterns of bile acid synthesis and bile flow have been described in rats, where the highest period of activity was observed at midnight and the nadir of activity at noon, during the sleep period (Duane et al., 1979; Gilberstadt et al., 1991, Bélanger et al., 2010). In short, the higher load of bile acids secreted into the small intestine and the greater rate of gastric emptying and intestinal motility during the morning hours parallel the time increase in the absorption parameters of most drugs in humans.

Cardiac output to the gastrointestinal tract, blood flow, and blood pressure follow a pattern of circadian rhythmic change which support the variations of different degrees of absorption of drugs from the gastrointestinal tract. In studies with rats it was observed that peak concentrations of radioactivity in the gastrointestinal tract were at their maximum at 03.00 hr and 21.00 hr (activity periods) and their nadir during the daytime (Labrecque et al., 1988; Bélanger et al., 2010; Larsen et al., 1991).

Internal Clock-Driven Rhythms and Distribution

Once a material is absorbed, it is then distributed to and throughout the body. This dissemination of any test article is affected by a number of different parameters, including body composition (bone, muscle, fat, liquid), body size, blood flow, and cardiac output to the various organs.

The plasma protein binding of a drug is also a very significant parameter in the distribution of a drug, because the release of test article from plasma proteins controls the amount of drug that is free in the plasma and can diffuse into the tissue compartment and target site (Bélanger, 1993). Plasma protein levels affect the degree of test article binding and accordingly the availability of many drugs and has been found to be subject to the influence of circadian oscillations (Nakano et al., 1984).

There are two plasma proteins that are most commonly involved in drug binding: α-1-glycoprotein and albumin. It is widely reported in the literature that α-1-glycoprotein and albumin and other plasma proteins are subject to the influence of circadian oscillations (Kanabrocki et al., 1988; Reinberg et al., 1977; Bélanger et al., 2010; Focan et al., 1988). Typically, the lowest levels of plasma proteins occur at night and peak levels occur in the first part of the afternoon. Reports in the literature have correlated the degree of drug binding and the amount of total plasma proteins in time (Hecquet et al., 1984, Bélanger et al., 2010; Tillement et al., 1984; Lindup, 1987). Unfortunately, the periodicity over time of such parameters as affinity-binding constants and the number of drug binding sites have never been reported in the literature. Furthermore, the correlation of such factors as feeding schedule, sex, diet composition, health state, age on time dependency, and plasma protein levels has never been reported in the literature (Bélanger et al., 2010). The binding of prednisolone to corticosteroid-binding globulin has been reported to demonstrate a temporal variation that is inversely related to the circadian oscillations observed with serum cortisol (Angeli et al., 1978).

Diurnal Patterns and Excretion

Excretion is the process by which the end products of the metabolism of a drug are eliminated from the body. The primary organs involved are the kidney for those drugs that are excreted into the urine and the liver for those drugs that are excreted via the biliary system. The kidney in addition to its role in the elimination of nitrogenous wastes such as urea and uric acid plays a vital role in maintaining salt and water balance in the body. Both the kidney and liver play important roles in the excretion and elimination of the end products of xenobiotic metabolism. In addition to the hepatic and renal routes of excretion, there are pulmonary routes, alimentary routes, and other minor routes of excretion and elimination. Essentially all chemicals are eliminated from the body either as the unchanged parent molecule or as metabolic degradation products of the parent. Typically, the kidney is considered the most important organ for elimination. With regard to urinary excretion, it is considered that there are generally three processes involved in the excretion of drugs and other xenobiotics. One is energy-dependent tubular transport, another is passive reabsorption and the third is glomerular filtration (Waterhouse and Minors, 1989). Circadian oscillations of significant degree have been described and reported in the literature with regard to renal plasma flow, the generation of urinary volume, urinary pH, urinary osmolality, urine specific gravity, glomerular filtration, and the urinary excretion and elimination of many electrolytes and other endogenous chemicals in rodents (Pons et al., 1994; Cambar et al., 1979). Similar correlations have been reported to exist in humans (Kanabrocki et al., 1988; Wisser and Breuer, 1981; Araki et al., 1983; Bélanger et al., 2010). For all of these oscillations, peak values occur during periods of activity and nadir values correlate with periods of rest. It should be obvious that these variations in time can and do affect the elimination of drugs that are excreted and eliminated into the urine predominantly as unchanged parent molecules. When one looks at the excretion of weakly acidic drugs and weakly basic drugs, the rhythm of fluctuations of urinary pH over a 24-hour period is very important in the process of urinary excretion (Elliot et al., 1959). In humans the value of urinary pH has been found to be typically lower during the sleep period than during the daytime activity hours. The relationship of urinary excretion of amphetamines and similar compounds and the temporal oscillations of urinary pH has been well reported in the literature (Beckett and

Rowland, 1964; Wilkinson and Beckett, 1968). Basic amines are more strongly ionized at lower pH values and accordingly readily excreted in the urine, but as values of pH increase the nonionized fraction of amine increases, causing a stronger degree of reabsorption of the compound and a decreased level of excretion and elimination. If the value of urine pH is maintained at a level of about 4.5–5.0 by the co-administration of a substance such as ammonium chloride, then the ionized forms of amphetamine and norephedrine are readily excreted independently of the urine output and the rate of excretion versus time is determined to be very similar to that of the plasma levels of drug (Bélanger et al., 2010). However, if the value of urine pH is alkalinized to a level of approximately 8.0 in order to drive the formation to the nonionized form of the drug and reabsorption, the rate of excretion of norephedrine fluctuates as a function of urinary output independently of pH. Other drugs such as salicylates and sulfonamides have been reported in the literature to exhibit temporal oscillations in their urinary excretion correlating with pH values (Dettli and Spring, 1966; Reinberg et al., 1967; Markiewicz and Semenowicz, 1979). Interestingly the time-related nephrotoxicity of such drugs as amikacin, gentamicin, tobramycin, and cisplatin have been correlated with time-dependent variations in renal excretion in both humans and rodents (Hosokawa et al., 1993; Cal et al., 1985; Lin et al., 1994; Hrusheski et al., 1982).

Internal Clock and the Renal System

The kidneys excrete a variety of waste products produced by metabolism into the urine. These include the nitrogenous wastes urea, from protein catabolism, and uric acid, from nucleic acid metabolism. The ability of mammals and some birds to concentrate wastes into a volume of urine much smaller than the volume of blood from which the wastes were extracted is dependent on an elaborate countercurrent multiplication mechanism. This requires several independent nephron characteristics to operate: a tight hairpin configuration of the tubules, water and ion permeability in the descending limb of the loop, water impermeability in the ascending loop, and active ion transport out of most of the ascending limb. In addition, passive countercurrent exchange by the vessels carrying the blood supply to the nephron is essential for enabling this function.

Glucose at normal plasma levels is completely reabsorbed in the proximal tubule. The mechanism for this is the Na^+/glucose cotransporter. A plasma level of 350 mg/dL will fully saturate the transporters and glucose will be lost in the urine. A plasma glucose level of approximately 160 is sufficient to allow glucosuria, which is an important clinical clue to diabetes mellitus.

Amino acids are reabsorbed by sodium-dependent transporters in the proximal tubule. Hartnup disease is a deficiency of the tryptophan amino acid transporter, which results in pellagra. The microscopic anatomy reabsorption functions of the kidney are outlined in Table 7.4 (Schnellman, 2013). Substances maintained in the body in a homeostatic balance by the kidney are outlined in Table 7.5 (Schnellman, 2013).

Two organ systems, the kidneys and lungs, maintain acid-base homeostasis, which is the maintenance of pH around a relatively stable value. The lungs contribute to acid-base homeostasis by regulating carbon dioxide (CO_2) concentration. The kidneys have two very important roles in maintaining the acid-base balance: to reabsorb and regenerate bicarbonate from urine, and to excrete hydrogen ions and fixed acids (anions of acids) into urine.

Any significant rise in plasma osmolality is detected by the hypothalamus, which communicates directly with the posterior pituitary gland. An increase in osmolality causes the gland to secrete antidiuretic hormone (ADH), resulting in water reabsorption by the kidney and an increase in urine concentration. The two factors work together to return the plasma osmolality to its normal levels.

ADH binds to principal cells in the collecting duct that translocate aquaporins to the membrane, allowing water to leave the normally impermeable membrane and be reabsorbed into the body by the vasa recta, thus increasing the plasma volume of the body.

There are two systems that create a hyperosmotic medulla and thus increase the body plasma volume: urea recycling and the "single effect."

TABLE 7.4

Microscopic Anatomy and the Function of Reabsorption in the Kidney

Site of Reabsorption	Reabsorbed Material	Comments
Early proximal tubule	Glucose (100%), amino acids (100%), bicarbonate (90%), Na^+ (65%), Cl^-, phosphate, and water (65%)	• PTH will inhibit phosphate reabsorption. • Angiotensin II stimulates Na^+, H_2O and HCO_3^- reabsorption.
Thin descending loop of Henle	Water	• Reabsorbs via medullary hypertonicity and makes urine hypertonic.
Thick ascending loop of Henle	Na^+ (10–20%), K^+, Cl^-; indirectly induces para cellular reabsorption of Mg^{2+}, Ca^{2+}	• This section is impermeable to water and urine becomes less concentrated as it ascends.
Early distal convoluted tubule	Na^+, Cl^-	• PTH causes Ca^{2+} reabsorption.
Collecting tubules	Na^+(3–5%), water	• Na^+ is reabsorbed in exchange for K^+ and H^+, which is regulated by aldosterone. • ADH acts on the V2 receptor and inserts aquaporins on the luminal side.

TABLE 7.5

Substances Maintained in a Homeostatic State by the Kidney

Substance	Description	Proximal Tubule	Loop of Henle	Distal Tubule	Collecting Duct
Bicarbonate	Helps maintain acid-base balance.	Reabsorption (80–90%)	Reabsorption (thick ascending)	–	Reabsorption (intercalated cells, via band 3 and pendrin)
Calcium	Uses calcium ATPase, sodium-calcium exchanger.	Reabsorption	Reabsorption (thick ascending) via passive transport	Reabsorption in response to PTH and ↑ reabsorption with thiazide diuretics	–
Carboxylate		Reabsorption (100%) via carboxylate transporters	–	–	–
Chloride	Usually follows sodium. Active (transcellular) and passive (paracellular).	Reabsorption	Reabsorption (thin ascending, thick ascending, Na-K-2Cl symporter)	Reabsorption (sodium-chloride symporter)	–
Glucose	If glucose is not reabsorbed by the kidney, it appears in the urine, in a condition known as glycosuria. This is associated with diabetes mellitus.	Reabsorption (almost 100%) via sodium-glucose transport proteins (apical) and GLUT (basolateral)	–	–	–
Magnesium	Calcium and magnesium compete, and an excess of one can lead to excretion of the other.	Reabsorption	Reabsorption (thick ascending)	Reabsorption	–

(Continued)

TABLE 7.5 (*Continued*)

Substances Maintained in a Homeostatic State by the Kidney

Substance	Description	Proximal Tubule	Loop of Henle	Distal Tubule	Collecting Duct
Oligopeptides, proteins, and amino acids	All are reabsorbed nearly completely.	Reabsorption	–	–	–
Phosphate	Excreted as titratable acid.	Reabsorption (85%) via sodium/phosphate cotransporter. Inhibited by parathyroid hormone.	–	–	–
Potassium	Varies upon dietary needs.	Reabsorption (65%)	Reabsorption (20%, thick ascending, Na-K-2Cl symporter)	–	Secretion (common, via Na+/K+-ATPase, increased by aldosterone), or reabsorption (rare, hydrogen potassium ATPase)
Protons	Uses vacuolar H+ATPase.	–	–	–	Secretion (intercalated cells)
Sodium	Uses Na-H antiport, Na-glucose symport, sodium ion channels (minor).	Reabsorption (65%, isosmotic)	Reabsorption (25%, thick ascending, Na-K-2Cl symporter)	Reabsorption (5%, sodium-chloride symporter)	Reabsorption (5%, principal cells), stimulated by aldosterone via ENaC
Urea	Regulation of osmolality. Varies with ADH.	Reabsorption (50%) via passive transport	Secretion	–	Reabsorption in medullary collecting ducts
Water	Uses aquaporin water channels. See also diuretic.	Absorbed osmotically along with solutes	Reabsorption (descending)	–	Reabsorption (regulated by ADH, via arginine vasopressin receptor 2)

Urea is usually excreted as a waste product from the kidneys. However, when plasma blood volume is low and ADH is released, the aquaporins that are opened are also permeable to urea. This allows urea to leave the collecting duct into the medulla creating a hyperosmotic solution that "attracts" water. Urea can then reenter the nephron and be excreted or recycled again depending on whether ADH is still present or not.

The "single effect" describes the fact that the ascending thick limb of the loop of Henle is not permeable to water but is permeable to NaCl. This allows for a countercurrent exchange system whereby the medulla becomes increasingly concentrated, but at the same time sets up an osmotic gradient for water to follow should the aquaporins of the collecting duct be opened by ADH.

Although the kidney cannot directly sense blood, long-term regulation of blood pressure predominantly depends upon the kidney. This primarily occurs through maintenance of the extracellular fluid compartment, the size of which depends on the plasma sodium concentration. Renin is the first in a series of important chemical messengers that make up the renin–angiotensin system. Changes in renin ultimately alter the output of this system, principally the hormones angiotensin II and aldosterone. Each hormone acts via multiple mechanisms, but both increase the kidney's absorption of sodium chloride, thereby expanding the extracellular fluid compartment and raising blood pressure. When renin levels are elevated, the concentrations of angiotensin II and aldosterone increase, leading to increased sodium chloride reabsorption, expansion of the extracellular fluid compartment and an increase in blood pressure. Conversely, when renin levels are low, angiotensin II and aldosterone levels decrease, contracting the extracellular fluid compartment, and decreasing blood pressure.

The kidneys secrete a variety of hormones, including erythropoietin, and the enzyme renin. Erythropoietin is released in response to hypoxia (low levels of oxygen at tissue level) in the renal circulation. It stimulates erythropoiesis (production of red blood cells) in the bone marrow. Calcitriol, the activated form of vitamin D, promotes intestinal absorption of calcium and the renal reabsorption of phosphate. Part of the renin–angiotensin–aldosterone system, renin is an enzyme involved in the regulation of aldosterone levels.

Unfortunately damage to the kidneys is a very common event as a result of exposure to drugs, toxins, and various xenobiotics (Batalla et al., 1994; Malmary et al., 1994; Cal et al. 1985; Dorian et al., 1988; Lévi and Schibler, 2007). This should be no surprise because approximately 20–25% of total cardiac output flows through the kidneys, which are a major route of excretion and elimination of many substances. Moieties are filtered through the glomerular barrier and enter the tubular lumen to reside in the tubular fluid. The structure and function of membrane structures in the kidney can be significantly altered as a result of exposure to aminoglycoside antibiotics, impairing their elimination processes. Xenobiotics may indeed reach very high levels of intracellular concentrations and therefore produce renal cell injury by one of three different metabolic pathways. Substances, if stable enough and whether metabolized or not, may inflict direct toxicity to the kidney. Secondly, a substance may enter renal tubular cells and interfere directly with cellular functions. Third, materials may be metabolized within renal cells to highly reactive metabolites that may bind covalently to structural or functional proteins or even initiate lipid peroxidation.

The kidneys are a prime target for heavy metals, because these substances are chiefly concentrated in the renal cortex (Cal et al., 1985; Dorian et al., 1988). This accumulation leads to the development of well-characterized pathological changes in the proximal tubules. The initial response of rat kidneys to the administration of a single dose of a heavy metal is the development of a focal necrosis in the mid-third of the proximal tubule (*pars recta*). If mercury is used as an example, within 3 hours after exposure, tubular cell brush borders become fragmented, and by 6 hours post exposure the epithelial cells of the *pars recta* begin to swell and protrude into the tubular lumen, creating tubular blockage (Cal et al., 1985; Dorian et al., 1988). The prevalence of intracellular vacuoles in the cytoplasm substantially increases. At 12 hours post exposure, a significantly increased number of lysosomal structures are appearing within the cells and the mitochondria are swelling. Eventually, cell necrosis results with the release of whole cells and cell parts into the tubular lumen. The exposure of rodents to cisplatin was monitored via histopathological evaluation and the determination of blood levels of serum urea nitrogen and N-acetyl-glucosamidase (NAG). Using these methods of evaluation, it was determined that rats exposed to cisplatin in the middle of the night period demonstrated trivial changes from that found in control kidneys and renal damage was minimal. However, a fourfold increase in NAG as well as a significant increase in serum urea was correlated with exposure to cisplatin during the daytime hours. Mercuric chloride's renal toxicity is well known and highly characterized. Single, sublethal doses of mercuric chloride administered via the subcutaneous route to rats at four different time points during a 24-hour period were evaluated for toxicity by measuring the levels of various tubular nephrotoxity biomarker enzymes. It was discovered that if the mercuric chloride was administered in the middle of the daytime or light period, urinary levels of the biomarkers approached increases of almost 2000%, but if the administration was performed in the middle of the dark period (high activity period of rats) the elevation of these biomarkers was only approximately 1000%. When one looks at the location of mercury or cisplatin in the kidney after exposure to cisplatin or mercuric chloride, if the exposure was performed at the end of the light period, mercury or cisplatin concentrations were found to be highest in the renal tissue and lowest in the urine. Simply put, these chemicals were excreted in larger amounts when animals were exposed during their active period as opposed to exposure during their rest phases (daytime) (Cal et al., 1985; Dorian et al., 1988). Similar results have been found in mice. Over the years, cadmium, platinum, and mercury have been popular heavy metal renal toxins for studies, which populate the literature (Cal et al., 1985; Dorian et al., 1988). It is now accepted that the lethality of heavy metals is highly dependent upon the timing of exposure (Cal et al., 1985; Dorian et al., 1988).

If administrations of anti-neoplastics for advanced-stage colon and rectal cancers were modulated over time rather than just constantly administered at a specific time each day, a significantly greater degree of effectiveness was observed. Similar effects or expressions of toxicity are observed with the administration of a variety of antibiotics cleared through the kidneys and many anti-neoplastic agents that are also cleared through the kidneys (Antoch and Kondratov, 2013; Roenneberg et al., 2013; Ortiz-Tudela et al., 2013)

Indeed, it has also been found that mice better tolerate the administration of adriamycin if exposure was during the latter half of the animal's rest period (daytime) and that cisplatin was best tolerated when administered in the latter half of the animal's activity period (nighttime) (Lévi et al., 1988; Lėvi, 2010). Shifting now to the clinical arena, when patients were exposed to adriamycin in the morning and cisplatin in the evening or adriamycin in the evening and cisplatin in the morning, complications and side effects were significantly worse for patients receiving adriamycin in the evening and cisplatin in the morning (Antoch and Kondratov, 2013; Roenneberg et al., 2013; Ortiz-Tudela et al., 2013). Similar results have been found in studies using human subjects exposed to 5-fluorouracil, cisplatin, or vindesine (Antoch and Kondratov, 2013; Roenneberg et al., 2013; Ortiz-Tudela et al., 2013).

Biological Rhythms and Mixed-Function Oxidases

Drug metabolism is the metabolic breakdown of drugs by living organisms, usually through specialized enzymatic systems. More generally, xenobiotic metabolism (from the Greek *xenos*, "stranger" and *biotic*, "related to living beings") is the set of metabolic pathways that modify the chemical structure of xenobiotics, which are compounds foreign to an organism's normal biochemistry, such any drug or poison. These pathways are a form of biotransformation present in all major groups of organisms, and are considered to be of ancient origin. These reactions often act to detoxify poisonous compounds (although in some cases the intermediates in xenobiotic metabolism can themselves cause toxic effects). The study of drug metabolism is called pharmacokinetics.

The metabolism of pharmaceutical drugs is an important aspect of pharmacology and medicine. For example, the rate of metabolism determines the duration and intensity of a drug's pharmacologic action. Drug metabolism also affects multidrug resistance in infectious diseases and in chemotherapy for cancer, and the actions of some drugs as substrates or inhibitors of enzymes involved in xenobiotic metabolism are a common reason for hazardous drug interactions. These pathways are also important in environmental science, with the xenobiotic metabolism of microorganisms determining whether a pollutant will be broken down during bioremediation, or persist in the environment. The enzymes of xenobiotic metabolism, particularly the glutathione S-transferases (GSTs), are also important in agriculture, since they may produce resistance to pesticides and herbicides.

Drug metabolism is divided into three phases. In phase I, enzymes such as cytochrome P450 oxidases introduce reactive or polar groups into xenobiotics. These modified compounds are then conjugated to polar compounds in phase II reactions. These reactions are catalyzed by transferase enzymes such as GSTs. Finally, in phase III, the conjugated xenobiotics may be further processed, before being recognized by efflux transporters and pumped out of cells. Drug metabolism often converts lipophilic compounds into hydrophilic products that are more readily excreted.

Most xenobiotics and drugs are eliminated from the body by drug metabolism, which involves a variety of different chemical reactions catalyzed by a wide spectrum of enzymatic systems (e.g., P-450 monooxygenases) that are for the most part located in the mammalian liver. The enzymes performing these metabolic changes exist in a variety of different isomeric forms, each with different kinetic properties and different substrate requirements. There are a number of reviews in the literature that describe the circadian oscillations of hepatic metabolism (Bélanger et al., 1991, 2010; Bélanger, 1987, 1993; Bélanger and Labrecque, 1989).

Phase I reactions are those that introduce a polar reactive group into lipophilic drugs or xenobiotics. In most cases this group becomes the site for conjugation during phase II reactions. Such reactions include microsomal mono-oxygenations, cytosolic and mitochondrial oxidations, co-oxidations in prostaglandin synthetase reactions, reductions, hydrolysis reactions, and epoxide hydration. The products of phase I reactions may be potent electrophiles that can be conjugated and detoxified in phase II reactions or which may react with nucleophilic groups or cellular constituents, thereby showing greater toxicity than the parent compound. The epoxidation of polycyclic aromatic hydrocarbons is one example of a typical phase I reaction.

Cytochromes P450 (CYPs) are proteins of the superfamily containing the heme group as a cofactor and, therefore, are hemoproteins. CYPs use a variety of small and large molecules as substrates in enzymatic reactions. They are, in general, the terminal oxidase enzymes in electron transfer chains, broadly categorized as P450-containing systems. The term "P450" is derived from the spectrophotometric peak at the wavelength of the absorption maximum of the enzyme (450 nm) when it is in the reduced state and complexed with carbon monoxide. CYP enzymes have been identified in all domains of life: animals, plants, fungi, protists, bacteria, archaea, and even in viruses. However, they are not omnipresent; for example, they have not been found in *Escherichia coli*. More than 21,000 distinct CYP proteins are known. Most CYPs require a protein partner to deliver one or more electrons to reduce the iron (and eventually molecular oxygen). Based on the nature of the electron transfer proteins, CYPs can be classified into several groups:

Microsomal P450 systems, in which electrons are transferred from NADPH via cytochrome P450 reductase (variously CPR, POR, or CYPOR). Cytochrome b5 (cyb5) can also contribute reducing power to this system after being reduced by cytochrome b5 reductase (CYB5R).

- Mitochondrial P450 systems, which employ adrenodoxin reductase and adrenodoxin to transfer electrons from NADPH to P450.
- Bacterial P450 systems, which employ a ferredoxin reductase and a ferredoxin to transfer electrons to P450.
- CYB5R/cyb5/P450 systems, in which both electrons required by the CYP come from cytochrome b5.
- FMN/Fd/P450 systems, originally found in Rhodococcus species, in which a FMN-domain-containing reductase is fused to the CYP.
- P450–only systems, which do not require external reducing power. Notable ones include thromboxane synthase (CYP5), prostacyclin synthase (CYP8), and CYP74A (allene oxide synthase).

Humans have 57 genes and more than 59 pseudogenes divided among 18 families of cytochrome P450 genes and 43 subfamilies. A summary of the human cytochrome P450 genes and the proteins that they encode is listed in Table 7.6 (Parkinson et al., 2013).

Of these enzymes and enzymatic systems referred to above, oxidative reactions are the most common, the most frequently studied, and the most important. The spectrum of substrates involved and the biological consequences of the products are wide. Indeed, oxidation reactions are in a large number of cases the rate-limiting process in the clearance of a drug from the body and they are the central point of consideration for many drug interactions that are of interest from a clinical perspective (Bélanger et al., 2010). Initial reports regarding the circadian oscillations of hepatic microsomal metabolism in the mouse and rat appeared in the late 1960s (Nair and Casper, 1969; Radzialowski and Bousquet, 1967). In these reports in the literature the values of oxidase activities were compared for the time periods of 22.00 hr–02.00 hr and 10.00 hr–14.00 hr; the highest levels of activities were found in the period 22.00 hr–02.00 hr and the lowest values were found in the period 10.00–14.00 hr. Utilizing purified microsomes isolated from rat livers, the oxidase activities for different substrates were determined; it was determined that the activities of aminopyrene de-ethylases and aniline hydroxylase were found to be higher at 21.00 hr than at 09.00 hr (Belanger and Lalande, 1988). However these researchers also found that the activity levels of testosterone 7α– and 6β–hydroxylases and the P-450 reductase were lower in the evening hours. When results for activity levels for testosterone 16α-hydroxylase were investigated, no diurnal rhythm was found and there was also no circadian effect on total

TABLE 7.6

Summary of the Most Common Human Cytochrome P-450 Genes and the Proteins That They Encode

Family	Members	Names	Function
CYP1	3 subfamilies, 3 genes, 1 pseudogene	CYP1A1, CYP1A2, CYP1B1	Drug and steroid (especially estrogen) metabolism, benzo[a] pyrene (forming (+)-benzo[a] pyrene-7,8-dihydrodiol-9,10-epoxide)
CYP2	13 subfamilies, 16 genes, 16 pseudogenes	CYP2A6, CYP2A7, CYP2A13, CYP2B6, CYP2C8, CYP2C9, CYP2C18, CYP2C19, CYP2D6, CYP2E1, CYP2F1, CYP2J2, CYP2R1, CYP2S1, CYP2U1, CYP2W1	Drug and steroid metabolism
CYP3	1 subfamily, 4 genes, 2 pseudogenes	CYP3A4, CYP3A5, CYP3A7, CYP3A43	Drug and steroid (including testosterone) metabolism
CYP4	6 subfamilies, 12 genes, 10 pseudogenes	CYP4A11, CYP4A22, CYP4B1, CYP4F2, CYP4F3, CYP4F8, CYP4F11, CYP4F12, CYP4F22, CYP4V2, CYP4X1, CYP4Z1	Arachidonic acid or fatty acid metabolism
CYP5	1 subfamily, 1 gene	CYP5A1	Thromboxane A2 synthase
CYP7	2 subfamilies, 2 genes	CYP7A1, CYP7B1	Bile acid biosynthesis 7-alpha hydroxylase of steroid nucleus
CYP8	2 subfamilies, 2 genes	CYP8A1 (prostacyclin synthase), CYP8B1 (bile acid biosynthesis)	Various multiple
CYP11	2 subfamilies, 3 genes	CYP11A1, CYP11B1, CYP11B2	Steroid biosynthesis
CYP17	1 subfamily, 1 gene	CYP17A1	Steroid biosynthesis, 17-alpha hydroxylase
CYP19	1 subfamily, 1 gene	CYP19A1	Steroid biosynthesis: aromatase synthesizes estrogen
CYP20	1 subfamily, 1 gene	CYP20A1	Unknown function
CYP21	2 subfamilies, 1 gene, 1 pseudogene	CYP21A2	Steroid biosynthesis
CYP24	1 subfamily, 1 gene	CYP24A1	Vitamin D degradation
CYP26	3 subfamilies, 3 genes	CYP26A1, CYP26B1, CYP26C1	Retinoic acid hydroxylase
CYP27	3 subfamilies, 3 genes	CYP27A1 (bile acid biosynthesis), CYP27B1 (vitamin D3 1-alpha hydroxylase, activates vitamin D3), CYP27C1 (unknown function)	Various multiple

(Continued)

TABLE 7.6 (*Continued*)

Summary of the Most Common Human Cytochrome p-450 Genes and the Proteins That They Encode

Family	Members	Names	Function
CYP39	1 subfamily, 1 gene	CYP39A1	7-alpha hydroxylation of 24-hydroxycholesterol
CYP46	1 subfamily, 1 gene	CYP46A1	Cholesterol 24-hydroxylase
CYP51	1 subfamily, 1 gene, 3 pseudogenes	CYP51A1 (lanosterol 14-alpha demethylase)	Cholesterol biosynthesis

amounts of cytochrome P-450 and protein. Interestingly, a very significant difference was found when daytime versus nighttime levels of the microsomal concentration of fatty acids were compared. At 09.00 in the morning the levels of various lipids were found to be much higher than in the evening, with the exception that arachidonic acid levels exhibited measurable amounts at 21.00 hr. The activity of each oxidase studied is mediated selectively by at least one cytochrome P-450 isoenzyme (Guengerich, 1987). The changes in activities of the various cytochromes P-450 in time with exposure to differing substrates support the contention that there are indeed diurnal variations in enzymatic kinetic parameters as well as amounts of some P-450 isoenzymes. The cytochromes P-450 isozymes IIB1, IIC11, and IIc6 were examined for temporal oscillatory behavior in rat liver microsomes using the Western blotting technique, selectivity for enantiomers, and the regiospecificity of testosterone hydroxylases at the 2α, 6β, 16α, and 16β positions (Miyazaki et al., 1990). Significant differences were observed in the temporal rhythms of total P-450 amounts for the various isoenzymes. Similarly, typical variations in time were found in the rates of various hydroxylations of testosterone at different sites on the core molecule (e.g. carbons 2, 6, and 16) or on the same carbon atom with products of differing conformation being generated (2α or 2β and 16α or 16β). However, correlation was found to be poor between oxidase activities and the total amount of microsomes. A lot of work remains to be performed in this area.

It is a simple and well-known fact that most drugs are metabolized to variously different products via oxidation by hepatic cytochrome P-450 isoenzymes that demonstrate differing temporal oscillations. This fact is undoubtedly why a large number of studies published in the literature have failed to demonstrate temporal changes in elimination parameters such as plasma half-life and non-renal clearance. The *in vivo* characterization of selected cytochrome P-450 isoenzymes is and has been studied in both animals and humans very adequately, using drugs that are metabolized to one or two major metabolites in reactions catalyzed by one specific family of cytochrome P-450 isoenzymes. The only conditions for these types of studies are that the drug has to demonstrate a low hepatic extraction ratio (Wilkinson and Shand, 1975). Examples of these types of compounds include phenylbutazone, tolbutamide, phenobarbital, antipyrine, and phenytoin (Rowland and Tozer, 1980). An approximate 40% decrease in the serum half-life of antipyrine in rats was observed when the drug was administered at 08.00 hr and an equivalent increase in the metabolic clearance of the drug when it was administered at 20.00 hr. Phenylbutazone and its major hydroxylated metabolite have been shown to demonstrate higher plasma levels at 20.00 hr and 01.00 hr than at 07.00 hr and 13.00 hr (Loubaris et al., 1984). With regard to the anti-inflammatory drug sundilac, higher plasma levels of the sulfone metabolite in humans have been observed when the drug was administered at 09.00 hr as opposed to 21.00 hr (Swanson et al., 1982). Circadian oscillations in the pharmacokinetic profile for aminopyrene demonstrated a 50% increase in the half-life and a 20% decrease in the clearance value when administered at 20.00 hr when compared to an administration at 08.00 hr (Poley et al., 1978). Circadian changes or rhythms in various pharmacokinetic parameters have been reported in studies in humans administered carbamazepine, phenytoin, methotrexate, and caffeine (Koren et al., 1992; Hartley et al., 1991; Garrettson and Jusko, 1974; Levy et al., 1984). These and other reports in the literature demonstrate that the hepatic microsomal metabolism of drugs and xenobiotics in both animals and humans are subject to diurnal variations in their pharmacokinetic profiles (Bélanger et al., 2010).

Hepatic Conjugation Reactions and Diurnal Variations

In subsequent phase II reactions, these activated xenobiotic metabolites are conjugated with charged species such as glutathione (GSH), sulfate, glycine, or glucuronic acid. Sites on drugs where conjugation reactions occur include carboxyl (-COOH), hydroxyl (-OH), amino (NH_2), and sulfhydryl (-SH) groups. Products of conjugation reactions have increased molecular weight and tend to be less active than their substrates, unlike phase I reactions, which often produce active metabolites. The addition of large anionic groups (such as GSH) detoxifies reactive electrophiles and produces more polar metabolites that cannot diffuse across membranes, and may, therefore, be actively transported.

These reactions are catalyzed by a large group of broad-specificity transferases, which in combination can metabolize almost any hydrophobic compound that contains nucleophilic or electrophilic groups. One of the most important classes of this group is that of the GSTs. Some of the most common phase II reactions are outlined in Table 7.7 (Parkinson et al., 2013).

After phase II reactions, the xenobiotic conjugates may be further metabolized via what is termed phase III modification. A common example is the

TABLE 7.7

Summary of Some of the Most Common Phase II Conjugation Reactions

Enzyme	Location	Mechanism of Action	Enzymatic Cofactor
Methyltransferase	liver, kidney, lung, CNS	methylation	S-adenosyl-L-methionine
Sulfotransferases	liver, kidney, intestine	sulfation	3′-phosphoadenosine-5′-phosphosulfate
N-acetyltransferases bile acid-CoA:amino acid N-acyltransferases	liver, lung, spleen, gastric mucosa, RBCs, lymphocytes	acetylation	acetyl coenzyme A
UDP-Glucuronosyltransferases	liver, kidney, intestine, lung, skin, prostate, brain	glucuronidation	UDP-glucuronic acid
Glutathione S-transferases	liver, kidney	glutathione conjugation	glutathione
Acetyl co-enzyme As	liver, kidney	glycine conjugation	glycine

processing of glutathione conjugates to acetylcysteine (mercapturic acid) conjugates. Here, the γ-glutamate and glycine residues in the glutathione molecule are removed by gamma-glutamyl transpeptidase and dipeptidases. In the final step, the cystine residue in the conjugate is acetylated.

Conjugates and their metabolites can be excreted from cells in phase III of their metabolism, with the anionic groups acting as affinity tags for a variety of membrane transporters of the multidrug resistance protein (MRP) family. These proteins are members of the family of ATP-binding cassette transporters and can catalyse the ATP-dependent transport of a huge variety of hydrophobic anions, and thus act to remove phase II products to the extracellular medium, where they may be further metabolised or excreted.

Phase II reactions are those that involve the conjugation of products generated from phase I and other xenobiotics containing functional groups such as hydroxyl, amino, carboxyl, epoxide, or halogens with endogenous metabolites. These endogenous metabolites include sugars, amino acids, glutathione, and sulfate. The conjugation products, with rare exceptions, are more polar, less toxic, and more readily excreted than their parent compounds. Conjugation reactions usually involve high-energy intermediates and have typically been classified into two general types: type I (glycoside and sulfate formation), in which an activated conjugating agent combines with the substrate to yield a conjugated product; and type II (amino acid conjugation), in which the substrate is activated and then combines with an amino acid to yield a conjugated product. The most important conjugation reactions are glycoside formation (especially glucuronidation), sulfate formation, glutathione conjugation, and mercapturic acid formation and acylation (especially amino acid conjugation).

The literature is sparsely populated with reports on the circadian oscillations observed in non-oxidative reactions involved in drug metabolism (Bélanger, 1987; Bélanger et al., 2010). Indeed, only a couple of publications report on the circadian oscillation behavior in rat liver homogenate of the conjugation of glucuronic acid, sulfate, and acetate (Bélanger et al., 2010; Bélanger, 1985, 1989). In this work, the rates of glucuronidation and acetylation were found to be significantly elevated at 21.00 hr, when compared to the levels observed at 09.00 hr. However, sulfation was found to follow a reverse pattern, with activity being twice as high at 09.00 hr when compared to 21.00 hr. Notably, these changes were found to be independent of protein concentration used in the assays. Interestingly, sulfotransferase activity was found to correlate with a fourfold decrease in the affinity of the substrate (Bélanger et al., 2010). The increased clearance of isoniazid was found to correlate well with a chronological pattern as regards the activity of N-acetyl transferase at 09.00 hr and 21.00 hr, with clearance being significantly higher at 21.00 hr (Bélanger et al., 2010, 1989). The circadian oscillatory pharmacokinetic behavior of hepatic sulfation and glucuronidation in both humans and rats may explain the diurnal variation in clearance of acetaminophen in humans and rats (Bélanger et al., 1987, 2010; Shiveley and Vessel, 1975). The plasma half-life of acetaminophen in humans was found to be approximately 20% longer at 06.00 hr than at 14.00 hr. Furthermore, the average ratio of glucuronide-acetaminophen conjugate to unaltered parent acetaminophen excreted in the first 3.5 hr urine sample varied from a value of 5.2 at 06.00 hr to a value of 7.8 at 14.00 hr (Shiveley and Vessel, 1975; Bélanger et al., 2010). However, in the rat, it was observed that clearance and metabolism of acetaminophen determined as the extraction ratio were both greater in the evening than in the morning (Bélanger et al., 2010, 1987).

A significant source of sulfhydryl groups within the cell is the tripeptide compound glutathione (Meister and Anderson, 1983; Kaplowitz et al., 1985). This compound in its reduced form has two main functions. Glutathione in its reduced form is a substrate for peroxidase redox system, which removes peroxides formed as a result of other enzymatically driven metabolic reactions. Additionally, glutathione in its reduced form and possibly most importantly forms adducts with electrophiles, free radicals, or reactive intermediates of toxic drugs that are produced by the cytochrome P-450 monooxygenases (Desgagné et al., 1988; Bélanger et al., 2010). Succinctly put, the conjugation to glutathione is a process that protects vital cellular components from reactive oxygen species of xenobiotics (Lavigne et al., 1983; Bélanger et al., 2010). Prominent circadian oscillations in the hepatotoxicity of chloroform have been reported to be inversely related to the diurnal variation of the hepatic concentration of glutathione (Bélanger et al., 2010, 1991; Desgagné et al., 1988). The lowest levels of glutathione in the rat were found at 21.00 hr and the highest levels of glutathione were found at 09.00 hr, which correlates very well to the maximal and minimal levels of hepatotoxicity.

Similar results and correlations were found as regards the hepatotoxicity of 1,1-dichloroethylene, styrene, acetaminophen, and allyl alcohol (Jaeger et al., 1973; Schnell et al., 1983; Desgagné and Bélanger, 1986; Bélanger et al., 2010; Hanson and Andres, 1978).

Biological Rhythms and the Otic System

Hearing function has been evaluated in rats exposed to kanamycin (Fisch et al., 1984). Kanamycin was administered in a study to female Sprague-Dawley rats four times a day via a subcutaneous route at 02:00 hr, 08:00 hr, 14:00 hr, and 20:00 hr of each day under light-synchronized conditions for six weeks. Observations at 2 weeks post dosing revealed that the 08:00 hr dose group showed significant hearing loss, while there were minimal changes recorded in the other animals and dose groups. However, it was then later discovered that after 6 weeks exposure, the 08:00 hr and 14:00 hr groups both demonstrated similar dramatic hearing loss, with other animals revealing minimal changes. The conclusion of the study was that it appeared that kanamycin-induced auditory damage is more dramatic during the day (rest period) than during the night (activity period). The improved nighttime tolerance was discovered also to be correlated with the peak of body temperature – another possible rhythm marker in aminoglycoside-induced toxicity.

Timed Oscillation Patterns Associated with Exposures to Substances Other Than Drugs

It is important to appreciate that it is not just chemical agents that can be associated with time-dependent toxicological effects. These same diurnal patterns of timing have also been associated with exposures to viruses, fungi, venoms, toxins, and bacteria (Bélanger, 1987; Cambar et al., 1992; Azou et al., 1992). The toxins released by these agents as part of their pathophysiologic activities can affect cell mobidity, mortality, and damage, the severity of which can be strongly related to the time of exposure. Mice exposed to *Brucella* endotoxin exposed at nighttime lived for approximately 75 hours, but those exposed during the daytime survived only 37 hours. A time-dependent change in locomotor activity in hamsters has been observed with exposure to the protein synthesis inhibitor anisomycine. The locomotor phase change is delayed if the anisomycine is administered at midnight, but accelerated if administered at midday. Interestingly, the induction of seizures by the presence of white noise (100 dB for 60 seconds) in mice has been shown to be circadian-dependent and X-ray irradiation has been similarly demonstrated to be circadian-dependent as measured by the mortalities observed in both *Drosophila* and mice (Hayes and Morgan, 1988). With regard to the irradiation, a dose of 555 r administered at night resulted in 100% mortality, but if done during the day the exposure was not lethal at all. Mice injected with *Escherichia coli* endotoxin at 0:00 hr and 16:00 hr showed survival rates of 85% and 20%, respectively. Tetrodotoxin (*Spheroides rubripes*) exposure results in significant changes in vasopressin release from the suprachiasmatic nucleus in the rat (Earnest et al., 1991). However, if tetrodotoxin is perfused into the rat during nighttime (12 hr period), no alterations of vasopressin release were observed, but when tetrodotoxin was perfused into rats during the daytime (12 hr), the levels of vasopressin were almost suppressed. Scorpion (*Heterometrus fulvipes*) venom has demonstrated a time-dependent toxicity in tropical mice (*Mus booduga*) as evaluated with the measurement of the blood levels of acetylcholine and acetylcholinesterase.

Patterns in Time and Other Physiologic Phenomena

There are other considerations that can alter the pharmacokinetic profile of a drug, such as the first-pass effect and blood flow (Bélanger and Labrecque, 1992). These have been found to exhibit circadian rhythms.

Blood flow through the liver is a significant factor in the clearance of drugs after intravenous administration, especially those with a high hepatic extraction ratio (Wilkinson and Shand, 1975). Hepatic blood flows have been demonstrated to be strongly correlated with extraction and rates of biotransformation (Bélanger and Labrecque, 1992). Elevated C_{max} +/– shorter T_{max} values can be observed as a result of differences in blood flow in the morning versus evening ingestion. Changes in blood flow in time in humans were found to be highest at 02.00–8.00 hr and lowest at 14.00 hr (Lemmer and Nold, 1991). Propanolol was found to have the shortest plasma half-life at 08.00 hr and the longest plasma half-life at 20.00 hr (Langner and Lemmer, 1988). The distribution of labelled microspheres in the rat in a variety of organs demonstrated that the lowest distribution of the microspheres in liver occurred at 15.00 hr and the highest distribution of microspheres occurred at 21.00 hr (Labrecque et al., 1988).

The first-pass effect is a process in which a substance is removed from the blood by the liver before reaching the systemic circulation. It occurs because, anatomically, blood from the upper portion of the gastrointestinal tract passes through the liver before reaching the general venous system. As a result, the measured systemic bioavailability may be less than the fraction of the dose absorbed from the gastrointestinal tract. The first-pass effect (also known as first-pass metabolism or presystemic metabolism) is a phenomenon of drug metabolism whereby the concentration of a drug is greatly reduced before it reaches the systemic circulation. Notable drugs that experience a significant first-pass effect are imipramine, morphine, propranolol, buprenorphine, diazepam, midazolam, pethidine, cimetidine, lidocaine, and nitroglycerin.

After a drug is swallowed, it is absorbed by the gastrointestinal tract and enters the hepatic portal system. From there, it is carried through the portal vein into the liver before it reaches the rest of the body. The liver metabolizes many chemicals, sometimes to such an extent that only a small amount of active drug emerges from the liver to the rest of the circulatory system. This *first pass* through the liver thus greatly reduces the bioavailability of the drug. The four primary systems that affect the first-pass effect of a drug are the enzymes of the gastrointestinal lumen, gut wall enzymes, bacterial enzymes, and hepatic enzymes (Bélanger et al., 2010). In drug design, candidates may have good drug-like properties, but fail on first-pass metabolism considerations and behavior.

Alternative routes of administration like suppository, intravenous, intramuscular, inhalational aerosol, transdermal, and sublingual avoid the first-pass effect because they allow drugs to be absorbed directly into the systemic circulation. Note that the intravenous route also avoids the absorption phase.

Drugs with high first-pass effect have a considerably higher oral dose than sublingual or parenteral dose. There is marked individual variation in the oral dose due to differences in the extent of first-pass metabolism. Oral bioavailability is apparently increased in patients with severe liver diseases like cirrhosis. It is also increased if another drug competing with it in first-pass metabolism is given concurrently, such as propranolol and chlorpromazine.

Subsequently, if a drug is totally absorbed and then eliminated by biotransformation, clearance of the drug from the body after oral administration is related to the drug-metabolizing capability independent of its hepatic extraction ratio (Wilkinson and Shand, 1975; Bélanger et al., 2010). The first-pass effect is significant only for drugs that exhibit a high hepatic extraction ratio. If one assumes complete oral absorption, the temporal variations in the first-pass effect and hence bioavailability have been demonstrated to be significant for many drugs (Bélanger et al., 2010). In the rat, isoniazid and acetaminophen demonstrated a first-pass effect of 70% at 09.00 hr and 40% at 21.00 hr (Bélanger et al., 1987, 1989). Surprisingly, the oral bioavailabilities of isoniazid and acetaminophen differed significantly as a function of time. Similar profiles have been generated and reported in humans for nifedipine, propranolol, and prednisone (English et al., 1983; Langner and Lemmer, 1988; Lemmer and Nold, 1991). However, in all cases, the AUC values of the plasma drug concentrations determined after oral ingestion were always higher in the morning hours, 08.00 hr–12.00 hr, than at any other time of the day (Bélanger et al., 2010). Accordingly, the oral bioavailability of drugs undergoing a significant first-pass effect is likely to show a temporal oscillation in their plasma levels.

Red blood cells can be viewed as a tissue compartment into which free or unbound test article present in the plasma can diffuse. The circadian oscillations of the binding of drug to red blood cells for local anesthetics and other drugs have been investigated (Bruguerolle and Jadot, 1983; Bruguerolle and Prat, 1987, 1988, 1989, 1990). For the case of lidocaine, it has been reported that lidocaine administered to rats at 22.00 hr demonstrated a ratio of drug concentration in erythrocytes to total plasma concentration of 0.74 but a ratio of 0.48 when the lidocaine was administered at 10.00 hr (Bruguerolle and Jadot, 1983).

Internal Clock and Mathematical Modeling

Computer-controlled mathematical models of medical outcomes are becoming increasingly prevalent in the literature. What is less common is an understanding of the methods used to construct such models, leaving the potential users in biology, medical, and pharmacologic research to just accept the interpretations as presented. A basic knowledge of the concepts used to generate models should provide the researcher with the insight needed to critically evaluate data appearing in the literature that is based on mathematical models.

The development of computerized mathematical models used to simulate medical outcomes is a growing area of specialization (Chambers, 2000). A search of articles using mathematical models to generate data yielded 43,764 articles dating from 1966. The majority (97%) of the manuscripts, including mathematical models, have been published since 1990. Since 1999, 9,219 articles have been published.

Researchers are beginning to accept the conclusions drawn from modeling, often without realizing that the data are simulated. Regression methods, the most commonly seen in modeling, use some variation of the classical linear model, $y = mx + b$, according to a transformation or derivation that plots a math function closely describing the data (Chambers, 2000). This is not new to biometrics, but the historical use is to compare two groups by the parameters of their lines from measured observations. Using the derived regression to predict outcomes from the same population has always been an accepted application of regression. Making the jump to using mathematical modeling to generate data from simulated population is a more difficult stretch.

However, skepticism at believing something that was not only unseen and unmeasured by the reader, but was also unseen and unmeasured by the math modelers, continues to be a barricade.

The suprachiasmatic nucleus (SCN) (see the section "Circadian Clock") in mammals is believed to be the central circadian oscillator. Neurons in the SCN are coupled via various neurotransmitters and gap junctions (Slat et al., 2013; Bordyugov et al., 2013). These neurons work together in synchrony to produce very precise controlling rhythms, which coordinate peripheral organs through neuronal and chemical pathways (Bordyugov et al., 2013). In mammals, the circadian clock can be viewed as an interconnected collection of simple oscillators. Transcriptional feedback loops in cells generate rhythms with a period of approximately 24 hours (Minami et al., 2013; Buhr and Takahashi, 2013). A number of different "clock" genes exist in a wide variety of tissues (Brown and Azzi, 2013; Keller et al., 2009; Bordyugov et al., 2013). Overall, this molecular clock directs the timing of a wide array of physiological and metabolic processes within the organism (Hastings et al., 2003). The SCN clock is itself modified by external light and dark and temperature cycles. The consumption of food is another modifying influence that can affect the circadian rhythms in the liver (Stokkan et al., 2001). Knowledge of these rhythms can be useful in the optimization of the timing of pharmacologic and therapeutic intervention in a variety of medical conditions (Lévi et al., 1997). The overall complexity of these systems has driven the development of systems-based biological approaches and mathematical modeling applied to determine the usefulness of circadian rhythms and chronotherapy (Ukai and Ueda, 2010). Various attempts at the mathematical modeling of circadian rhythms have already been made (Bordyugov et al., 2013; Daan and Berde, 1978; Winfree, 1980). Indeed, the most popular of these models is the Hodgkin–Huxley model, established in 1952 (Hodgkin and Huxley, 1952). However, this model is very cumbersome to use. Computers should permit accelerated development of these types of models and many more models yet to be developed. Mathematical models can deeply explore the functions of feedback loops, sensitivity to variations in the various parameters, noise, and the actual efficacy of various chronotherapies. These initial attempts at mathematically modeling the circadian oscillation process were created in an attempt to better understand seasonal variations and in general the synchronization of organisms to external rhythms and stimuli and to better understand the optimization of responses of circadian oscillators (Kronauer et al., 1982; Westermark et al., 2009; Bordyugov et al., 2011, 2013). Over time, detailed models of the core circadian clock have been developed (Leloup and Goldbeter, 2003; Forger and Peskin, 2003; Becker-Weimann et al., 2004). These models provide greater detail in the process describing nuclear translocation, complex formation, protein degradation, post-translational modifications, protein expression, and transcriptional regulation (Mirsky et al., 2009; Relógio et al., 2011). However, overall the models that have been developed have not yet taken researchers to where they want to be. In actuality, simple cell models still remain the best predictors for a simulation of the activity of coupled cells (Gonze et al., 2005). There have even been recent attempts to connect clock models to the process of cell proliferation in an attempt to simulate chronotherapy and its effectiveness (Lévi et al., 2008). At this current time, mathematical modeling in chronobiology focuses on three different areas: (1) chronotherapy and its optimization, (2) a negative feedback loop simple oscillator model, and (3) coupling of cells through the use of external stimuli (Bordyugov et al., 2013).

Conclusion

Toxicologists have many times taken issue with some results reported in the literature and the discrepancies that were found to exist. The data from this relatively new area of research mandate that it is necessary to critically evaluate and reevaluate toxicity, especially if mortality is involved with respect to exposures and the time of the day and even the time of the year. The examples provided above demonstrate that diurnal and timed patterns of other periodicity patterns do indeed exist and can significantly alter the susceptibility and response levels of living organisms. It is now fundamental for toxicology reports and studies to state not only the lighting schedule for animals on a given study, but also the exact hour of the day and the month of the year to be able to fairly and accurately compare toxicity results. The lack of adequate consideration of this timing parameter can result in significantly altered efficacy profiles, toxicity profiles, therapeutic indices, safety factors, and tolerance. These days, one should keep in mind the potential utility of using circadian oscillations to investigate or maximize the pharmacology, pharmacodynamics, and bioavailability of a test article in preclinical development. While not all animal profiles necessarily translate completely to the human, nevertheless the consideration of circadian rhythms can be something that is investigated in the drug development process as part of the formulation development component. The existence, importance, and roles of circadian rhythms have been reported and documented in the literature in development, aging, asthma, pharmacokinetics, cardiovascular disease, endocrine disease, sleep, hormone levels, treatment of cancer, function and synthesis of neurotransmitters, diabetes, noradrenaline synthesis, clinical depression, behavior, metabolism, the function of the hematopoietic system, the function of the hemostatic system, allergy, efficacy of local anesthetics, the pharmacology of H_1-receptor antagonists, analgesia, and the treatment of pain (Antoch and Kondratov, 2013; Belanger et al., 2010; Redfern and Lemmer, 2010).

References

Albrecht U. (2013) Circadian clocks and mood-related behaviors. In: A Kramer and M Merrow (Eds.) *Circadian Clocks, Handbook of Experimental Pharmacology, Circadian Control of Physiology and Behavior (Part II)*, Vol. 217. Germany: Springer-Verlag Berlin Heidelberg, pp. 227–242.

Angeli A, Frajria R, DePaoli R, Fonzo D, Ceresa F. (1978) Diurnal variation of prednisolone binding to serum corticosteroid binding-globulin in man. *Clin Pharmacol Ther*, 23:47–53.

Antoch MP, Kondratov RV. (2013) Pharmacological modulators of the circadian clock as potential therapeutic drugs: Focus on genotoxic/anticancer therapy. In: A Kramer and M Merrow (Eds.) *Circadian Clocks, Handbook of Experimental Pharmacology, Chronopharmacology and Chronotherapy (Part III)*, Vol. 217. Germany: Springer-Verlag Berlin Heidelberg, pp. 289–310.

Araki S, Murata K, Yokoyama K, Yanaghara S, Niinuma Y, Yamanoto R, Ishihara N. (1983) Circadian rhythms in the urinary excretion of metals and organic substances in healthy men. *Arch Environ Health*, 38:360–366.

Arendt J. (2010) The pineal gland, circadian rhythms and photoperiodism. In: PH Redfern and B Lemmer (Eds.) *Physiology and Pharmacology of Biological Rhythms, Handbook of Experimental Pharmacology*, Vol. 125. Germany: Springer-Verlag Berlin Heidelberg, pp. 375–414.

Aungst BJ, Fung H-L. (1981) Intestinal lead absorption in rats: Effects of circadian rhythm, food, undernourishment, and drugs which alter gastric emptying and GI motility. *Res Commun Chem Path Pharmacol*, 34:515–530.

Batalla A, Malmary MF, Cambar J, Labat C, Oustrin J. (1994) Dosing-time dependent nephrotoxicity of cyclosporin A during 21-day administration to Wistar rats. *Chronobiol Int*, 11(3):195.

Becker-Weimann S, Wolf J, Herzel H, Kramer A. (2004) Modeling feedback loops of the mammalian circadian oscillator. *Biophys J*, 87:3023–3034.

Beckett AH, Rowland M. (1964) Rhythmic urinary excretion of amphetamine in man. *Nature*, 204:1203–1204.

Bélanger PM. (1987) Chronobiological variation in the hepatic elimination of drugs and toxic chemical agents. *Ann Rev Chronopharmacol*, 4:1–46.

Bélanger PM. (1993) Chronopharmacology in drug research and therapy. *Adv Drug Res*, 24:1–80.

Bélanger PM, Desgagné M, Bruguerolle B. (1991) Temporal variations in microsomal lipid peroxidation and in glutathione concentration of rat liver. *Drug Metab Disp*, 19:241–244.

Bélanger PM, Labrecque G. (1989) Temporal aspects of drug metabolism. In: B Lemmer (Ed.) *Chronopharmacology: Cellular and Biochemical Interactions*, Chap. 2. Basel: Marcel Dekker, pp. 15–34.

Bélanger PM, Labrecque G. (1992) Biological rhythms in the hepatic drug metabolism. In: Y Touitou and E Haus (Eds.) *Biological Rhythms in Clinical and Laboratory Medicine*. New York, NY: Springer-Berlin Heidelberg, pp. 403–409.

Bélanger PM, Lalande M, Doré F, Labrecque G. (1987) Time-dependent variations in the organ extraction ratios of acetaminophen in rat. *J Pharmacokin Biopharm*, 15:133–143.

Bélanger PM, Lalande M, Doré F, Labrecque G. (1989) Temporal variations in the pharmacokinetics of isoniazid and N-acetylisoniazid in rats. *Drug Metab Disp*, 17:91–97.

Bélanger PM, Lalande M, Labrecque G, Doré F. (1985) Diurnal variations in the transferases and hydrolases involved in glucuronide and sulfate conjugation of rat liver. *Drug Metab Disp*, 13:386–389.

Bélanger PM, Bruguerolle B, Labrecque G. (2010) Rhythms in pharmacokinetics: Absorption, distribution, metabolism, and excretion. In: PH Redfern and B Lemmer (Eds.) *Physiology and Pharmacology of Biological Rhythms, Handbook of Experimental Pharmacology*, Vol. 125. Germany: Springer-Verlag Berlin Heidelberg, pp. 177–204.

Bélanger PM, Lalande M. (1988) Day-night variations in the activity and composition of the microsomal mixed-function oxidase of rat liver. *Ann Rev Chronopharmacol*, 5:219–222.

Bogentoft C, Carlson I, Ekenved G, Magnusson A. (1978) Influence of food on the absorption of acetylsalicylic acid from enteric-coated dosage forms. *Eur J Clin Pharmacol*, 15:351–355.

Bordyugov G, Granada A, Herzel H. (2011) How coupling determines the entrainment of circadian clocks. *Eur Phys J B*, 82:227–234.

Bordyugov G, Westermark PO, Korenčič A, Bernard S, Herzel H. (2013) Mathematical modeling in chronobiology. In: A Kramer and M Merrow (Eds.) *Circadian Clocks, Handbook of Experimental Pharmacology, Systems Biology of Circadian Clocks (Part IV)*, Vol. 217. Germany: Springer-Verlag Berlin Heidelberg, pp. 335–358.

Brown SA, Azzi A. (2013) Peripheral circadian oscillators in mammals. In: A Kramer and M Merrow (Eds.) *Circadian Clocks, Handbook of Experimental Pharmacology, Molecular and Cellular Basis of Circadian Clocks (Part I)*, Vol. 217. Germany: Springer-Verlag Berlin Heidelberg, pp. 45–66.

Bruckner JV, Luthra R, Lakatua D, Sackett-Lunden L. (1984) Influence of time of exposure to carbon tetrachloride on toxic liver injury. *Ann Rev Chronopharmacol*, 1:373–376.

Bruguerolle B, Jadot G. (1983) Influence of the hour of administration of lidocaine on its intraerythrocytic passage in the rat. *Chronobiologia*, 10:295–297.

Bruguerolle B, Lemmer B. (1993) Recent advances in chronopharmacokinetics: Methodological problems. *Life Sci*, 52:1809–1824.

Bruguerolle B, Prat M. (1987) Temporal changes in bupivacaine kinetics. *J Pharm Pharmacol*, 39:148–150.

Bruguerolle B, Prat M. (1988) Circadian phase-dependent pharmacokinetics and acute toxicity of mepivacaine. *J Pharm Pharmacol*, 40:592–594.

Bruguerolle B, Prat M. (1989) Temporal variations in the erythrocyte permeability to bupivacaine, etidocaine and mepivacaine in mice. *Life Sci*, 45:2587–2589.

Bruguerolle B, Prat M. (1990) Circadian phase-dependent acute toxicity and pharmacokinetics of etidocaine in serum and brain of mice. *J Pharm Pharmacol*, 42:201–202.

Buhr ED, Takahashi JS. (2013) Molecular components of the mammalian circadian clock. In: A Kramer and M Merrow (Eds.) *Circadian Clocks, Handbook of Experimental Pharmacology, Molecular and Cellular Basis of Circadian Clocks (Part I)*, Vol. 217. Germany: Springer-Verlag Berlin Heidelberg, pp. 3–28.

Cal JC, Dorian C, Cambar J. (1985) Circadian and circannual changes in nephrotoxic effects of heavy metals and antibiotics. *Ann Rev Chronopharmacol*, 2:143–176.

Cambar J, Lemoigne F, Toussaint C. (1979) Etude des variations nychtémérales de la filtration glomérulaire chez le rat. *Experientia*, 35:1607–1609.

Cambar J, Pons M. (2010) New trends in chronotoxicology. In: PH Redfern and B Lemmer (Eds.) *Physiology and Pharmacology of Biological Rhythms, Handbook of Experimental Pharmacology*, Vol. 125. Germany: Springer-Verlag Berlin Heidelberg, pp. 557–588.

Cambar J, L'Azou B, Cal JC. (1992) Chronotoxicology. In: Y Touitou and E Haus (Eds.) *Biologic Rhythms in Clinical and Laboratory Medicine*. New York, NY: Springer Berlin Heidelberg, pp. 138–150.

Chambers RB. (October, 2000) The role of mathematical modeling in medical research: Research without patients? *The Ochsner Journal*, 4:218–223.

Daan S, Berde C. (1978) Two coupled oscillators: Simulations of the circadian pacemaker in mammalian activity rhythms. *J Theor Biol*, 70:297–313.

Denvis MJ. (1990) Absorption processes. In: C Hansch, PG Sammes, and JB Taylor (Eds.) *Comprehensive Medicinal Chemistry: Biopharmaceutics*, Vol. 5. New York, NY: Pergamon, pp. 1–43.

Desgagné M, Bélanger PM. (1986) Chronohepatotoxicity of styrene in rat. *Ann Rev Chronopharmacol*, 3:103–106.

Desgagné M, Boutet M, Bélanger PM. (1988) The mechanism of the chronohepatotoxicity of chloroform in rat: Correlation between binding of hepatic subcellular fractions and histologic changes. *Ann Rev Chronopharmacol*, 5:235–238.

Dettli L, Spring P. (1966) Diurnal variations in the elimination rate of a sulfonamide in man. *Helv Med Acta*, 33:291–306.

Dorian C, Catroux P, Cambar J. (1988) Chronobiological approach to aminoglycosides. *Arch Toxicol*, 12:151–157.

Duane WC, Gilberstadt ML, Weigand DM. (1979) Diurnal rhythms of bile acid production in the rat. *Amer J Physiol*, 236:R175–R179.

Earnest DJ, Digiorgio SM, Sladek CD. (1991) Effects of tetrodotoxine on the circadian pacemaker mechanism in suprachiasmatic explants in vitro. *Brain Res Bull*, 26:677–682.

Edmunds Jr., LN. (2010) Regulation of cell division cycles by circadian oscillators: Signal transduction between clocks. In: PH Redfern and B Lemmer (Eds.) *Physiology and Pharmacology of Biological Rhythms, Handbook of Experimental Pharmacology*, Vol. 125. Germany: Springer-Verlag Berlin Heidelberg, pp. 29–54.

Elliot JS, Sharp RF, Lewis L. (1959) Urinary pH. *J Urol*, 81:339–343.

English J, Dunne M, Marks V. (1983) Diurnal variation in prednisolone kinetics. *Clin Pharmacol Ther*, 33:381–385.

Fisch J, Yonovitz A, Smolensky M. (1984) Effects of circadian rhythm on kanamycin-induced hearing loss. *Ann Rev Chronopharmacol*, 1:385–388.

Fisher SP, Foster RG, Peirson SN. (2013) The circadian control of sleep. In: A Kramer and M Merrow (Eds.) *Circadian Clocks, Handbook of Experimental Pharmacology, Circadian Control of Physiology and Behavior (Part II)*, Vol. 217. Germany: Springer-Verlag Berlin Heidelberg, pp. 157–184.

Focan C, Bruguerolle B, Arnaud C, Levi F, Mazy V, Focan-Henrard D, Bouvenot G. (1988) Alteration of circadian time structure of plasma proteins in patients with inflammation. *Ann Rev Chronopharmacol*, 5:21–24.

Forger DB, Peskin CS. (2003) A detailed predictive model of the mammalian circadian clock. *Proc Natl Acad Sci USA*, 100:14806–14811.

Garrettson LK, Jusko WJ. (1974) Diphenylhydantoin elimination kinetics in overdosed children. *Clin Pharmacol Ther*, 17:481–491.
Gilberstadt ML, Bellinger LL, Lindblad S, Duane WC. (1991) Liver denervation does not alter the circadian rhythm of bile acid synthesis in rats. *Am J Physiol*, 261:G799–G802.
Goo RH, Moore JG, Greenberg E, Alazraki NP. (1987) Circadian variation in gastric emptying of meals in humans. *Gastroenterology*, 93:515–518.
Gonze D, Bernard S, Waltermann C, Kramer A and Herzel H. (2005) Spontaneous synchronization of coupled circadian oscillators. *Biophys J*, 89:120–129.
Guengerich FP. (1987) Enzymology of rat liver cytochromes p-450. In: FP Guengerich (Ed.) *Mammalian Cytochromes P-450*, Vol. 1, Chap. 1. Boca Raton: CRC Press, pp. 2–54.
Guissou P, Cuisinaud G, Llorca G, Lejeune E, Sassard J. (1983) Chronopharmacologic study of a prolonged release form of indomethacin. *Eur J Clin Pharmacol*, 24:667–672.
Hanson SK, Andres MW. (1978) The effect of diethyl maleate treatment, fasting and time of administration of allyl alcohol hepatotoxicity. *Toxicol Lett*, 1:301–305.
Hartley R, Forsythe WI, McLain B, Ng PC, Lucok MD. (1991) Daily variations in steady-state plasma concentrations of carbamazepine and its metabolite in epileptic children. *Clin Pharmacokin*, 20:237–244.
Hastings M, Reddy A, Maywood E. (2003) A clockwork web: circadian timing in brain and periphery, in health and disease. *Nat Rev Neurosci*, 4(8):649–661.
Hastings MH. (2010) The vertebrate clock: Localisation, connection and entrainment. In: PH Redfern and B Lemmer (Eds.) *Physiology and Pharmacology of Biological Rhythms, Handbook of Experimental Pharmacology*, Vol. 125. Germany: Springer-Verlag Berlin Heidelberg, pp. 1–28.
Hayes DK, Morgan NO. (1988) Insects as animal models for chronopharmacological research. *Ann Rev Chronopharmacol*, 5:243–246.
Hecquet B, Meynadier J, Bonneterre J, Adenis L. (1984) Circadian rhythm in cisplatin binding on plasma proteins. *Ann Rev Chronopharmacol*, 3:115–118.
Hodgkin A, Huxley A. (1952) A quantitative description of membrane current and its application to conduction and excitation in nerve. *J Physiol*, 117:500–544.
Hosokawa H, Nyu S, Nakamura N, Mifune K, Nakano S. (1993) Circadian variation in amikacin clearance and its effects on efficacy and toxicity in mice with and without immunosuppression. *Chronobiol Int*, 10:259–270.
Hrusheski WJM, Borsch R, Levi F. (1982) Circadian time dependence of cisplatin urinary kinetics. *Clin Pharmacol Ther*, 32:330–339.
The International Transporter Consortium. (2010) *Nat Rev Drug Discov*, 9(3): 215–236.
Jaeger RJ, Conolly B, Murphy ID. (1973) Diurnal variation of hepatic glutathione concentration and its correlation with 1,1-dichloroethylene inhalation toxicity in rats. *Res Commun Chem Pathol Pharmacol*, 6:465–671.
Jespersen CM, Fredericksen M, Fisher Hansen J, Klitgaard NA, Soerum C. (1989) Circadian variation in the pharmacokinetics of verapamil. *Eur J Clin Pharmacol*, 37:613–615.
Kanabrocki EL, Sothern RB, Scheving LE, Halberg F, Pauly JE, Greco J, Nemchausky BA, DeBartelo M, Kaplan E, McCormick JB, Olwin JH, Marks GE, Bird T, Redmond DP, Graeber RC, Ferrara A, Hrushesky WJM. (1988) Ten-year-replicated circadian profiles for 36 physiological, serological and urinary variables in healthy men. *Chronobiol Int*, 5:237–284.

Kaplowitz N, Aw TY, Ookhtens M. (1985) The regulation of hepatic glutathione. *Ann Rev Pharmacol Toxicol*, 25:715–744.

Keller M, Mazuch J, Abraham U, Eom G, Herzog E, Volk H, Kramer A, Maier B. (2009) A circadian clock in macrophages controls inflammatory immune responses. *Proc Natl Acad Sci USA*, 106:21407–21412.

Koren G, Ferrazzini G, Gohl H, Robieux I, Johnson D, Giesbrecht E. (1992) Chronopharmacology of methotrexate pharmacokinetics in childhood leukemia. *Chronobiol Int*, 9:434–438.

Kronauer RE, Czeisler CA, Pilato SF, Moore-Ede MC, Weitzman ED. (1982) Mathematical model of the human circadian system with two interacting oscillators. *Am J Physiol*, 242:R3–17.

Kumar D, Wingate D, Ruckebusch Y. (1986) Circadian variation in the propagating velocity of the migrating motor complex. *Gastroenterology*, 91:926–930.

Labrecque G, Bélanger PM. (1991) Biological rhythms in the absorption, distribution, metabolism and excretion of drugs. *Pharmacol Ther*, 52:95–107.

Labrecque G, Bélanger PM, Doré F, Lalande M. (1988) 24-hour variations in the distribution of labeled microspheres to the intestines, liver and kidneys. *Ann Rev Chronopharmacol*, 5:445–449.

Labrecque G, Karzazi M, Vanier M-C. (2010) Biological rhythms in pain analgesi. In: PH Redfern and B Lemmer (Eds.) *Physiology and Pharmacology of Biological Rhythm, Handbook of Experimental Pharmacology*, Vol. 125. Germany: Springer-Verlag Berlin Heidelberg, pp. 619–650.

Langner B, Lemmer B. (1988) Circadian changes in the pharmacokinetics and cardiovascular effects of oral propranolol in healthy subjects. *Eur J Clin Pharmacol*, 33:619–624.

Larsen KR, Doyton MT, Moore JG. (1991) Circadian rhythm in gastric mucosal blood flow in fasting rat stomach. *J Surg Res*, 51:275–280.

Lavigne JG, Bélanger PM, Doré F, Labrecque G. (1983) Temporal variations in chloroform-induced hepatotoxicity in rats. *Toxicology*, 26:267–273.

Lehman-McKeeman LD. (2013) Absorption, distribution, and excretion of toxicants. In: CD Klaassen (Ed.) *Casarett and Doull's Toxicology, The Basic Science of Poisons*, Chap. 5. New York: McGraw-Hill Education, pp. 153–183.

Leloup JC, Goldbeter A. (2003) Toward a detailed computational model for the mammalian circadian clock. *Proc Natl Acad Sci USA*, 100:7051–7056.

Lemmer B, Bruguerolle B. (1994) Chronopharmacokinetics: Are they clinically relevant? *Clin Pharmacokinet*, 26:419–427.

Lemmer B, Nold G. (1991) Circadian changes in estimated hepatic blood flow in healthy subjects. *Br J Clin Pharmacol*, 32:627–629.

Lėvi F. (2010) Chronopharmacology of anticancer agents. In: PH Redfern and B Lemmer (Eds.) *Physiology and Pharmacology of Biological Rhythms, Handbook of Experimental Pharmacology*, Vol. 125. Germany: Springer-Verlag Berlin Heidelberg, pp. 299–332.

Lévi F, Altinok A, Clairambault J, Goldbeter A. (2008) Implications of circadian clocks for the rhythmic delivery of cancer therapeutics. *Philos Trans R Soc A*, 366:3575–3598.

Lévi F, Boughattas NA, Blazsek I. (1988) Comparative murine chronotoxicity of anticancer agents and related mechanisms. *Ann Rev Chronopharmacol*, 4:283–331.

Lévi F, Schibler U. (2007) Circadian rhythms: Mechanisms and therapeutic implications. *Annu Rev Pharmacol Toxicol*, 47:593–628.

Lévi F, Zidani R, Misset J. (1997) Randomised multicentre trial of chronotherapy with oxaliplatin, fluorouracil, and folinic acid in metastatic colorectal cancer. *Lancet*, 350:681–686.

Levy M, Granet L, Zylber-Katz E. (1984) Chronopharmacokinetics of caffeine in healthy volunteers. *Ann Rev Chronopharmacol*, 1:97–100.

Lin L, Grenier L, Bergeron Y, Simard M, Bergeron MG, Labrecque G, Beauchamp D. (1994) Temporal changes of pharmacokinetics, nephrotoxicity, and subcellular distribution of tobramycin in rats. *Antimicrob Agents Chemother*, 38:54–60.

Lindup WE. (1987) Plasma protein binding of drugs- some basic and clinical aspects. *Prog Drug Metab*, 10:142–185.

Loubaris N, Michel A, Cros G, Serrano JJ, Katz S, Bouchard M. (1984) Circadian changes in carrageenan-induced edema: the anti-inflammatory effect and bioavailability of phenylbutazone in rats. *Life Sci*, 34:2379–2384.

Marcheva B, Ramsey KM, Peek CB, Affinati A, Maury E, Bass J. (2013) Circadian clocks and metabolism. In: A Kramer and M Merrow (Eds.) *Circadian Clocks, Handbook of Experimental Pharmacology, Circadian Control of Physiology and Behavior (Part II)*, Vol. 217. Germany: Springer-Verlag Berlin Heidelberg, pp. 127–156.

Markiewicz A, Gomoluch T, Marek E, Boldys H. (1981) Circadian absorption of vitamin B_{12}. *Scand J Gastroenterol*, 16:541–544.

Markiewicz A, Senenowicz K. (1979) Time-dependent changes in the pharmacokinetics of aspirin. *J Clin Pharmac Biopharm*, 17:409–411.

Meister A, Anderson ME. (1983) Glutathione. *Ann Rev Biochem*, 52:711–760.

Minami Y, Ode KL, Ueda HR. (2013) Mammalian circadian clock: The roles of transcriptional repression and delay. In: A Kramer, M Merrow (Eds.) *Circadian Clocks, Handbook of Experimental Pharmacology*, Vol. 217. Heidelberg: Springer.

Mirsky H, Liu A, Welsh D, Kay S, Doyle F. (2009) A model of the cell-autonomous mammalian circadian clock. *Proc Natl Acad Sci*, 106:11107–11112.

Miyazaki Y, Yatagai M, Imaoka S, Funae Y, Motohashi Y, Kobayashi Y. (1990) Temporal variations in hepatic cytochromes P-450 isoenzymes in rats. *Ann Rev Chronopharmacol*, 7:149–153.

Moore JG, Datz FL, Christian PE. (1990) Exercise increases solid food gastric emptying rates in men. *Dig Dis Sci*, 35(4):448–452.

Moore JG, Halberg F. (1986) Circadian rhythm of gastric acid secretion in men with active duodenal ulcer. *Dig Dis Sci*, 31:1185–1191.

Moore JG, Merki H. (2010) Gastrointestinal tract. In: PH Redfern and B Lemmer (Eds.) *Physiology and Pharmacology of Biological Rhythms, Handbook of Experimental Pharmacology*, Vol. 125. Germany: Springer-Verlag Berlin Heidelberg, pp. 351–374.

Moore RY. (2010) Chemical neuroanatomy of the mammalian circadian system. In: PH Redfern and B Lemmer (Eds.) *Physiology and Pharmacology of Biological Rhythms, Handbook of Experimental Pharmacology*, Vol. 125. Germany: Springer-Verlag Berlin Heidelberg, pp. 79–94.

Motahashi Y, Miyazaki Y. (1990) Temporal variations in acute neurotoxicity of trichloroethylene in the rat. *Ann Rev Chronopharmacol*, 7:177–180.

Nair V, Casper R. (1969) The influence of light on daily rhythm in hepatic drug metabolizing enzymes in rat. *Life Sci*, 8:1291–1298.

Nakano S, Watanabe H, Nagai K, Ogawa N. (1984) Circadian stage-dependent changes in diazepam kinetics. *Clin Pharmacol Ther*, 36:271–277.

Ortiz-Tudela E, Mteyrek A, Ballesta A, Innominato PF, Lévi F. (2013) Cancer chronotherapeutics: Experimental, theoretical, and clinical aspects. In: A Kramer and M Merrow (Eds.) *Circadian Clocks, Handbook of Experimental Pharmacology, Chronopharmacology and Chronotherapy (Part III)*, Vol. 217. Germany: Springer-Verlag Berlin Heidelberg, pp. 261–288.

Parkinson A, Ogilvie BW, Buckley DB, Kazmi F, Czerwinski M and Parkinson O. (2013) Biotransformation of xenobiotics. In: Klaassen CD, PhD. (Ed.) *Casarett and Doull's Toxicology The Basic Science of Poisons*, Chap 6, Edition 8, McGraw-Hill Education, New York, pp 185–366.

Patel IH, Venkataramanan R, Levy RH, Viswanathan CT, Ojemann LM. (1982) Diurnal oscillations in plasma protein binding of valproic acid. *Epilepsia*, 32:282–290.

Poley GE, Shiveley CH, Vesell ES. (1978) Diurnal rhythms of aminopyrine metabolism: Failure of sleep deprivation to affect them. *Clin Pharmacol Ther*, 24:726–732.

Pons M, Tranchot J, L'Azou B, Cambar J. (1994) Circadian rhythms of renal hemodynamics in unanesthetized, unrestrained rats. *Chronobiol Int*, 11:301–308.

Pooler PA, Duane WC. (1988) Effects of bile acid administration on bile acid synthesis and its circadian rhythm in man. *Hepatology*, 8:1140–1146.

Prat M, Bruguerolle B. (1988) Chronotoxicity and chronokinetics of two local anaesthetic agents, bupivacaine and mepivacaine, in mice. *Ann Rev Chronopharmacol*, 5:263.

Radzialowski FM, Bousquet WF. (1967) Circadian rhythm in hepatic drug metabolizing activity of the rat. *Life Sci*, 6:2545–2548.

Reinberg A, Lévi F, Smolensky MH, Labrecque G, Ollagnier M, Decousus H, Bruguerolle B. (1990) Chronokinetics. In: C Hansch, PG Sammes, and JB Taylor (Eds.) *Comprehensive Medicinal Chemistry 5: Biopharmaceutics*. New York, NY: Pergamon, pp. 280–295.

Reinberg A, Levi F, Touitou Y, Le Liboux A, Simon J, Frydman A, Bicakova-Rocher A, Bruguerolle B. (1986) Clinical chronokinetic changes in a sustained release preparation of ketoprofen. *Ann Rev Chonopharmacol*, 3:317–320.

Reinberg A, Schuller E, Deslanerie N, Clench J, Helary M. (1977) Rythmes circadiens et circannuels des leucocytes, protéines totales, immunoglobulines A, G et M. Etude chez neuf adultes jeunes et sains. *Nouv Prese Med*, 6:3819–3823.

Reinberg A, Zagulla-Mally ZW, Ghate J, Halberg F. (1967) Circadian rhythm in duration of sodium salicylate excretion referred to phase of excretory rhythms and routine. *Proc Soc Exp Biol Med*, 124:826–832.

Relógio A, Westermark P, Wallach T, Schellenberg K, Kramer A, Herzel H. (2011) Tuning the mammalian circadian clock: Robust synergy of two loops. *PLoS Comput Biol*, 7:e1002309.

Rensing L. (2010) Genetics and molecular biology of circadian clocks. In: PH Redfern and B Lemmer (Eds.) *Physiology and Pharmacology of Biological Rhythms, Handbook of Experimental Pharmacology*, Vol. 125. Germany: Springer-Verlag Berlin Heidelberg, pp. 55–78.

Roenneberg T, Kantermann T, Juda M, Vetter C, Allebrandt KV. (2013) Light and the human circadian clock. In: A Kramer and M Merrow (Eds.) *Circadian Clocks, Handbook of Experimental Pharmacology, Chronopharmacology and Chronotherapy (Part III)*, Vol. 217. Germany: Springer-Verlag Berlin Heidelberg, pp. 311–334.

Rowland M, Tozer TN. (1980) *Clinical Pharmacokinetics. Concepts and Applications*. Philadelphia, PA: Lea and Febiger, pp. 65–76.

Sahar S, Sassone-Corsi P. (2013) The epigenetic language of circadian clocks. In: A Kramer and M Merrow (Eds.) *Circadian Clocks, Handbook of Experimental Pharmacology, Molecular and Cellular Basis of Circadian Clocks (Part I)*, Vol. 217. Germany: Springer-Verlag Berlin Heidelberg, pp. 29–44.

Schnell RC, Bozigian HP, Davies MH, Merrick BA, Johnson KL. (1983) Circadian rhythm in acetaminophen toxicity: Role of nonprotein sulfhydryls. *Tox Appl Pharmacol*, 71:353–361.

Schnellmann RG. (2013) Toxic Responses of the Kidney. In: Klaassen CD, PhD. (Ed.) *Casarett and Doull's Toxicology The Basic Science of Poisons*, Chap 14, Edition 8, McGraw-Hill Education, pp. 665–690.

Shiga T, Fujimura A, Tateishi T, Ohaski K, Ebihara A. (1993) Differences of chronopharmacokinetic profiles between propranolol and atenolol in hypertensive subjects. *J Clin Pharmacol*, 33:756–761.

Shiveley CA, Vessel ES. (1975) Temporal variations in acetaminophen and phenacetin half-life in man. *Clin Pharmacol Ther*, 18:413–424.

Slat E, Freeman GM, Herzog ED. (2013) The clock in the brain: Neurons, glia and networks in daily rhythms. In: A Kramer and M Merrow (Eds.) *Circadian Clocks, Handbook of Experimental Pharmacology*, Vol. 217. Germany: Springer, Heidelberg.

Smolensky M. (1989) Chronopharmacology of theophylline and beta-sympathomimetics. In: B Lemmer (Ed.) *Chronopharmacology: Cellular and Biochemical Interactions*, Chap. 5. Basel: Marcel Dekker, pp. 65–113.

Stokkan K, Yamazaki S, Tei H, Sakaki Y, Menaker M. (2001) Entrainment of the circadian clock in the liver by feeding. *Science*, 291:490–439.

Swanson BN, Boppana VK, Vlasses PH, Holmes GI, Monsell K, Ferguson RK. (1982) Sulindac disposition when given once and twice daily. *Clin Pharmacol Ther*, 32:397–403.

Taggart AJ, McElnay JC, Kerr B, Passmore P. (1987) The chronopharmacokinetics of indomethacin suppositories in healthy volunteers. *Eur J Clin Pharmacol*, 31:617–619.

Tarquini B, Romano S, De Scalzi M, de Leonardis V, Benvenuti F, Chegai E, Comparini T, Moretti R, Cognoni M. (1979) Circadian variation of iron adsorption in healthy human subjects. *Adv Biosci*, 19:347–354.

Tillement JP, Houin G, Zini R, Urien S, Albengres E, Barré J, Lecompte M, D'Athis P, Sébille B. (1984) The binding of drugs to blood plasma macromolecules: Recent advances and therapeutic significance. *Adv Drug Res*, 13:59–94.

Ukai H, Ueda HR. (2010) Systems biology of mammalian circadian clocks. *Annu Rev Physiol*, 72:579–603.

van Berge Henegower GP, Hoffmann AF. (1978) Nocturnal gallbladder storage and emptying in gallstone patients and healthy subjects. *Gastroenterology*, 75:879–885.

Vener KJ, Moore JG. (1987) Chronobiologic properties of the alimentary canal affecting xenobiotic absorption. *Ann Rev Chronopharmacol*, 4:257–281.

Walker CA, Soliman MR, Soliman KF. (1983) Chronopharmacology and chronotoxicology of CNS drugs: Interrelationships with neuromodulators. *Int J Toxicol*, 2(6):359–370.

Waterhouse JM, Minors DS (1989) Temporal aspects of renal drug elimination. In: B Lemmer (Ed.) *Chronopharmacology: Cellular and Biochemical Interactions*, Chap. 3. Basel: Marcel Dekker, pp. 35–50.

Westermark PO, Welsh DK, Okamura H, Herzel H. (2009) Quantification of circadian rhythms in single cells. *PLoS Comput Biol*, 5:e1000580.

Wilkinson GR, Beckett AH. (1968) Absorption, metabolism and excretion of the ephedrines in man. I. The influence of urinary pH and urine volume output. *J Pharmacol Exp Ther*, 162:139–147.

Wilkinson GR, Shand DG. (1975) A physiological approach to hepatic drug clearance. *Clin Pharmacol Ther*, 18:377–390.

Winfree A. (1980) *The Geometry of Biological Time*. New York, NY: Springer.

Wisser H and Breuer H. (1981) Circadian changes of clinical, chemical and endocrinological parameters. *J Clin Chem Clin Biochem*, 19:323–337.

Wrobel J, Nagel G. (1979) Diurnal rhythm of active calcium transport in ratintestine. *Experientia*, 35:1581–1582.

8

Fundamentals of Nonclinical Formulation: Chemistry and Biological Behavior

The perfect drug would be along the lines of Paul Ehrlich's "magic bullet" – as illustrated in Figure 8.1, a drug molecule is readily administered, completely absorbed, moves to the desired therapeutic target site (receptor), does what it is supposed to, and is completely eliminated. The most pressing (and rewarding, if successful) area for drug development currently is optimizing the drug for a therapeutic target delivery as part of this process. One of the key steps in the nonclinical and clinical formulation of the drug is the selection of vehicles and of the inactive ingredients (excipients). Excipients are essential components of drug products in the United States and one must adequately address the safety of the proposed exposure to the excipients in those products. The specific safety data that may be needed will vary depending upon the clinical situation, including such factors as the duration, level, and route of exposure (that is, actual means of clinical drug administration).

Many guidances exist to aid in the development of pharmaceutical drugs, but very few guidances exist to aid in the formulation of drugs for nonclinical safety evaluation, or for the assessment of pharmaceutical excipient safety. The FDA/CDER adopted, in 2005, the guidance for industry "Nonclinical Studies for Development of Pharmaceutical Excipients," which focuses on the development of safety profiles to support use of new excipients as components of drug or biological products.

A similar guidance was published in 1995 by the IPEC (International Pharmaceutical Excipient Council) association: "Excipient Safety Evaluation Guidance" (updated in IPEC, 2012). These guidelines are presented in a tiered approach of recommended data that should be available on an excipient to provide a pharmaceutical formulator with a rational basis for including a new excipient in a drug formulation. The objective of the current proposal is to codify existing excipients and provide a logical and rational approach to qualifying new excipients. The final aim of these Safety Evaluation Guidelines for Excipients is to provide an important element in the acceptability of a new excipient by regulators, independent of the approval of a specific drug formulation.

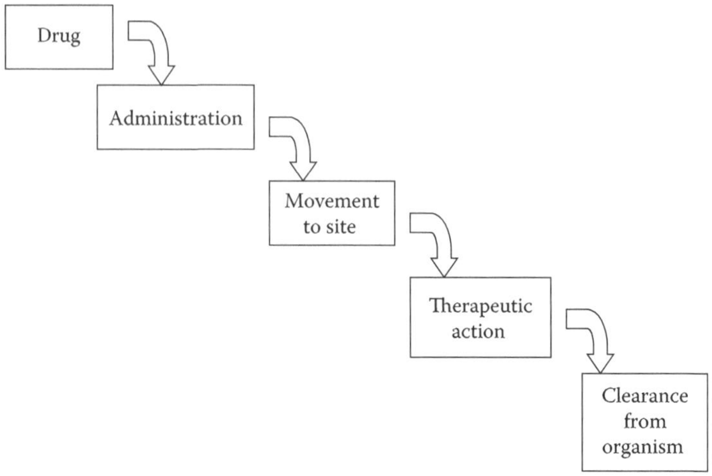

FIGURE 8.1
The magic bullet concept.

The three essential requirements of active pharmaceutical ingredients (API) principles are compared with those of excipients. Fundamental to both are quality and safety. The requirement of therapeutic efficacy for drugs is replaced by that of functionality for the excipient, defined as "the physical, physicochemical and biopharmaceutical properties" of the same.

Throughout the development process for pharmaceuticals, formulation development is proceeding with several objectives in mind. The importance of each of these factors changes over time (Monkhouse and Rhodes, 1998; Strickley, 2008) as illustrated in Figure 8.2. First is optimizing the bioavailability of the therapeutic agent at the target organ site by the intended clinical route. Clinical route(s) are selected on a number of grounds (nature of the drug, patient acceptance, issues of safety, marketability, competition). Second is minimizing any safety concerns. This means not just the systemic toxicity but also local tissue tolerance at the site of application. Third is optimizing the stability of the drug active ingredient. Its activity and integrity must be maintained for a period of sufficient duration to be made effectively available to patients. Early on in preclinical development, simplicity and maximized bioavailability are essential. Early single-dose studies in animals are the starting place, and usually bear no relationship to what is used later.

Formulations used to administer potential drugs undergoing development occupy an unusual place in pharmaceutical safety assessment compared to the rest of industrial toxicology. Eventually, a separate function in the pharmaceutical company developing a drug will develop a specific formulation that is to be administered to people – a formulation that optimizes the conditions of absorption and stability for the drug entity (Racy, 1989). The final formulation will need to be assessed to see if it presents any unique local or short-term hazards, but as long as its nonactive constituents are drawn from the approved

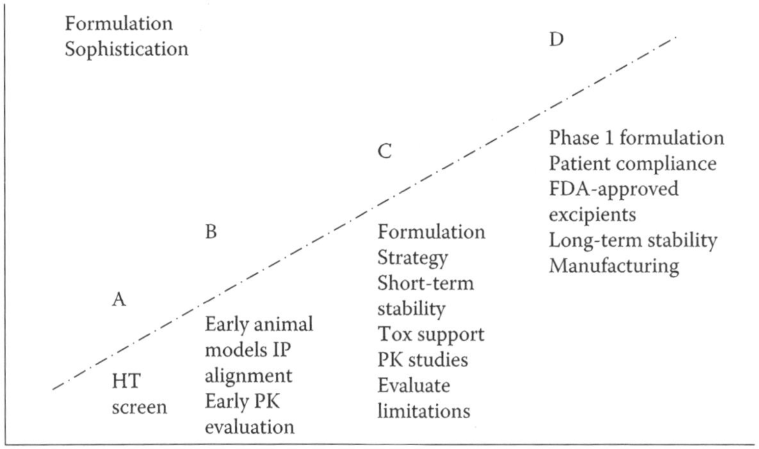

FIGURE 8.2
Evolution of formulations through phase 1.

formulary lists, no significant separate evaluation of their safety is required preclinically. They can, of course, alter the toxicity of the drug under study.

Simultaneous with this development of an optimized clinical formulation, however, preclinical evaluations of the safety of the drug moiety must by performed. Separate preclinical formulations (which generally are less complex than the clinical ones) are developed, sometimes by a formulation group and other times by a toxicology group itself. These preclinical formulations will frequently include much higher concentrations of the drug moiety being tested than do any clinical formulations. These formulations have to support a high and wide range of doses in animals (3- to 100-fold or greater over the clinical dose). The preclinical formulations are developed and evaluated with the aim of reproducibly delivering the drug (if at all possible by the route intended in man). Maintaining drug stability through an optimum period of time, and occluding the observed effects of the drug with vehicle effects to the minimum extent possible. These preclinical formulations are not restricted to materials that will or could be used in final clinical formulations (Gad, 2008b; Boersen et al., 2014).

In pivotal studies, the actual blood levels of an active moiety that are achieved will be determined so that correlations to later clinical studies can be made.

The formulations that are developed and used for preclinical studies are sometimes specific for the test species to be employed, but their development always starts with consideration of the route of exposure that is to be used clinically and, if possible, in accordance with a specified regimen of treatment (mirroring the intended clinical protocol as much as possible). One aspect of both nonclinical and clinical formulation and testing which prevents an important but often overlooked aspect of pharmaceutical safety assessment is the special field of excipients. These will be considered at the end of this chapter.

Among the cardinal principles of both toxicology and pharmacology is that the means by which an agent comes in contact with or enters the body (i.e., the route of exposure or administration) does much to determine the nature and magnitude of an effect. However a rigorous understanding of formulations, routes, and their implications to the design and analysis of safety studies is not widespread. And in the day-to-day operations of performing studies in animals, such an understanding of routes, their manipulation, means and pitfalls of achieving them, and the art and science of vehicles and formulations is essential to the sound and efficient conduct of a study. Newer approaches to formulations include nanosuspensions and self-emulsifying drug delivery systems (SEDDS).

As presented in Table 8.1, there are at least 26 potential routes of administration, of which 10 are commonly used in safety assessment and, therefore, are addressed here.

TABLE 8.1

Potential Routes of Administration

A. Oral routes
1. Oral (PO)[a]
2. Inhalation[a]
3. Sublingual
4. Buccal
B. Placed into a natural orifice in the body other than the mouth
1. Intranasal
2. Intraauricular
3. Rectal
4. Intravaginal
5. Intrauterine
6. Intraurethral
C. Parenteral (injected into the body or placed under the skin)
1. Intravenous (IV)[a]
2. Subcutaneous (SC)[a]
3. Intramuscular (IM)[a]
4. Intra-arterial
5. Intradermal (ID)[a]
6. Intralesional
7. Epidural
8. Intrathecal
9. Intracisternal
10. Intracardial
11. Intraventricular
12. Intraocular
13. Intraperitoneal (IP)[a]
D. Topical routes
1. Cutaneous[a]
2. Transdermal (also called percutaneous)[a]
3. Ophthalmic[a]

[a] Commonly used in safety assessment

Mechanisms

There are three primary sets of reasons why differences in formulations and the route of administration are critical in determining the effect of an agent of the biological system. These are (1) local effects, (2) absorption and distribution, and (3) metabolism.

Local effects. Local effects are those that are peculiar to the first area or region of the body to which a test material gains entry or that it contacts. For the dermal route, these include irritation, corrosion, and sensitization. For the parenteral routes, these include irritation, pyrogenicity, sterility, and blood compatibility. In general, these same categories of possible adverse effects (irritation, immediate immune response, local tissue/cellular compatibility, and physicochemical interactions) are the mechanisms of, or basis for, concern.

In general, no matter what the route, certain characteristics will predispose a material to have local effects (and, by definition, if present, tend to limit the possibility of local effects). These factors include pH, redox potential, high molar concentration, and the low level of flexibility (malleability) and sharp edges of certain solids. These characteristics will increase the potential for irritation by any route and, subsequent to the initial irritation, other appropriate regional adaptive responses (for orally administered materials, for example, emesis and diarrhea).

Absorption and distribution. For a material to be toxic, it must be absorbed into the organism (local effects are largely not true toxicities by this definition).

There are characteristics that influence absorption by the different routes, and these need to be understood by any person trying to evaluate and/or predict the toxicities of different moieties. Some key characteristics and considerations are summarized by route in the list that follows.

Table 8.2 presents the normal pH ranges for human physiologic fluids. These need to be considered in terms of the impact on solubility and stability of a formulation and active drug.

TABLE 8.2

Normal pH Range for Human Physiologic Fluids

Medium	Normal pH Range
Tears	7.35–7.45
Saliva	6.0–8.0
Gastric juice	1.5–6.5
Intestinal juice	6.5–7.6
Blood	7.35–7.45
Skin (sweat)	4.0–6.8

1. Oral and rectal routes (GI tract)
    a. Lipid-soluble compounds (nonionized) are more readily absorbed than water-soluble compounds (ionized).
        i. Weak organic bases are in the nonionized, lipid-soluble form in the intestine and tend to be absorbed there.
        ii. Weak organic acids are in the nonionized, lipid-soluble form in the stomach, and one would suspect that they would be absorbed there, but absorption in the intestine is greater because of time and area of exposure.
    b. Specialized transport systems exist for some moieties: sugars, amino acids, pyrimidines, calcium, and sodium.
    c. Almost everything is absorbed – at least to a small extent (if it has a molecular weight below 10,000).
    d. Digestive fluids may modify the structure of a chemical.
    e. Dilution increases toxicity because of more rapid absorption from the intestine, unless stomach contents bind or degrade the moiety.
    f. Physical properties are important; for example, dissolution of metallic mercury is essential to allow its absorption.
    g. Age is important; for example, neonates have a poor intestinal barrier.
    h. Effect of fasting on absorption depends on the properties of the chemical of interest.
2. Inhalation (lungs)
    a. Aerosol deposition
        i. Nasopharyngeal –5 µm or larger in man, less in common laboratory animals
        ii. Tracheobronchial –1 to 5 µm
        iii. Alveolar – 1 µm
    b. If inhalant is a solid, mucociliary transport from lungs to GI tract may clear it out of the pulmonary system.
    c. Lungs are anatomically good for absorption.
        i. Large surface area (50–100 m2).
        ii. Blood flow is high.
        iii. Accessibility to blood (10 µm between gas media and blood).
    d. Absorption of gases is dependent on solubility of the gas in blood.
        i. Chloroform, for example, has high solubility and is all absorbed, though respiration is limited.

    ii. Ethylene has low solubility and only a small percentage is absorbed – blood flow limits absorption.

3. Parenteral routes
    a. Irritation at the site of injection is influenced by solubility, toxicity, temperature, osmolality, and pH of injected solution.
    b. Pyrogenicity and blood compatibility are major concerns for intravenously administered materials.
    c. Solubility of test material in an aqueous or modified aqueous solution is the chief limitation on how much material may be given intravenously.
4. Dermal routes
    d. In general, any factor that increases the absorption through the stratum corneum will also increase the severity of an intrinsic response. Unless this factor mirrors potential exposure conditions, it may, in turn, adversely affect the relevance of test results (Barry, 1983).
    e. The physical nature of solids must be carefully considered both before testing and in interpreting results. Shape (sharp edges), size (small particles may abrade the skin due to being rubbed back and forth under the occlusive wrap), and rigidity (stiff fibers or very hard particles will be physically irritating) of solids may all enhance an irritation response and alter absorption.
    f. The degree of occlusion (in fact, the tightness of the wrap over the test site) also alters percutaneous absorption and therefore irritation. One important quality control issue in the laboratory is achieving a reproducible degree of occlusion in dermal wrappings.
    g. Both the age of the test animal and the application site (saddle of the back versus flank) can markedly alter test outcome. Both of these factors are also operative in humans, of course, but in dermal irritation tests, the objective is to remove all such sources of variability. In general, as an animal ages, the sensitivity to irritation decreases. And the skin on the middle of the back (other than directly over the spine) tends to be thicker (and therefore less sensitive to irritations) than that on the flanks.
    h. The sex of the test animals can also alter study results, because both regional skin thickness and surface blood flow vary between males and females.

As a generalization, there is a pattern of relative absorption rates that characterizes the different routes that are commonly employed. This order of absorption (by rate from fastest to slowest and, in a less rigorous manner, by degree of absorption from most to least, is IV > inhalation > IM > IP > SC > oral > ID > other.

Metabolism. Metabolism is directly influenced by both the region of the body onto which or into which a material is initially absorbed, and by the resultant distribution (both the rate and the pattern). Rate determines whether the primary enzyme systems will handle the entire xenobiotic dose, or if these are excessively saturated and overwhelmed. The pattern determines which routes of metabolism are operative.

Absorption (total amount and rate, distribution, metabolism, and species similarity in response) is the reason for selecting particular routes in toxicology in general. In the safety assessment of pharmaceuticals, however, the route is usually dictated by the intended clinical route and dosing regimen. If this intended route of human exposure is uncertain, or if there is the potential for a number of other routes where the human absorption rate and pattern are greater than the most common route or the route with the greater degree of human absorption rate and more desirable pattern, the most common plan forward becomes that of the most conservative approach. This approach stresses maximizing potential absorption in the animal species (within the limits of practicality) and selecting from among those routes commonly used in the laboratory those that get the most material into the animal's system as quickly and completely as possible to evaluate the potential toxicity. Under this approach, many compounds are administered intraperitoneally in acute testing, though there is little or no real potential for human exposure by this route.

Assuming that a material is absorbed, distribution of a compound in early preclinical studies is usually of limited interest. In so-called heavy acute studies (Gad and Chengelis, 1998) where acute systemic toxicity is intensive and evaluated to the point of identifying target organs, or in range-finder-type study results, for refining the design of longer-term studies, distribution would be of interest. Some factors that alter distribution are listed in Table 8.3.

The first special case is the parenteral route, where the systemic circulation presents a peak level of the moiety of interest to the body at one time, tempered only by the results of a first pass through the liver.

The second special case arises in cases of inhalation exposures. Because of the arrangement of the circulatory system, inhaled compounds (and those administered via the buccal route) enter the full range of systemic circulation without any "first-pass" metabolism by the liver. Kerberle (1971) and O'Reilly (1972) have previously published reviews of absorption, distribution, and metabolism that are relevant to acute testing.

One of the areas that is overlooked by virtually everyone in toxicology testing and research, yet is of crucial importance, is the need for formulation of candidate drugs and the use of vehicles and excipients in the formulation of test chemicals for administration to test animals (Strickley, 2008). For a number of reasons, a drug of interest is rarely administered or applied as is ("neat"). Rather, it must be put in a form that can be accurately

TABLE 8.3

Selected Factors That May Affect Chemical Distribution to Various Tissues

A. Factors relating to the chemical and its administration:
1. Degree of binding of chemical to plasma proteins (i.e., agent affinity for proteins) and tissues.
2. Chelation to calcium, which is deposited in growing bones and teeth (e.g., tetracyclines in young children).
3. Whether the chemical distributes evenly throughout the body (one-compartment model) or differentially between different compartments (models of two or more compartments).
4. Ability of chemical to cross the blood–brain barrier.
5. Diffusion of chemical into the tissues or organs and degree of binding to receptors that are and are not responsible for the drug's beneficial effects.
6. Quantity of chemical given.
7. Route of administration or exposure.
8. Partition coefficients (nonpolar chemicals are distributed more readily to fat tissues than are polar chemicals).
9. Interactions with other chemicals that may occupy receptors and prevent the drug from attaching to the receptor, inhibit active transport, or otherwise interfere with a drug's activity.
10. Molecular weight of the chemical.

B. Factors relating to the test subject:
1. Body size.
2. Fat content (e.g., obesity affects the distribution of drugs that are highly soluble in fats).
3. Permeability of membranes.
4. Active transport for chemicals carried across cell membranes by active processes.
5. Amount of proteins in blood, especially albumin.
6. Pathology or altered homeostasis that affects any of the other factors (e.g., cardiac failure, renal failure).
7. Presence of competitive binding substances (e.g., specific receptor sites in tissues bind drugs).
8. pH of blood and body tissues.
9. pH of urine.[a]
10. Blood flow to various tissues or organs (e.g., well-perfused organs usually tend to accumulate more chemical than less well-perfused organs).

[a] The pH of urine is usually more important than the pH of blood.

given to animals in such a way that it will be absorbed and not be too irritating. Most laboratory toxicologists come to understand vehicles and formulation, but to the knowledge of the author, guidance on the subject is limited to a short chapter on formulations by Fitzgerald et al. (1983). There is also a very helpful text on veterinary dosage forms by Hardee and Baggot (1998).

Table 8.4 presents an overview of typical forms of excipients for oral dosage forms. The entire process of drug development – even during the preclinical and nonclinical phase – includes a continuous development (with increased sophistication) of formulation, the use of nanoparticles (actually, developed as seeking even finer *micronized* particles in dosing formulations – it has been around for decades) being the latest approach.

TABLE 8.4

Nonactive Formulation Components in Drug Candidate Oral Delivery

Type	Purpose	Examples
Tablets, sachets, and capsules		
Binder	Facilitates agglomeration of powder into granules	Povidones, starches
Capsule shell	Contains powders or liquids	Gelatin, hypromellose
Coating agent	May mask unpleasant tastes or odors, improve ingestion or appearance, protect ingredients from the environment, or modify release of the active ingredient	Shellac, hypromellose
Colorant	Produces a distinctive appearance and may protect light-sensitive ingredients	FD&C colors, titanium dioxide
Disintegrant	Promotes rapid disintegration to allow a drug to dissolve faster	Sodium starch glycolate, crospovidones
Enteric coatings	Protects from dissolution in stomach ("gastro resistant")	HPMC, methylacrylate copolymers
Filler or diluent	Increases volume or weight	Calcium phosphate, lactose
Glidant or anticaking agent	Promostes powder flow and reduces caking or clumping	Talc, colloidal silicon dioxide
Lubricant	Reduces friction between particles themselves and between particles and manufacturing equipment	Magnesium stearate, glycerides
Release modifier	Provides extended-release capability	Ethylcellulose, guar gum
Solvent/vehicle	Improves stability and bioavailability	Water++ (see Gad et al., 2016)
Surfactant	Solubilizing agent	Irganox, SLS
Oral liquids		
Antimicrobial preservative	Prevents growth of bacteria, yeast, and mold	Glycerin, benzyl alcohol
Antioxidant	Reduces oxidative reactions that could alter ingredients	Ascorbic acid, butylated hydroxyanisole
Chelating or complexing agent	Stabilizes ions	Ethylenediaminetetraacetic acid salts, cyclodextrins
Liposome preparations	Improve bioavailability and protect from first-pass metabolism	Special vehicle with at least one lipid bilayer
pH modifier	Controls pH to improve drug stability or avoid irritation when consumed	Citric acid and its salts, salts of phosphoric acid
Surfactants and solubilizing agent	Promotes dissolution of insoluble ingredients	Sodium lauryl sulfate, polysorbates
Sweetening agent	Improves palatability	Sucrose, saccharin

Sources: Gad SC, *Pharmaceutical Manufacturing Handbook: Production and Processes*, Wiley, Hoboken, NJ, 2008a; Gad SC, *Drug Safety Evaluation*, 3rd Ed., Wiley, Hoboken, NJ, 2016.

Regulatory toxicology in the United States can be said to have arisen, due to the problem of vehicles and formulation, in the late 1930s, when attempts were made to formulate the new drug sulphanilamide. This drug is not very soluble in water, and a U.S. firm called Massengill produced a clear, syrupy elixir formulation that was easy to take orally. The figures illustrate how easy it is to be misled. The drug sulphanilamide is not very soluble in glycerol, which has an LD_{50} in mice of 31.5 g kg^{-1}, but there are other glycols that have the characteristic sweet taste and a much higher solvent capacity. Ethylene glycol has an LD_{50} of 13.7 g kg^{-1} in mice and 8.5 g kg^{-1} in rats, making it slightly more toxic than diethylene glycol, which has an LD_{50} in rats of 20.8 g kg^{-1}, similar to that for glycerol. The drug itself, which is inherently toxic, was marketed in a 75% aqueous diethylene glycol-flavored elixir. Then early in 1937 appeared the first reports of deaths, but the situation remained obscure for about six months until it became clear that the toxic ingredient in the elixir was the diethylene glycol. Even as late as March 1937, Haag and Ambrose were reporting that the glycol was excreted substantially unchanged in dogs, suggesting that it was likely to be safe (Hagenbusch, 1937). Within a few weeks, Holick (1937) confirmed that a low concentration of diethylene in drinking water was fatal to a number of species. Hagenbusch (1937) found that the results of necropsies performed on patients who had been taking 60–70 mL of the solvent per day were similar to those of rats, rabbits, and dogs taking the same dose of solvent with or without the drug. This clearly implicated the solvent, although some authors considered that the solvent was simply potentiating the toxicity of the drug. Some idea of the magnitude of this disaster may be found in the paper of Calvary and Klump (1939), who reviewed 105 deaths and a further 2,560 survivors who were affected to varying degrees, usually with progressive failure of the renal system. It is easy to be wise after the event, but the formulator fell into a classic trap, in that the difference between acute and chronic toxicity had not been adequately considered. In passing, the widespread use of ethylene glycol itself as an antifreeze has led to a number of accidental deaths, which suggests that the lethal dose in humans is around 1.4 mL kg^{-1}, or a volume of about 100 mL. In the preface to the first United States Pharmacopoeia, published in 1820, there is a the statement that "It is the object of the Pharmacopoeia to select from among substances which possess medical power, those, the utility of which is most fully established and best understood; and to form from them preparations and compositions, in which their powers may be exerted to the greatest advantage." This statement suggests that the influence that formulation and preparation may have on the biological activity of a drug (and on nonpharmaceutical chemicals) has been appreciated for a considerable time.

Available and commonly used vehicles and formulating agents are reviewed, along with basic information on their characteristics and usages, in Appendix at the end of this book. There is a general presumption that those excipients and formulating agents listed in the Pharmacopoeia (USP, 2014) or in the *Inactive Ingredient Guide* prepared by the FDA are safe to use and without biological effect. This may not always be the case in either experimental scenarios or humans (see Weiner and Bernstein, 1989), either directly or in how they alter absorption of and response to the active ingredient.

There are some basic principles to be observed in developing and preparing test material formulations. These are presented in Table 8.5. A start to all of this should be preformulation – characterization of the chemical, physical, and physical chemistry aspects of the drug molecule (Neervannan, 2006). Parts of this are identifying the optimal phase of the drug for formulation (Palucki et al., 2010) and useful salts (Stahl and Wermuth, 2011).

Preformulation

While advances in molecular biology and genomics have produced a flood of molecules with vastly improved target receptor specificity, these molecules have frequently turned out to be very, very difficult to get absorbed and to the desired target tissue site.

Lipinski's rule of five (RO5, Lipinski et al., 2001; Lipinski, 2004) predates these recent target advances in specificity, but not the problems. In its original form, the RO5 proposed four guiding principles:

- No more than 5 hydrogen bond donors (the total number of nitrogen-hydrogen and oxygen-hydrogen bonds)
- Not more than 10 hydrogen bond acceptors (all nitrogen or oxygen atoms)

TABLE 8.5

Basic Principles to be Observed in Developing and Preparing Test Material Formulations

A. Preparation of the formulation should not involve heating of the test material anywhere near to the point where its chemical or physical characteristics are altered.
B. If the material is a solid and it is to be assessed for dermal effects, its shape and particle size should be preserved. If intended for use in humans, topical studies should be conducted with the closest possible formulation to that to be used on humans.
C. Multicomponent test materials (mixtures) should be formulated so that the administered form accurately represents the original mixture (i.e., components should not be selectively suspended or taken into solution).
D. Formulation should preserve the chemical stability and identity of the test material.
E. The formulation should be such as to minimize total test volumes. Use just enough solvent or vehicle.
F. The formulation should be easy to administer accurately.
G. pH of dosing formulations should be between 5 and 9, if possible.
H. Acids or bases should not be used to divide the test material (for both humane reasons and to avoid pH partitioning in either the gut or the renal tubule).
I. If a parental route is to be employed, final solutions should be as nearly isotonic as possible. Do not assume a solution will remain such upon injection into the bloodstream. It is usually a good idea to verify that the drug stays in solution upon injection by placing some drops into plasma.
J. Particularly if use is to be more than a single injection, steps (such as filtration) should be taken to ensure suitable sterility.

- A molecular mass less than 500 daltons
- An octanol-water partition coefficient[5] log *P* not greater than 5

While Lipinski *prisma facie* applies to oral route drugs, it also is useful for other routes.

For clinically useful drugs (and therefore for drugs proceeding through preclinical and nonclinical evaluation and development), there are a number of desirable attributes:

- A simple structure
- Aimple and efficient synthesis
- Non-hygroscopic
- Avoid chiral centers
- Lack of mutagenicity
- Crystalline with solid-state stability
- No strong odors, colors, or (of oral) tastes
- Compatible with standard excipient
- Stable at ambient temperatures and at physiologic pHs (Niazi, 2007; Gibson, 2009; Adeyeye and Brittain, 2008)

Bioavailability is defined as the fraction of the dose reaching either the therapeutic target organ or tissue or the systemic circulation as unchanged compound following administration by any route. For an agent administered orally, bioavailability may be less than unity, for several reasons. The molecule may be incompletely absorbed. It may be metabolized in the gut, the gut wall, the portal blood, or the liver prior to entry into the systemic circulation (see Figure 8.3). It may undergo enterohepatic cycling with incomplete reabsorption following elimination into the bile. Biotransformation of some chemicals in the liver following oral administration is an important factor in the pharmacokinetic profile, as will be discussed further. Bioavailability measures following oral administration are generally given as the percentage of the dose available to the systemic circulation.

As the components of a mixture may have various physiochemical characteristics (solubility, vapor pressure, density, etc.), great care must be taken in preparing and administering any mixture so that what is actually tested is the mixture of interest.

Examples of such procedures are making dilutions (not all components of the mixture may be equally soluble or miscible with the vehicle) and generating either vapors or respirable aerosols (not all the components may have equivalent volatility or surface tension, leading to a test atmosphere that contains only a portion of the components of the mixture).

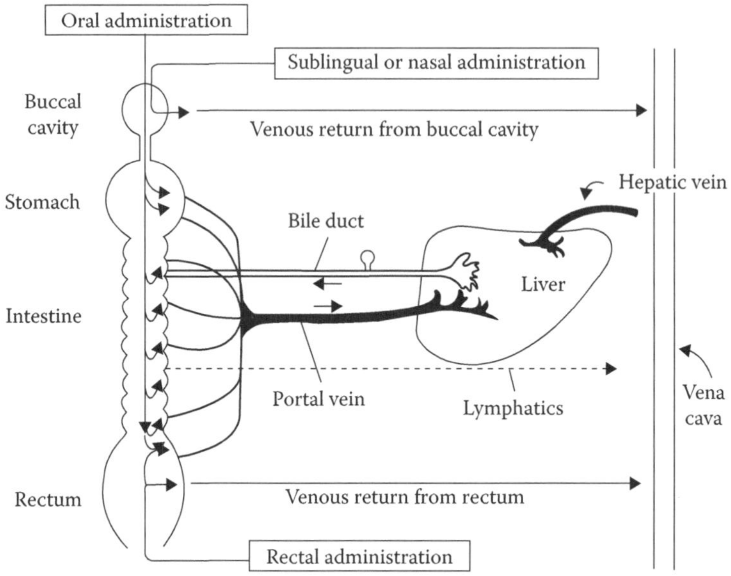

FIGURE 8.3
Path of drugs through the body after absorption by one of three routes of administration.

By increasing or decreasing the viscosity of a formulation, the absorption of a toxicant can be altered (Ritschel et al., 1974). Conversely, the use of absorbents to diminish absorption has been used as an antidote therapy for some forms of intoxication. Using the knowledge that rats cannot vomit, there have been serious attempts at making rodenticides safer to nontarget animals by incorporating emetics into the formulations, but this has had only a limited degree of success. Gaines used *in vivo* liver perfusion techniques to investigate the apparent anomaly that the carbamate Isolan was more toxic when administered to rats percutaneously than when administered orally (Gaines, 1960). It has been shown that these results, a manifestation of different formulations, have been used for the two routes of exposure (oral and percutaneous) in estimating the LD_{50} values using a common solvent, *n*-octanol. It was found that Isolan was significantly more toxic by the oral route than by the percutaneous route; by regression analysis it was found that at no level of lethal dose values was the reverse correct.

Although the oral route is the most convenient for most therapeutic uses, there are numerous factors that make it unpredictable, particularly for drug molecules that have very limited water solubility (Liu, 2008). Absorption by his route is subject to significant variation from animal to animal, and even in the same individual animal at different times. Considerable effort has been spent by the pharmaceutical industry to develop drug formulations with absorption characteristics that are both effective and dependable. Protective enteric coatings for pharmaceuticals were introduced long ago to retard the action of gastric fluids and then disintegrate and dissolve after passage of a tablet into the human intestine. The purposes of these coatings for drugs are to protect the active ingredient, which would be degraded in the stomach; to prevent nausea and vomiting caused by local gastric irritation (also a big problem in rodent studies, where over a long time period gastric irritation frequently leads to forestomach hyperplasia); to obtain higher local concentrations of the active ingredient intended to act locally in the intestinal tract; and to produce a delayed biological effect, or to deliver the active ingredient to the intestinal tract for optimal absorption there. Such coatings are generally fats, fatty acids, waxes, or other such agents, and all of these intended purposes for drug delivery can readily be made to apply for some toxicity studies. Their major drawback, however, is the marked variability in time for a substance to be passed through the stomach. In humans, this gastric emptying time can range from minutes to as long as 12 h. One would expect the same for animals, as the limited available data suggest is the case. Similar coating systems, including microencapsulation (see Melnick et al., 1987), are available for, and currently used in, animal toxicity studies.

The drug molecule is unlikely to be absorbed or excreted unless it is first released from its formulation. It is this stage of the process that is the first and most critical step for the activity of many chemicals. If the formulation does not release the chemical, the rest of the process becomes somewhat pointless.

It might be argued that the simplest way around the formulation problem is to administer any test as a solution in water, thereby avoiding the difficulties altogether. However, since multiple, small, accurately measured doses of a chemical are required repeatedly, reproducible dilutions must be used. Also, the water itself is to be regarded as the formulation vehicle, and the test substance must be water soluble and stable in solution, and many are not. If one takes into account this need for accuracy, stability, and optimum performance *in vivo*, the whole problem becomes very complex.

It is uncommon for direct connections between observed toxicity and formulation components to be made. Indeed, it is usually assumed that vehicles and other nontest chemical components are innocuous or have only transitory pharmacological effects. Historically, however, this has certainly not been the case. Even lactose may have marked toxicity in individual test animals (or humans) who are genetically incapable of tolerating it.

The initial stage of drug release from the formulation, both in terms of the amount and the rate of release, may exercise considerable influence at the clinical response level. A close consideration of the formulation parameters of any chemical is therefore essential during the development of any new drug, and, indeed, there are examples where formulations of established drugs also appear to require additional investigation.

The effects of formulation additives on chemical bioavailability from oral solutions and suspensions have been previously reviewed by Swarbrick (2006). He pointed out how the presence of sugars in a formulation may increase the viscosity of the vehicle. However, sugar solutions alone may delay stomach-emptying time considerably when compared to solutions of the same viscosity prepared with celluloses, which may be due to sugar's effect on osmotic pressure. Sugars of different types may also have an effect on fluid uptake by tissues and this, in turn, correlates with the effect of sugars such as glucose and mannitol on drug transport.

Surfactants have been explored widely for their effects on drug absorption, in particular using experimental animals (Gibaldi and Feldman, 1970; Gibaldi, 1976). Surfactants alter dissolution rates (of lipid materials), surface areas of particles and droplets, and membrane characteristics, all of which affect absorption.

Surfactants may increase the solubility of the drug via micelle formation, but the amounts of material required to increase solubility significantly are such that, at least orally, the laxative effects are likely to be unacceptable. The competition between the surfactant micelles and the absorption sites is also likely to reduce any useful effect and make any prediction of net overall effect difficult. However, if a surfactant has any effect at all, it is likely to be in the realm of agents that help disperse suspensions of insoluble materials and make them available for solution. Natural surfactants, in particular bile salts, may enhance absorption of poorly soluble materials.

The effective surface area of an ingested chemical is usually much smaller than the specific surface area that is an idealized *in vitro* measurement. Many drugs whose dissolution characteristics could be improved by particle-size reduction are extremely hydrophobic and may resist wetting by gastrointestinal fluids. Therefore, the gastrointestinal fluids may come in intimate contact with only a fraction of the potentially available surface area. The effective surface area of a hydrophobic particle can often be increased by the addition of a surface-active agent to the formulation, which reduces the contact angle between the solid and the gastrointestinal fluids, thereby increasing effective surface area and dissolution rate.

Formulations for administering dermally applied toxicants present different considerations and problems. The extent of penetration and speed with which a biologically active substance penetrates the skin or other biological membrane depend on the effect that the three factors – vehicle, membrane, and chemical – exert on the diffusion process. It is now accepted that they together represent a functional unit that controls the penetration and location of the externally applied chemicals in the deeper layers of the skin or membrane layer. The importance of the vehicle for the absorption process has been neglected until recently. One of the few requirements demanded of the vehicle has been that it acts as an inert medium that incorporates the test chemical in the most homogeneous distribution possible. In addition, chemical stability and good cosmetic appearance have been desirable characteristics. Most formulations in toxicology are based on empirical experience.

The chemical incorporated in a vehicle should reach the surface of the skin at a suitable rate and concentration. If the site of action lies in the deeper layers of the epidermis or below, the substance must cross the stratum corneum, if the skin is intact. Both processes, diffusion from the dosage form and diffusion through the skin barriers, are inextricably linked. They should be considered simultaneously and can be influenced by the choice of formulation.

The thesis that all lipid-soluble compounds basically penetrate faster than water-soluble ones cannot be supported in this absolute form. A lipophilic agent can penetrate faster or slower or at the same rate as a hydrophilic agent, depending on the vehicle used.

Disregarding such chemical-specific properties as dissociation constants (in the case of ionic compounds), particle size, and polymorphism, as well as side effects of viscosity, binding to vehicle components, complex formulation, and the like, the following formulation principles arise:

1. Optimization of the concentration of chemical capable of diffusion by testing its maximum solubility
2. Reduction of the proportion of solvent to a degree that is adequate to keep the test material still in solution
3. Use of vehicle components that reduce the permeability barriers

These principles lead to the conclusion that each test substance requires an individual formulation. Sometimes different ingredients will be required for different concentrations to obtain the maximum rate of release. No universal vehicle is available for any route, but a number of primary approaches are. Any dosage preparation lab should be equipped with glassware, a stirring hot plate, a sonicator, a good homogenizer, and a stock of the basic formulating material, as detailed at the end of this chapter.

Dermal Formulations

Preparing formulations for application to the skin has special considerations associated with it, which, in the case of human pharmaceuticals, has even led to a separate book (Barry, 1983).

The physical state of the skin is considerably affected by external factors such as relative humidity, temperature, and air movement at the skin surface. If this contact is broken (for example, by external applications of ointments or creams), it is reasonable to assume that the new skin will change in some way, sometimes to an extent that creates new conditions of permeability for the test material. This would be the case, for example, if the stratum corneum becomes more hydrated than normal due to the topical delivery form. Temperature might also have an effect, as is the case when any constituents of the vehicle affect the inner structure of the skin through interactions with endogenous skin substances. Often several of these processes occur together. Figure 8.4 shows the relationship between water content (%) and relative humidity (%).

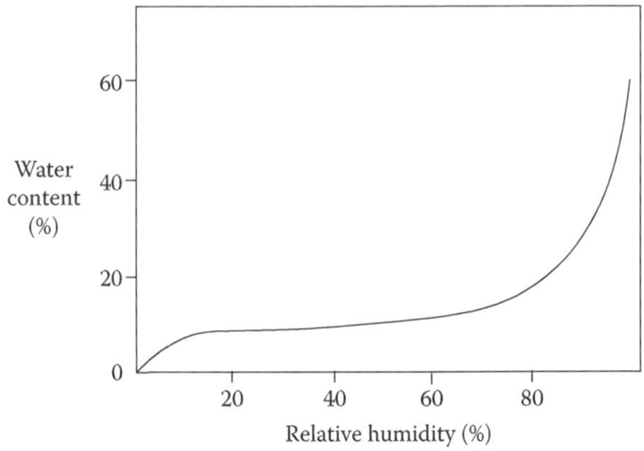

FIGURE 8.4
Course of moisture absorption of skin.

Since this is a question of interactions between the vehicle and the skin (and the latter cannot be viewed as an inert medium), the composition of the vehicle itself may be altered (e.g., by incorporation into skin constituents or through loss of volatile components).

The first contact between vehicle and skin occurs on the skin's surface. The first phase of interaction undoubtedly begins with the lipid mantle, in the case of so-called normal skin. If the skin has been damaged by wounds, the surface can form a moist milieu of serous exudate, resulting in abnormal wetting properties. Normally it is impregnated with oily sebaceous secretions and horny fat, presenting a hydrophobic surface layer. Water will not spread out as a film but will form droplets, while bases with a high affinity to the skin surface constituents spread spontaneously into a film and can wet. In the case of a base, low viscosity, the degree of wetting can often be determined by measuring the angle of contact. If the preparation wets the skin surface, is drawn by capillary action from the visible area into the large inner surface of the stratum corneum, and is transported away into the interior, then it is said that the ointment or cream penetrates well. Spreading and wetting are purely surface phenomena, not penetration in the strict sense. If the skin shows a high content of its own lipids, spreading is limited. It is also reduced if the value of the surface tension of the skin (o_s) decreases compared to the value of the interfacial force between the skin and subject liquid ($y_{s/1}$) and the surface tension of the subject liquid (0_1), as is the case with aqueous bases. Addition of amphiphilic compounds decreases o_1 and $y_{s/1}$ and thus spreadability increases.

How much the endogenous emulsifying substances of the fatty film, such as cholesterol esters, and fatty acid salts, affect this spreading process is not clear. They can probably promote the emulsification of hydrophobic substances with water. Whether the sebaceous and epidermal lipids alone are sufficient to emulsify water and so form a type of emulsive film remains controversial. However, it is assumed that they, together with appropriate vehicle components, improve the spreading of the applied vehicle and that this effect can be potentiated by mechanical means such as intensive rubbing in. A good spreadability ensures that the active ingredient is distributed over a large area.

High local concentrations are avoided and, at the same time, close contact is made between the chemical and the upper layers of the skin.

In grossly simplified terms, hydrogels, suspensions, and water emulsions behave on the skin surface similarly to aqueous solutions. By contrast, pastes and water-oil emulsions act like oil. The ability of an organic solvent to stick or wet depends on its specific properties (e.g., its viscosity and its surface tension).

At present, the information concerning alterations in vehicle composition on the skin surface is sparse. However, two possible extremes are conceivable. On the one hand, if the vehicle has a high vapor pressure, it often completely evaporates shortly after application. On the other hand, the vehicle may remain on the skin surface in an almost completely unchanged composition (e.g., highly viscous Vaseline or similar thick covering systems). Between these two extremes lie the remaining types of vehicles.

The first situation applies for the short-chain alcohols, acetone, or ether. After their evaporation, the drug remains finely dispersed on or in the skin at 100% concentration.

If individual components evaporate, the structure of the vehicle changes and, under certain circumstances, also the effective drug concentration. Oil-water emulsions lose water rapidly, giving rise to the well-known cooling effect. If evaporation continues, the dispersed oil phase coalesces and forms a more or less occlusive film on the skin, together with the emulsifier and the drug. Of course, it is possible that a certain hydrophilic proportion of the drug is then present in suspended form or at least can react with charged molecules and is thus removed from the diffusion process at the start; at the same time, it is to be expected that soluble constituents of the skin are incorporated so that a new system can be formed on the surface and the adjoining layers of skin. Comparable transformations probably also occur after application of water-oil emulsions, providing one realizes that the water evaporates more slowly, the cooling effect is less strong, and, due to the water-oil character of the molecule, the occlusive effect can be more marked because of the affinity of the oily components for the skin.

By contrast, Vaseline and similar highly viscous, lipid bases from the outset form an impenetrable layer, virtually unaffected by external factors or effects emanating from the skin itself. Interactions with the skin lipids are likely only at the boundary between ointment and skin.

The evaporation of the water from the skin into the atmosphere is a continuous process. It can be increased or decreased by the use of suitable vehicles. An evaporation increase will always occur if the water vapor from the vehicle is taken away more quickly than water can diffuse from the deeper layers into the stratum corneum. This applies in principle to all hydrophilic bases, particularly for systems with an oil-water character. After loss of most of their own water, hydrophilic bases develop a true draining effect that can lead to the drying out of the underlying tissue. How much the penetration of hydrophilic drugs can be proved with the help of oil-water systems depends on the solution properties of the rest of the components in the skin. Generally, such compounds can only seldom reach deeper layers. It is equally difficult to show an adequate release of water from hydrophilic systems to a dry skin. If any such effects do occur, they are short term, and are quickly overtaken by opposing processes. The same seems to apply to most of the traditional moisturizers such as glycerin and propylene glycol (Powers and Fox, 1957; Rieger and Deems, 1974). They can also cause a large rise in the rate of evaporation, depending on the relative humidity, and thus increase the transepidermal loss of water. It is probably impossible to prevent this drying out without preparations having some occlusive properties.

In contrast, vehicles that are immiscible with water and those with a high proportion of oils have occlusive effects. They reduce both insensible perspiration and the release of sweat. The sweat collects as droplets at the opening of the glands, but does not spread as a film between the hydrophobic skin surface and the lipophilic base because the free surface energy of the vehicle–skin interface is smaller than that between water and skin. If a lipophilic layer of vehicle is present, this is not spontaneously replaced by the water–skin layer if sweat is secreted.

The horny layer consists of about 10% extracellular components such as lipids, proteins, and mucopolysaccharides. Around 5% of the protein and lipids form the cell wall. The majority of the remainder is present in the highly organized cell contents, predominantly as keratin fibers, which are generally assigned an α-helical structure. They are embedded in a sulphur-rich amorphous matrix, enclosed by lipids that probably lie perpendicular to the protein axis. Since the stratum corneum is able to take up considerably more water than the amount that corresponds to its volume, it is assumed that this absorbed fluid volume is mainly located in the region of these keratin structures.

Some insight into where on the relative humidity continuum water molecules are absorbed can be gained from equilibrium isotherms (Ziegenmeyer, 1982) (Figure 8.4), which show a characteristic sigmoidal shape. At low relative humidity, water is first absorbed at specific skin sites, probably in the region of the peptide compounds and the various polar side chains. At higher moisture content, layers of water form on the skin. By using Zimm–Lundberg cluster theory (Zimm and Lundberg, 1956), additional information can be obtained about the nature of the absorbed water.

Because of thick intertwining protein fibers in the cell and in the area of the cell membrane, cell structure is rigid and remains so, but is altered by the osmotic effect of the penetrating water. The uptake of water entails a continual shifting of the cell matrix, which gradually develops elastic opposing forces that increasingly resist further expansion. Eventually an equilibrium is reached if both forces balance each other. In the case of water, it takes quite a long time to completely hydrate the cell. This process can, however, be shortened if there are components present with a solvent effect diffusing out of a basic vehicle. The duration and degree of swelling depends on the affinity of all the dissolved substances for the tissue, and on the size of the maximum possible elastic reaction, which stabilizes cell structure.

Interactions between Skin, Vehicle, and Test Chemical

The diffusion coefficient of the hydrated stratum corneum is larger than that of dry skin. Therefore, hydration increases the rate of passage of all substances that penetrate the skin. If the hydrated keratin complex is represented by a biphasic system, then it can be considered to exist as a continuous region covered with layers of water and intervening layers of lipids. Nonpolar compounds are predominantly dissolved in the nonpolar lipid matrix and diffuse through it. Polar substances, by contrast, pass through the aqueous layers. The diffusion of water and low-molecular-weight hydrophilic molecules through these layers of water is more difficult than a corresponding free diffusion in an aqueous solution. This could, under certain circumstances, be due to a higher degree of organization of water in the protein structures (than in plasma or the free state), in the sense that this water is available as the driving force of the diffusion process only to a limited degree.

The degree of hydration can be controlled by the choice of vehicle. Lipophilic paraffin bases are available, but vehicles such as water-oil emulsions are more acceptable since they are less occlusive and offer ease of formulation.

In principle, temperature can also have an effect on penetration, which may be exerted on the basic vehicle if it contains temperature-sensitive components (e.g., nonionic entities or thermotherapeutic agents) (Groves 1966). Room and body temperatures can be enough to change the hydrophilic-lipophilic balance and thus possibly change the entire system. It has long been known that increasing temperature can considerably reduce diffusional resistance and thereby increase the rate of penetration of substances. In practice, however, this effect is of no importance. Of course, skin temperature will be increased a few degrees by occlusion because of the prevention of sweating and restriction of heat radiation. However, compared to the increase in penetration achieved by the simultaneous hydration process, this effect is insignificant.

Additives aimed at accelerating penetration always attempt to enable diffusion of pharmacologically active compounds into or through the stratum corneum without damaging it and without causing undesirable systemic effects. Although attempts have been made to limit these effects, this goal has not been achieved as yet. There are numerous substances that decrease the diffusional resistance of the skin, such as propylene glycol, tensides, parotic substances such as urea, DMSO, DMF, and various other organic solvents, mostly of medium chain length. They all improve the penetration of dissolved agents, but only at the cost of the integrity of skin structure, raising the question of the degree of damage and reversibility.

If the substances have passed the stratum corneum, they also generally diffuse into the living part of the epidermis, reach the circulation, and then have systemic effects depending on the amount absorbed. Because these are often constituents of formulations, one generally expects them to have little direct influence on skin penetration. However, their amphiphilic properties allow them to form new systems with the body's constituents and even to change the physical state of water in the skin. By this means, a pathway is cleared for other hydrophilic substances to gain entry into the general circulation.

Most of a permeability enhancer (such as tenside) is bound to the stratum corneum. It is assumed that the underlying mechanism of the process involves interactions with keratin structures. Positively and negatively charged ionic groups of proteins have been suggested as binding sites for ionic substances. Ion pairs can also form. On the other hand, hydrophobic areas are present that can bind with the uncharged part of the enhancers. The total free binding energy of molecules to keratin is made up of the contributions arising from electrostatic and nonpolar interactions. Nonpolar interactions increase with chain length of the molecule. This would be the reason why predominantly anionic molecules of medium chain length exert stronger effects on the keratin structure than those of shorter chain length (Dominguez et al., 1977).

In order to reach the interior of the tightly enmeshed keratin, the molecule must overcome the elastic energy of the polypeptide matrix. The energy necessary to do this is proportional to the volume of the penetrating molecule. The larger the volume, the more difficult it will be for the molecule to approach the various binding sites of proteins in the interior of the keratin complex. Thus, the size of the penetrating molecule is subject to certain limitations. If more molecules are present than can become bound, it is possible that a few of them will reach the living layers of the epidermis, as has been described for several anionic, mostly medium-chain enhancer molecules such as tensides. It remains unclear whether this is a consequence of pure saturation or if other interactions are involved (e.g., with structural lipids or hydrophilic materials from the intercellular lipids).

The extent to which the vehicle can affect the entire diffusion process can be shown by an example. In a four-component system of 40% oil, 40% water, and 20% of an emulsifying agent and co-emulsifier, alteration of only the proportion of emulsifier to co-emulsifier leads to systems of completely different colloidal-chemical structures, which can be labeled as either creams, gels, or microemulsions.

Dermal administration presents fewer logistic difficulties than oral administration. Liquids can be administered as supplied and powders or solids can be moistened with saline to form a thick paste or slurry, or can be applied dry and moistened with saline. Solid materials (sheets or plastic, fabric, etc.) can also be administered dermally. Liquid materials or slurries are applied directly to the skin, taking care to spread the material evenly over the entire area or as much of the area as can reasonably be covered, and then covering with a strip of gauze. If a large amount of material is being administered and the abdominal skin will be exposed, it is sometimes necessary to apply material to the gauze and to the skin. Dry materials are weighed out, then placed on the gauze strip and moistened with physiological saline (generally 15 mL) so that they will adhere to the gauze. The gauze is then wrapped around the animal. This porous gauze dressing is then held in place by an additional wrapping, generally of an impervious material, to create an "occlusive" covering. This occlusion enhances penetration and prevents ingestion or evaporation of the test material.

Another recently developed approach is the use of plastic containment capsules (modified Hilltop Chambers) for administration of well-measured doses in a moisturized microenvironment (Derelanko et al., 1987).

Finally, it should be noted that for some agents (contrary to the general rule), decreasing the concentration of chemical in a vehicle may increase its apparent intrinsic toxicity.

Oral Formulations

The physical form of a material destined for oral administration often presents unique challenges. Liquids can be administered as supplied or diluted with an appropriate vehicle, and powders or particulates can often be dissolved or suspended in an appropriate vehicle. However, selection of an appropriate vehicle is often difficult. Water and oil (such as the vegetable oils) are used most commonly. Materials that are not readily soluble in either water or oil can frequently be suspended in a 1% aqueous mixture of methylcellulose. Occasionally, a more concentrated methylcellulose suspension (up to 5%) may be necessary. Materials for which appropriate solutions or suspensions cannot be prepared using one of these three vehicles often present major difficulties.

Limited solubility or suspendability of a material often dictates preparation of dilute mixtures that may require large volumes to be administered. The total volume of liquid dosing solution or suspension that can be administered to a rodent is limited by the size of its stomach. However, because rats lack a gagging reflex and have no emetic mechanism, any material administered will be retained. Guidelines for maximum amounts to be administered are given in Table 8.6.

TABLE 8.6

Volume Guidelines for Administration of Compounds by Route of Administration to Laboratory Animals

Species	Route: Intranasal µL/nostril	Gavage mL/kg[a]		IV Bolus[b] mL/kg[a]		IP[c] mL/kg[a]		SC[d] mL/kg[a]		IM mL/kg[a]		ID mL/site[a]		IV Infusion[b] mL/kg		Continuous Infusion		Intra Vaginal mL	Intra Vitreal µL/eye	Intra Articular mL/joint		Perivenous/ Periarterial mL/ear
	Vol.	Ideal	Max	Ideal	Max	Ideal	Max	Ideal	Max	Ideal	Max	Ideal	Max	Vol.	Rate mL/min	Max[s] mL/kg/hr	Cath. Maint. Rate mL/hr	Max	Vol.	Joint	Vol.	Max
Mouse	25	10	40[f]	5	25	5-10	50	1-5	20[k]	0.1	1[l,m]	0.1	0.1	50	1	NIA[e]		NIA	2	NIA	NIA	NIA
Rat	50	10	20[g]	1	20	5-10	20	1	20[k]	0.1	10[p]	0.1	0.1	50	1	5	0.1-0.5	0.35	5	Stifle Tarsal	0.1 0.05	NIA
Guinea pig	100	10	30	1	5	1-5	20	1-5	10[k]	0.1	0.5-1	0.1	0.1	10	1	NIA	NIA	NIA	NIA	NIA	NIA	NIA
Rabbit	200	1	20[h]	1[i]	10[i]	3	5[j]	1-2.5	10[k]	0.1-0.5	1	0.1	0.1	20	1	NIA	NIA	2.0	100[r]	Stifle	0.5	0.5
Dog	500	5	20	1	10	3	5	0.5	2[k]	0.1-0.25	1[n,q]	0.1	0.1	20	5	5	1-2	2.0	100	Stifle	1	NIA
Non-human primate	200	5	10	1	10	3	5	0.5	2[k]	0.1-0.5	11[n,q]	0.1	0.1	20	1	5	1-2	NIA	50	NIA	NIA	NIA

(Continued)

TABLE 8.6 (*Continued*)

Volume Guidelines for Administration of Compounds by Route of Administration to Laboratory Animals

Species	Route: Intranasal µL/nostril	Gavage mL/kg[a]		IV Bolus[b] mL/kg[a]		IP[c] mL/kg[a]		SC[d] mL/kg[a]		IM mL/kg[a]		ID mL/site[a]		IV Infusion[b] mL/kg		Continuous Infusion		Intra Vaginal mL	Intra Vitreal µL/eye	Intra Articular mL/joint		Perivenous/ Periarterial mL/ear
	Vol.	Ideal	Max	Ideal	Max	Ideal	Max	Ideal	Max	Ideal	Max	Ideal	Max	Vol.	Rate mL/min	Max[s] mL/kg/hr	Cath. Maint. Rate mL/hr	Max	Vol.	Joint	Vol.	Max
Mini-swine	NIA	NIA	NIA	1	10	1	5	1	3[k]	0.25	0.5[o]	0.1	0.1	10	1	5	1-2	NIA	NIA	NIA	NIA	NIA

Source: Baker HJ et al., *The Laboratory Rat*, Academic Press, New York, 1979; Spector WS., *Handbook of Biological Data*, W. B. Saunders, Philadelphia, 1956; Swenson MJ., *Duke's Physiology of Domestic Animals*, Comstock, Ithaca, 1977; Adapted from Gad et al. 2016.

[a] Single dose per day except where noted otherwise.

[b] Solution properties such as tonicity, pH, etc. need to be taken into account when approaching the volume limits or determining the volume to be infused IV. The recommended working range for pH is 4.5–8.0. The order of degree of tolerance of pH for different dosing routes is oral > intravenous > intramuscular > subcutaneous > intraperitoneal. Animal health must also be taken into consideration, such as kidney function and cardiovascular function. These systems must be normal to handle increased fluid volumes.

[c] When administering a solution IP, the viscosity, concentration, tonicity, and pH of the solution need to be taken into account.

[d] When administering a solution SC, the concentration, tonicity, and pH of the solution must be taken into account.

[e] NIA – no information currently available.

[f] To accommodate a larger volume, the dose may be divided over time (e.g., 20mL/kg administered four times per day to reach a total of 80mL/kg in a 24-hour period).

[g] To accommodate a larger volume, the dose may be divided over time (e.g., 10mL/kg administered four times per day to reach a total; of 40 mL/kg in a 24-hour period).

[h] Rabbits should not be fed prior to administration. Rabbits should be fed after the completion of dose administration.

[i] These volumes may also be used for intra-arterial injection.

[j] Not often used.

[k] If volumes greater than those cited above are used, the volume must be divided over multiple sites.

[l] May be used if divided over multiple sites and alternating legs, maximum of 5 sites per leg. Final volume not to exceed 0.10mL.

[m] 0.05 mL total volume limit per site.

[n] 3 mL total volume limit per site.

[o] 5 mL total volume limit per site.

[p] Up to 20 mL/kg if divided over multiple sites.

[q] 3 mL total volume limit.

[r] Repeat dose 67 µl in the rabbit.

[s] Solution properties such as tonicity, pH, etc., need to be taken into account when determining the volume that may be infused IV. Animal health must also be taken into consideration, such as kidney function and cardiovascular function. These systems need to be normal to handle increased fluid volumes.

Limitations on total volume, therefore, present difficulties for materials that cannot easily be dissolved or suspended. The most dilute solutions that can be administered for a limit-type test (5000 mg kg^{-1}), using the maximum volumes shown in Table 8.6, generally are 1% for aqueous mixtures and 50% for other vehicles.

Although vehicle control animals are not required for commonly used vehicles (e.g. water, oil, methylcellulose), most regulations require that the biological properties of a vehicle be known and/or that historical data be available. Unfortunately, the best solvents are generally toxic and, thus, cannot be used as vehicles. Ethanol and acetone can be tolerated in relatively high doses but produce effects that may complicate interpretation of toxicity associated with the test material alone. It is sometimes possible to dissolve a material in a small amount of one of these vehicles and then dilute the solution in water or in oil.

Gels and resins often present problems because of their viscosity at room temperature. Warming these materials in a water bath to a temperature of up to 50°C will frequently facilitate mixing and dosing. However, it is important to ascertain that no thermal degradation occurs and that actually administered formulations be at or near body temperature.

Other possibilities for insoluble materials are to mix the desired amount of material with a small amount of the animal's diet or to use capsules. The difficulty with the diet approach is the likelihood that the animal will not consume all of the treated diet or that it may selectively not consume chunks of test material. Use of capsules, meanwhile, is labor intensive. In rare cases, if all of these approaches fail, it may not be possible to test a material by oral administration. In capsules, particle size is generally inversely related to solubility and bioavailability. However, milling of solids may adversely affect their chemical nature and/or pose issues of safety.

If necessary, the test substance should be dissolved or suspended as a suitable vehicle, preferably in water, saline, or an aqueous suspension such as 0.5% methyl cellulose in water. If a test substance cannot be dissolved or suspended in an aqueous medium to form a homogenous dosage preparation, corn oil or another solvent can be used. The animals in the vehicle control group should receive the same volume of vehicle given to animals in the highest-dose group.

The test substance can be administered to animals at a constant concentration across all dose levels (i.e., varying the dose volume) or at a constant dose volume (i.e., varying the dose concentration). However, the investigator should be aware that the toxicity observed by administration in a constant concentration may be different from that observed when given in a constant dose volume. For instance, when a large volume of corn oil is given orally, gastrointestinal motility is increased, causing diarrhea and decreasing the time available for absorption of the test substance in the GI tract. This situation is particularly true when a highly lipid-soluble chemical is tested.

If an organic solvent is used to dissolve the chemical, water should be added to reduce the dehydrating effect of the solvent within the gut lumen. The volume of water or solvent-water mixture used to dissolve the chemical should be kept low, since excess quantities may distend the stomach and cause rapid gastric emptying. In addition, large volumes of water may carry the chemical through membrane pores and increase the rate of absorption. Thus, if dose-dependent absorption is suspected, it is important that the different doses are given in the same volume of solution.

Larger volumes than those detailed earlier may be given, although nonlinear kinetics seen under such circumstances may be due to solvent-induced alteration of intestinal function. The use of water-immiscible solvents such as corn oil (which are sometimes used for gavage doses) should be avoided, since it is possible that mobilization from the vehicle may be rate limiting. Magnetic stirring bars or homogenizers can be used in preparing suspensions. Sometimes a small amount of a surfactant such as Tween 80, Span 20, or Span 60 is helpful in obtaining a homogenous suspension.

A large fraction of such a material may quickly pass through the gastrointestinal tract and remain unabsorbed. Local irritation by a test substance generally decreases when the material is diluted. If the objective of the study is to establish systemic toxicity, the test substance should be administered in a constant volume to minimize gastrointestinal irritation that may, in turn, affect its absorption. If, however, the objective is to assess the irritation potential of the test substance, then it should be administered undiluted.

Parenteral Formulations

Parenteral dose forms include aqueous, organic, and oily solutions, emulsions, suspensions, and solid forms for implantation (Spiegel and Noseworthy, 1963). These parenterals need to be sterile and pyrogen-free. Additionally they are, if possible, buffered as close to normal physiological pH and preferably are isotonic with the body fluids.

The preparation of parenteral dosage forms of approved and potential drugs for animals is the same as for humans. Turco and King (1974) provide a comprehensive review of the subject, which was written with human therapeutics in mind and contains very little that is not applicable to animals. Sterility, lack of pyrogenicity, blood compatibility, and low to no irritation at the point of injection are biological requirements; there are also a corresponding set of physicochemical requirements.

Parenteral products are usually given to humans when an immediate effect is needed, when a patient is unable to accept medication by the oral route, or when the drug will be ineffective by the oral route. These conditions apply to animals used in safety evaluation.

Parenteral products can be easily administered to confined or restrained animals, leaving no doubt that the animal received its medication.

To be acceptable, a subcutaneous or intramuscular formulation should cause only a minimum amount of irritation, no permanent damage to the tissues, and be systemically distributed and active when administered by this route. The ideal parenteral product is an aqueous solution isotonic with the body fluids with a pH between 7 and 8. When the drug lacks sufficient aqueous solubility, a suspension may be considered; however, in most cases, the bioavailability of the drug may be affected and encapsulation by the body at the site of injection is extremely likely. The solubility of the drug in water may be improved by the addition of cosolvents such as alcohol, propylene glycol, polyethylene glycol, dimethylacetamide, dimethylsulfoxide, or dimethylformamide. The resulting solution must have additional tolerance for water so that the drug will not precipitate at the site of injection when the solution is diluted by body fluids. If precipitation occurs at the site of injection, the absorption of the drug may be delayed or even completely inhibited.

Water-miscible solvents alone can be used when the drug is chemically unstable in the presence of any water. The number of solvents available for this purpose is extremely limited. The classic review of this subject was made in 1963 (Spiegel and Noseworthy), and some 30 years later, no additional solvents are available. This is unlikely to change in the near future due to the extensive effort necessary to determine the safety of a solvent used as a vehicle. When a nonaqueous vehicle is used, one can invariably expect some degree of pain upon injection, and subsequent tissue destruction is possible. This damage may be due, in part at least, to the heat of solution as the vehicle mixes with body fluids, it may be associated with tissues rejecting the solvent, or it may be an inherent property of the solvent.

Fixed oils of vegetable origin and their esters may be used as parenteral vehicles for some drugs, particularly steroidal hormones. While an oleaginous vehicle may delay or impair absorption of the drug, this characteristic has been used to advantage with some drugs where a small dose is desired over a long period of time. The formulator must know which species will receive the formulation and the type of equipment used in its administration. A product intended for a dog or primate is usually given to a single animal at a time. Conventional glass or disposable syringes will be used with a 20- or 22-gauge needle, which may impede the flow of the liquid, especially when an oleaginous vehicle is used. Impedance is usually compensated for by using small animals, since the volume of injection is small and no more than one injection is normally given at one time.

The viscosity of the solution will influence its acceptability when automatic injection equipment is used. If many animals are injected at one time, a viscous solution that requires a great deal of force to eject will rapidly tire the user. When the automatic injector is refilled from a reservoir, a viscous solution will be slow to fill the volumetric chamber. The subjective aspect of measuring the ease of expelling a dose can be eliminated by constructing an apparatus that will measure the pressure needed to expel a dose. An objective means of measuring ease will allow the formulator to vary the composition of the injection and measure any improvement in injectability. For example, the addition of a wetting agent can be investigated and, if improvement is seen, the level of use can be optimized.

A parenteral product in a multidose vial must contain a preservative to protect the contents of the vial against contamination during repeated withdrawal of dose aliquots.

Dosing Calculations

One of the first things a new technician (or graduate student) must learn is how to calculate dose. Generally, administered doses in systemic toxicity studies are based on the body weight of the animal (expressed as either weight or volume – for liquids – of the test substance per kilogram of body weight of the animal), although some would maintain that surface area may be a more appropriate basis on which to gauge individual dose. The weight (or dose) of the test substance is often expressed in milligrams or grams of active ingredient if the test substance is not pure (i.e., if it is not 100% active ingredient).

Ideally, only the 100% pure sample should be tested; however, impurity-free samples are difficult to obtain and preparation of formulations (as previously discussed) is frequently essential. The toxicity of impurities or formulation components should be examined separately if the investigator feels that they may contribute significantly to the toxicity of the test substance.

If the test substance contains only 75% active ingredient and the investigator chooses a constant dose volume of 10 mL kg^{-1} body weight across all dose levels, it will be more convenient to prepare a stock solution such that when 10 mL kg^{-1} of this stock solution is given to the animal, the dose will be the desired one (say 500 mg kg^{-1} of active ingredient). The concentration of this stock solution would be (500 mg/10 mL)/0.75 = 66.7 mg of the test substance per milliliter of diluent.

Aliquots of the test substance for other dose levels can then be prepared by dilution of the stock solution. For example, the solution concentration for a 250 mg kg^{-1} dose level is (200 mg/10 mL)/0.75 = 26.7 mg of the test substance per milliliter of diluent.

This solution can be prepared by diluting the stock solution 25 times; that is, for each mL of the 26.7 mg mL^{-1} solution to be prepared,

$$\frac{\left(26.7\ \text{mg mL}^{-1}\right)(1\ \text{mL})}{66.7\ \text{mg mL}^{-1}} = 0.400\ \text{mL of the stock solution}$$

This amount should be diluted to a final volume of 1 mL with the vehicle.

The other way to express a relative dose in animals or humans is to do so in terms of body surface area. There are many reasons for believing that the surface area approach is more accurate for relating doses between species (Schmidt-Nielsen, 1984) – and especially between test animals and humans – but this is still a less common approach in safety assessment, although it is the currently accepted norm in a couple of areas: carcinogenesis and chemotherapy, for example.

Calculating Material Requirements

One of the essential basic skills for the efficient design and conduct of safety assessment studies is to be able to accurately project compound requirements for the conduct of a study. In theory, this simply requires plugging numbers into a formula such as

$$(A \times B \times C \times D)\ 1.1 = \text{total compound requirement}$$

where

A = number of animals in each study group

B = the *sum* of doses of the dose groups (such as 0.1 + 0.3 + 1.0 mg kg^{-1}) = 1.4 mg kg^{-1}

C = the number of doses to be delivered (usually the length of the study in days)

D = the average body weight per animal (assuming dosing is done on a per body weight basis)

1.1 = a safety factor (in effect, 10%) to allow for spillage, etc.

As an example of this approach, consider a study that calls for 10 dogs/sex/group (A = 10 × 2 = 20) to receive 0, 10, 50, or 150 mg kg^{-1} day^{-1} (B = 10 + 50 + 150 = 210 mg kg^{-1}) for 30 days (C = 30). On average, the dogs of the age range used weigh 10 kg (D = 10 kg). Our compound need is then (20 × 210 mg kg^{-1} × 30 × 10 kg) 1.1 = 1.386 kg.

The real-life situation is a bit more complicated, since animal weights change over time, diet studies have doses dependent on daily diet consumption, the material may be a salt but dosage should be calculated on the basis of the parent compound, and not all animals may be carried through the entire study.

TABLE 8.7

Standardized Total Compound Requirements for Rodent Diet Studies[a]

	Total Compound Requirement (g) per Dose (mg kg^{-1} day^{-1})					
Length of Study (Days)	**1**	**3**	**10**	**30**	**100**	**300**
Rat[b]						
2 wks	0.2	0.4	1.2	4	10.6	32
4 wks	0.43	0.7	2.5	7.5	25	75
13 wks	0.8	2.6	8.5	25.5	85	260
52 wks	7	21	70	210	0.7[c]	2.1[c]
2 yrs	15	45	150	450	1.5[c]	4.5[c]
Mouse						
2 wks	0.03	0.06	0.22	0.65	2.2	6.4
4 wks	0.08	0.14	0.8	1.4	8	14
13 wks	0.14	0.42	1.4	4.2	14	42
18 mos	0.85	2.5	8.5	25	85	250

Source: Katdare A, Chaubal MV., *Excipient Development for Pharmaceutical Biotechnology and Drug Delivery Systems*, Informa Healthcare, New York, 2006.

[a] Based on 10 animals per sex per group for the length of the study that are 6–8 weeks old at study initiation. Animals are weighed to determine body weights.

[b] Sprague-Dawley rats (body weights and compound requirements for Fischers would be less).

[c] In kilograms.

For rats and mice (where weight change is most dramatic and diet studies most common), Table 8.7 presents some reliable planning values for compound requirements during diet studies.

Excipients

Excipients are usually thought of as being inert substances (such as gum arabic and starch) that form the vehicle or bulk of the dosage form of a drug. They are, of course, both much more complicated than this and not necessarily inert. A better definition would be that of the USP and National Formulary, which defined excipients as any component, other than the active substances (that is, drug substances or DS) intentionally added to the formulation of a dosage form. These substances serve a wide variety of purposes: enhancing stability, adding bulking, increasing and/or controlling absorption, providing or masking flavor, coloring, and serving as a lubricant in the manufacturing process. They are, in fact, essential for the production and delivery of marketed drug products. As will soon be made clear, they are regulated both directly and as part of the drug product (DP). For the pharmaceutical manufacturers, the use of established and accepted excipients (such as can be found in Smolinske, 1992 or USP, 2012 – though these lists are not complete) is much preferred. However, both pharmaceutical manufacturers and the companies which supply excipients must from time to time utilize (and therefore develop, evaluate for safety, and get approved) new excipients.

In the last 10 years, the use of nanosuspensions as components of formulations has presented a new set opportunities – and potential problems (Rabinow, 2004).

While for the purpose of nonclinical formulation, our concerns are generally limited to vehicles, their formulation components can be important. Table 8.4 lists examples of this.

Regulation of Excipients

Table 8.8 lists the relevant sections of CFR 21 which govern excipients. Under Section 201(g)(1) of the Federal Food, Drug, and Cosmetic Act (FD&C Act; 1), the term *drug* is defined as:

> (A) articles recognized in the official *United States Pharmacopeia*, official *Homeopathic Pharmacopeia of the United States*, or official *National Formulary*, or any supplement to any of them; and (B) Articles intended for use in the diagnosis, cure, mitigation, treatment, or prevention of disease in man or other animals; and (C) Articles (other than food) intended to affect the structure of any function of the body of man or other animals; and (D) Articles intended for use as a component of any articles specified in clause (A), (B), or (C).

An excipient meets the definitions as listed in (A) and (D) above.

In 21 CFR § 210.3(b)(2), an "inactive ingredient means any component other than an active ingredient." According to the CFR, the term inactive ingredient includes materials in addition to excipients. 21 CFR § 201.117 states the following:

> Inactive ingredients: A harmless drug that is ordinarily used as an inactive ingredient, such as a coloring, emulsifier, excipient, flavoring, lubricant, preservative, or solvent in the preparation of other drugs shall be exempt from Section 502(f)(1) of the Act. This exemption shall not apply to any substance intended for a use which results in the preparation of a new drug, unless an approved new-drug application provides for such use.

Excipients also meet the definition of component in the Good Manufacturing Practice (GMP) regulations in 21 CFR § 210.3(b)(3): "Component means any ingredient intended for use in the manufacture of a drug product, including those that may not appear in such drug product."

The *NF* Admissions Policy in the *United States Pharmacopeia 30/National Formulary 25* defines the word *excipient*: "An excipient is any component other than the active substance(s), intentionally added to the formulation of a dosage form. It is not defined as an inert commodity or an inert component of a dosage form."

TABLE 8.8

U.S. Code of Federal Regulations References to Excipients

Subject	Reference	Content
General	21 CFR § 210.3(b)(8)	Definitions
	21 CFR § 201.117	Inactive ingredients
	21 CFR § 210.3(b)(3)	Definitions
Over-the-counter drug products	21 CFR § 330.1(e)	General conditions for general recognition as safe, effective, and not misbranded
	21 CFR § 328	Over-the-counter drug products intended for oral ingestion that contain alcohol
Drug Master Files	21 CFR § 314.420	Drug master files
Investigational New Drug Application	21 CFR § 312.23(a)(7)	IND content and format
New Drug Application	21 CFR § 312.31	Information amendments
	21 CFR § 314.50(d)(1)(ii)(a)	Content and format of an application
	21 CFR § 314.70	Supplements and other changes to an approved application
Abbreviated New Drug Application	21 CFR § 314.94(a)(9)	Content and format of an abbreviated application
	21 CFR § 314.127	Refusal to approve an abbreviated new drug application
	21 CFR § 314.127(a)(8)	Refusal to approve an abbreviated new drug application
Current Good Manufacturing Practice	21 CFR § 211.84(d)	Testing an approval or rejection of components, drug product containers and closures
	21 CFR § 211.165	Testing and release for distribution
	21 CFR § 211.180(b)	General requirements
	21 CFR § 211.80	General requirements
	21 CFR § 211.137	Expiration dating
Listing of drugs	21 CFR § 207	Registration of procedures of drugs and listing of drugs in commercial distribution
	21 CFR § 207.31(b)	Additional drug listing information
	21 CFR § 207.10(e)	Exceptions for domestic establishments
Labeling	21 CFR § 201.100(b)(5)	Prescription drugs for human use
	21 CFR § 201.20	Declaration of presence of FD&C Yellow No. 5 and/or FD&C Yellow No. 6 in certain drugs for human use
	21 CFR § 201.21	Declaration of presence of phenylalanine as a component of aspartame in over-the-counter and prescription drugs for human use
	21 CFR § 201.22	Prescription drugs containing sulfites; required warning statements

Similar to all other drugs, excipients must comply with the adulteration and misbranding provisions of the FD&C Act. Under Section 501(a), an excipient shall be deemed to be adulterated if it consists in whole or in part of any filthy, putrid, or decomposed substance, or if it has been prepared, packed, or held under unsanitary conditions whereby it may have been contaminated with filth, or whereby it may have been rendered injurious to health. An excipient is adulterated if the methods used in, or the facilities or controls used for its manufacture, processing, packing, or holding do not conform to or are not operated or administered in conformity with current Good Manufacturing Practice to assure that such drug meets the requirements of the act as to safety and has the identity and strength, and meets the quality and purity characteristics which it purports or is represented to possess. In addition, under Section 501(b), an excipient shall be deemed to be adulterated if it purports to be or is represented as a drug, the name of which is recognized in an official compendium, and its strength differs from, or its quality or purity falls below, the standards set forth in such compendium.

In 2005, the US Food and Drug Administration (FDA) promulgated new guidance on the selection and use of excipients in nonclinical and clinical studies. US FDA compliance officials require the use of inactive ingredients that meet compendial standards when standards exist and either have previous use in US FDA–approved pharmaceuticals or that they be qualified as "novel" excipients (with studies as summarized in Table 8.9) (FDA, 2005). The FDA Center for Drug Evaluation and Research maintains an Inactive Ingredient Committee whose charter includes the evaluation of the safety of inactive ingredients on an as-needed basis, preparation of recommendations concerning the types of data needed for excipients to be declared safe for inclusion in a drug product, and other related functions.

From a regulatory standpoint, the FDA's concern regarding safety involves the toxicity, degradants, and impurities of excipients, as discussed in other chapters in this book. In addition, other chapters of this book address types of toxicity concerns, toxicity testing strategies, and exposure and risk assessment of excipients.

Excipients must be safe for their intended use. Under 21 CFR § 330.1(e), over-the-counter (OTC) human drugs that are generally recognized as safe and effective and not misbranded may only contain inactive ingredients if they are suitable and if the amounts administered are safe and do not interfere with the effectiveness of the drug or with required tests or assays. Color additives may be used in accordance with the provisions of the FD&C Act and the regulations of 21 CFR Parts 70–82. The FDA proposed that to make it clear that, to be considered as suitable within the meaning of 21 CFR § 330.1(e), each inactive ingredient in an OTC human drug product should perform a specific function (5). The proposed regulation defined *safe and suitable*

TABLE 8.9

Summary of Toxicological Studies Recommended for New Pharmaceutical Excipients Based on Route of Exposure

Tests	Oral	Mucosal	Trans-Dermal	Dermal/Topical	Paren-Teral	Inhalation/Intranasal	Ocular
Appendix 1-base set	R	R	R	R	R	R	R
Acute oral toxicity	R	R	R	R	R	R	R
Acute dermal toxicity	C	C	C	C	C	R	C
Acute inhalation toxicity	R	R	R	R	R	R	R
Eye irritation	R	R	R	R	R	R	R
Skin irritation	R	R	R	R	R	R	R
Skin sensitization	R	R	R	R	R	R	R
Acute parenteral toxicity	–	–	–	–	R	–	–
Application site evaluation	–	–	R	R	R	R	–
Pulmonary sensitization	–	–	–	–	–	R	–
Phototoxicity/photoallergy	–	–	R	R	–	–	–
Ames test	R	R	R	R	R	R	R
Micronucleus test	R	R	R	R	R	R	R
ADME-intended route	R	R	R	R	R	R	R
28-day toxicity (2 species)-intended route	R	R	R	R	R	R	R
Appendix 2							
90-day toxicity (most appropriate species)	R	R	R	R	R	R	R
Developmental toxicity (rat and rabbit)	R	R	R	R	R	R	R
Additional assays	C	C	C	C	C	C	C
Genotoxicity assays	R	R	R	R	R	R	R
Appendix 3							
Chronic toxicity (rodent, nonrodent)	C	C	C	C	C	C	C
Photocarcinogenicity	–	–	C	C	–	–	–
Carcinogenicity	C	C	C	C	C	C	–

R, required; C, conditionally required.

Note: In the author's experience, novel excipients used for true single administration drugs may require only expanded acute studies (as performed for "Phase 0 INDs).

to mean that the inactive ingredient meets various conditions as mentioned in the foregoing. OTC drug manufacturers are responsible for assuring that these conditions are met. There is no formal approval mechanism.

In the United States, the safety and suitability of excipients used in new drugs are considered as part of the New Drug Application (NDA) process. There is no separate and independent review and approval system for excipients. There are no specific regulations or guidelines that specify the requirements needed to gain approval of a new drug that contains a new excipient. Generally, pharmaceutical companies choose excipients that previously have been approved for commercial use in other NDAs. The FDA's Inactive Ingredient Guide (FDA [2017]) contains a listing of inactive ingredients present in approved drug products. There is currently no way of gaining a listing for an excipient in the guide independent of the NDA route. The FDA reviews the status of an excipient in food as information to support its use in drug products. Factors relative to the use of an excipient, such as dosing regimen and route of administration, are also reviewed. Advances in excipient technology and drug dosage from technology have created a need for a separate regulatory approval process for new excipients. The *USP* published IPEC's Excipient Safety Evaluation Guidelines as Information Chapter <1074> Excipient Biological Safety Evaluation Guideline.

Information on existing or new excipients can be described and provided to the FDA in an NDA directly. Alternatively, the manufacturers of excipients may prepare and submit type IV Drug Master Files (DMF) to support the use of an excipient in one or more NDAs. The DMFs are discussed in the FDA's regulations under 21 CFR § 314.420 and the FDA-issued Guidance for Drug Master Files (8). When authorized by the DMF submitter (i.e., the excipient manufacturer) and cross-referenced by an NDA submitter, the FDA reviews the DMF to make determinations on the safety, manufacture, and quality of the excipient use in the new drug that is the subject of the then pending NDA. The DMF becomes active when reviewed in conjunction with the review and approval of an NDA.

The *USP/NF* provides a listing of excipients by categories in a table according to the function of the excipient in a dosage form, such as tablet binder, disintegrant, and such. An excellent reference for excipient information is the *Handbook of Pharmaceutical Excipients* (Rowe et al., 2012). Additionally, Gad et al. (2016) provide an excellent and extensive database of nonclinical formulation components and either acceptable maximum usage levels by species route and duration of study.

Excipients have historically not been subjected to extensive safety testing because they have been considered a priori to be biologically inactive and therefore nontoxic. Many, if not most, excipients used are approved food ingredients, the safety of which has been assured by a documented history of safe use or appropriate animal testing. Some of the excipients are Generally Recognized as Safe (GRAS) food ingredients. The excipient is an integral component of the finished drug preparation and, in most countries, is evaluated as part of this preparation. There has been no apparent need to develop specific guidelines for the safety evaluation of excipients, and most developed countries do not have specific guidelines. However, as drug development has become more complex and/or new dosage forms have developed, improved drug bioavailability has become more important. It was noted that as the available excipients were often inadequate, new pharmaceutical excipients specifically designed to meet the challenges of delivering new drugs were needed, and these are being developed. The proper safety evaluation of new excipients has now become an integral part of drug safety evaluation.

In the absence of official regulatory guidelines, the Safety Committees of the International Pharmaceutical Excipients Council (IPEC) in the United States, Europe, and Japan developed guidelines for the proper safety evaluation of new pharmaceutical excipients (IPEC, 1997). The Committees critically evaluated guidelines for the safety evaluation of food ingredients, cosmetics, and other products, as well as textbooks and other appropriate materials. Before initiating a safety evaluation program for a new pharmaceutical excipient, it is advisable to address the following:

1. Chemical and physical properties and functional characterization of the test material (Hawley, 1971).
2. Analytical methods that are sensitive and specific for the test material and that can be used to analyze the test material in animal food used in feeding studies or in the vehicle used for other studies
3. Available biological, toxicological, and pharmacological information on the test material and related materials (which involves a thorough search of the scientific literature)
4. Intended conditions of use, including reasonable estimates of exposure
5. Potentially sensitive segments of the population

As pharmaceutical excipients are assumed to be biologically nonreactive, dose-response relations cannot always be established. An acceptable alternative is to use a maximum attainable or maximum feasible dose. This is the highest dose possible that will not compromise the nutritional or health status of the animal. Table 8.10 summarizes the maximum or limit doses for various types of studies by different routes of exposure. For example, 2000 mg/kg body weight of an orally administered test material is the maximum dose recommended for a testing strategy that has been developed for new pharmaceutical excipients that takes into consideration the physical-chemical nature of the product and the potential route(s) and duration of exposures, both through its intended use as part of a drug product and through workplace exposure during manufacturing. The number and types of studies recommended in this tiered approach are based on the duration and routes of

TABLE 8.10
Limit Doses for Toxicological Studies

Nature of Test	Species	Limit dose[a]
Acute oral	Rodent	2000 mg/kg bw
Acute dermal	Rabbit	2000 mg/kg bw
	Rat	
Acute inhalation[b]	Rat	5 mg/L air for 4 h or maximum attainable level under conditions of study
Dermal irritation	Rabbit	0.5 mL liquid
		0.5 g solid
Eye irritation	Rabbit	0.1 mL liquid
		100 mg solid
14-day/28-day oral repeated dosing; 90-day subchronic	Rodent, Non-rodent	1000 mg/kg bw/day
14-day/28-day oral repeated dosing; 90-day subchronic	Rat, rabbit	1000 mg/kg bw/day
Chronic toxicity, carcinogenicity	Rats, mice	5% maximum dietary concentration for nonnutrients
Reproduction	Rats	1000 mg/kg bw/day
Developmental toxicity (teratology)	Mice, rats, rabbits	1000 mg/kg bw/day

Source: Adapted from Weiner ML, Katkoskie LA., *Excipient Toxicity and Safety,* Marcel Dekker, New York, 1999.

[a] mg/kg bw, milligrams of test material dosed per kilogram of body weight to the test species.

[b] Acute inhalation guidelines that indicate this limit dose are US Environmental Protection Agency Toxic Substance Health Effect Test Guidelines, Oct. 1984; (PB82-232984) Acute Inhalation Toxicity Study; the OECD Guidelines of the Testing of Chemicals, Vol 2, Section 4; Health Effects, 403, Acute Inhalation Toxicity Study, May 12, 1982; and the Official Journal of the European Communities, L383A, Vol 35, Dec. 29, 1992, Part B.2.

potential human exposure. Thus, the longer the exposure to the new pharmaceutical excipient, the more studies are necessary to assure safety. Table 8.9 summarizes the entire set of toxicological studies recommended for new pharmaceutical excipients (Weiner and Katkoskie, 1999; IPEC, 2012).

Tests have been outlined for each exposure category to assure safe use of the time period designated. The tests for each exposure category assure the safe use of the new pharmaceutical excipient for the time frame specified for the specific exposure category. Additional tests are required for longer exposure times.

The base set required for all excipients is detailed in Table 8.11. These are sufficient, however, only for those excipients intended for use for up to 2 weeks in humans.

TABLE 8.11

Base Set Studies for a Single Dose in Humans (Exposure up to 2 Weeks)

Test	Purpose
Acute oral toxicity	To determine the potential acute toxicity-lethality following a single oral dose
Acute dermal toxicity	To determine the potential acute toxicity-lethality following a single dermal dose
Acute inhalation toxicity	To determine the potential acute toxicity-lethality following a single 4-h inhalation exposure to a test atmosphere containing the new pharmaceutical excipient (aerosol, vapor, or particles)
Eye irritation	To determine the potential to produce acute irritation or damage to the eye
Skin irritation	To determine the potential to produce acute irritation or damage to the skin
Skin sensitization	To determine the potential to induce skin sensitization reactions
Ames test	To evaluate potential mutagenic activity in a bacterial reverse mutation system with and without metabolic activation
Micronucleus test	To evaluate the clastogenic activity in mice using polychromatic erythrocytes
ADME – intended route	To determine the extent of absorption, distribution, metabolism, and excretion by the intended route of exposure following a single dose and repeated doses
28-day toxicity – intended route	To assess the repeated-dose toxicity in male and female animals of two species following dosing for 28 days by the intended route of exposure

TABLE 8.12

Appendix 3 Studies for Repeated Chronic Exposure in Humans

Test	Purpose
Chronic toxicity	To assess the toxicity following chronic (lifetime) exposure by the route of intended exposure
Oncogenicity	To assess the potential to induce tumors by the intended route of exposure
One-generation reproduction	To assess the potential reproductive and developmental toxicity in males and females by the intended route of exposure

If exposure to the new pharmaceutical excipient is expected to occur for longer than 2 but no more than 6 weeks, additional toxicological studies are required, as shown in Table 8.12. The longer the expected human exposure, the more extensive will be the toxicological studies to assure safety. A tiered approach assures that those tests necessary to ensure safety for the expected duration of human exposure are conducted. Thus, to assure safe use for greater than 2 weeks, but no more than 6 weeks in humans, subchronic toxicity and developmental toxicity studies are required. To assure safe use for greater than 6 continuous weeks, chronic or oncogenicity studies are conditionally required, as per Table 8.12. This means long-term studies should be considered for prolonged human exposures, but may not be absolutely required. A thorough, scientific review of the data generated in the base set and Appendix 2 studies should be undertaken. From a critical evaluation by a competent toxicologist, the results of the physical-chemical properties of the test material, the 28-day and 90-day tests, the ADME-PK acute and repeated-dose tests, and the developmental toxicity test(s), a final determination can be made on the value of chronic toxicity or oncogenicity studies.

For example, if no toxicity is observed at a limit dose of 1000 mg/kg body weight per day following the 90-day toxicity study, no genotoxicity was found, and the ADME-PK profile indicates that the material is not absorbed and is completed excreted unchanged in the feces, then it is likely that a chronic study is not necessary. The decision to conduct chronic studies should be determined on a case-by-case basis using scientific judgment. It will be interesting to observe how this scheme may change in light of ICH.

References

Adeyeye MC, Brittain HG. (Eds.) (2008) Preformulation in solid dosage form development. In: *Drugs and the Pharmaceutical Sciences*, Vol. 178. New York, NY: Informa Healthcare USA.

Baker HJ, Lindsey JR, Weisbroth SH. (1979) *The Laboratory Rat*, Vol. 1. New York, NY: Academic Press, pp. 411–412.

Barry BW. (1983) *Dermatological Formulations.* New York, NY: Marcel Dekker.

Boersen N, Lee T, Hui H-W. (2014) Development of preclinical formulations for toxicology studies. In: AS Faqi (Ed.) *A Comprehensive Guide to Toxicology in Preclinical Drug Development*. San Diego, CA: Academic Press, pp. 69–86.

Calvary HO, Klump TG. (1939) The toxicity for human beings for diethylene glycol with sulfanilimide. *South Med J*, 32:1105.

Derelanko MJ, Gad SC, Gavigan FA, Babich PC, Rinehart WE. (1987) Toxicity of hydroxylamine sulfate following dermal exposure: Variability with exposure method and species. *Fund Appl Toxicol*, 8(3):583–594.

Dominguez JG, Parra JL, Infante MR, Pelejero CM, Balaguer F, Sastre T. (1977) A new approach to the theory of adsorption and permeability of surfactants on keratinic proteins: the specific behaviour of certain Hydrophobie chains. *J Soc Cosmet Chem*, 28:165–182.

FDA (2017) *Inactive Ingredient* Database. Available at https://www.accessdata.fda.gov/scripts/cder/iig/index.cfm

FDA. (May 2005) *Guidance for Industry: Nonclinical Studies for the Safety Evaluation of Pharmaceutical Excipients.* Rockville, MD:FDA.

Fitzgerald JM, Boyd VF, Manus AG. (1983) Formulations of insoluble and immiscible test agents in liquid vehicles for toxicity testing. In: CW Jameson and DB Walters (Eds.) *Chemistry for Toxicity Testing.* Boston, MA: Butterworth, pp. 83–90.

Gad SC. (2008a) *Pharmaceutical Manufacturing Handbook: Production and Processes.* Hoboken, NJ: Wiley.

Gad SC. (Ed.) (2008b) *Preclinical Development Handbook: ADME and Biopharmaceutical Properties.* Hoboken, NJ: Wiley.

Gad SC. (2016) *Drug Safety Evaluation,* 3rd Ed. Hoboken, NJ: Wiley.

Gad SC, Chengelis CP. (1998) *Acute Toxicology Testing, 2nd Ed.* San Diego: Academic Press.

Gad SC, Spainhour CB, Shoemake C, Stackhouse Pallman DR, Stricker-Krongrad A, Downing PA, Seals RE, Eagle LA, Polhamus K, Daly J. (2016) Tolerable levels of nonclinical vehicles and formulations used in studies by multiple routes in multiple species with notes on methods to improve utility. *Int J Toxicol,* 35(2):95–178.

Gaines TB. (1960) The acute toxicity of pesticides to rats. *Toxicol Appl Pharmacol,* 2:88–99.

Gibaldi M. (1976) Biopharmaceutics. In: L Lachman, HA Lieberman, and JL Kanig (Eds.) *The Theory and Practice of Industrial Pharmacy,* 2nd Ed. Philadelphia, PA: Lea & Febiger.

Gibaldi M, Feldman S. (1970) Mechanisms of surfactant effects on drug absorption. *J Pharm Sci,* 59:579.

Gibson M. (Ed.) (2009) Pharmaceutical preformulation and formulation. In: *Drugs and the Pharmaceutical Sciences,* 2nd Ed, Vol. 199. New York, NY: Informa Healthcare USA.

Groves M. (1966) The influence of formulation upon the activity of thermotherapeutic agents. *Reps Progr Appl Chem,* 12:51–151.

Hagenbusch OE. (1937) Elixir of sulfanilamide massengill. *J Am Med Assoc,* 109:1531.

Hardee GE, Baggot JD. (Eds.) (1998) Development and formulation of veterinary dosage forms. In: *Drugs and the Pharmaceutical Sciences,* 2nd Ed, Vol. 88. New York, NY: CRC Press.

Hawley GG. (1971) *The Condensed Chemical Dictionary.* New York, NY: Van Nostrand Reinhold.

Holick HGO. (1937) Glycerine, ethylene glycol, propylene glycol and diethylene glycol. *J Am Med Assoc,* 109:1517.

IPEC (1997) *The Proposed Guidelines for the Safety Evaluation of New Excipients.* International Pharmaceutical Excipients Council (IPEC), November 1997, Berlin.

IPEC (2012) *The Joint IPEC–PQG Guide for Pharmaceutical Excipients.* London: Pharmaceutical Quality Group.

Katdare A, Chaubal MV. (2006) *Excipient Development for Pharmaceutical Biotechnology and Drug Delivery Systems.* New York, NY: Informa Healthcare.

Kemsley J. (February 3, 2014) Eyes on Excipients: Inactive but essential drug ingredients are getting checked out with modern rigor to ensure their quality. *Chem Eng News,* 92(5):9–11. Accessible at: http://cen.acs.org/articles/92/i5/Eyes-Excipients.html. Last accessed April 7, 2015.

Kerberle H. (1971) Physiocochemical factors of drugs affecting absorption, distribution and excretion. *Acta Pharmacol Toxicol,* 29(suppl 3):30–47.

Lipinski CA. (2004) Lead- and drug-like compounds: The rule-of-five revolution. *Drug Discov Today Technol*, 1(4):337–341.

Lipinski CA, Lombardo F, Dominy BW, Feeney PJ. (2001) Experimental and computational approaches to estimate solubility and permeability in drug discovery and development settings. *Adv Drug Deliv Rev*, 46(1–3):3–26.

Liu R. (Ed.) (2008) *Water-Insoluble Drug Formulation*, 2nd Ed. Boca Raton, FL: CRC Press.

Melnick RL, Jameson CW, Goehl TJ, Kuhn GO. (1987) Application of microencapsulation for toxicology studies. *Fund Appl Toxicol*, 11:42531.

Monkhouse DC, Rhodes CT. (1998) *Drug Products for Clinical Trials*. New York, NY: Marcel Dekker.

Neervannan S. (2006) Preclinical formulations for discovery and toxicology: Physicochemical challenges. *Expert Opin Drug Metab Toxicol*, 2(5):715–731.

Niazi SK. (2007) *Handbook of Preformulation: Chemical, Biological, and Botanical Drugs*. New York, NY: Informa Healthcare USA.

O'Reilly WJ. (1972) Pharmacokinetics in drug metabolism and toxicology. *Canad J Pharm Sci*, 7:6–77.

Palucki M, Higgins JD, Kwong E, Templeton AC. (2010) Strategies at the interface of drug discovery and development: Early optimization of the solid state phase and preclinical toxicology formulation for potential drug candidates. *J Med Chem*, 53(16):5897–5905.

Powers DH, Fox C. (1957) A study of the effect of cosmetic ingredients, creams and lotions on the rate of moisture loss. *Proc Sci Sect Toil Goods Assoc*, 28:21–26.

Rabinow BE. (2004) Nanosuspensions in drug delivery. *Nat Rev Drug Discov*, 3(9):785–796.

Racy I. (1989) *Drug Formulation*. New York, NY: John Wiley.

Rieger MM, Deems DE. (1974) Skin moisturizers. II. The effects of cosmetic ingredients on human stratum comeum. *J Soc Cosmet Chem*, 25:253.

Ritschel WA, Siegel EG, Ring PE. (1974) Biopharmaceutical factors influencing LD50. 1. Viscosity. *Arzneim-Forsch*, 24:907–910.

Rowe RC, Sheskey PJ, Cook WG, Fenton ME. (2012) *Handbook of Pharmaceutical Excipients*, 7th Ed. Washington, DC: American Pharmaceutical Association.

Schmidt-Nielsen K. (1984) *Scaling: Why Is Animal Size So Important?* New York, NY: Cambridge University Press.

Smolinske SC. (1992) *Handbook of Food, Drug and Cosmetic Excipients*. Boca Raton, FL: CRC Press.

Spector WS. (1956) *Handbook of Biological Data*. Philadelphia, PA: W. B. Saunders.

Spiegel AJ, Noseworthy MM. (1963) Use of nonaqueous solvents in parenteral products. *J Pharm Sci*, 52:917–927.

Stahl PH, Wermuth CG. (Eds.) (2011) *Handbook of Pharmaceutical Salts: Properties, Selection and Use*, 2nd Ed. Zurich: Wiley-VCH & VHCA.

Strickley RG. (2008) Formulation in drug discovery. In: A Wood and M Desai (Eds.) *Annual Report on Medicinal Chemistry*. Oxford, UK: Elsevier, pp. 419–451.

Swarbrick J. (Ed.) (2006) *Current Concepts in Pharmaceutical Sciences: Dosage Form, Design and Bioavailability*. Philadelphia, PA: Lea & Febiger.

Swenson MJ. (Ed.) (1977) *Duke's Physiology of Domestic Animals*. Ithaca, NY: Comstock, p. 178.

Turco S, King RE. (1974) *Sterile Dosage Forms*. Philadelphia, PA: Lea & Febiger.

USP. (2014) *United States Pharmacopiea and National Formulary. USP 37/NF 32.* Washington, DC: United States Pharmacopiea Convention.

Weiner M, Bernstein IL. (1989) *Adverse Reactions to Drug Formulation Agents.* New York, NY: Marcel Dekker.

Weiner ML, Katkoskie LA. (1999) *Excipient Toxicity and Safety.* New York, NY: Marcel Dekker.

Ziegenmeyer J. (1982) The influence of the vehicle on the absorption and permeation of drugs. In: R Brandan and P Reisen (Eds.) *Dermal and Transdermal Absorption.* Stuttgart: Wissenschaftliche, Verlagsgesellschaft, pp. 73–89.

Zimm BH, Lundberg JL. (1956) Sorption of vapors by high polymers. *J Phys Chem,* 60:425.

9

Vehicles and Excipients

For most toxicologists, pharmacologists, and early clinical drug developers, the fundamental approach to meeting (at least the first two) objectives targeted by this book are through formulation approaches. While the fundamentals of formulation theory were addressed in Chapter 8, the principal working components are vehicles ("solvents") and excipients – the materials comprising a formulation (for an oral example set, see Table 9.1). There are separate and specialized sets for each route. Since the primary objective for these nonclinical uses is to achieve having a stable solution for administration, vehicles are generally more the focus than the other types of excipients.

The Appendix provides a modified version of Gad et al. (2016), the most comprehensive summary of common nonclinical vehicles and excipients and their tolerated levels by route, species, and duration of use. The first version of such a summary appeared in Gad and Chengelis (1988). It should be remembered that (with the exception of systemic dermal studies) nonclinical study formulation components are *not* limited to those components that are allowed/approved for human (clinical) use.

For completeness, it should be noted that for medical devices, where systemic testing is done either by implantation of the device or biomaterial or by injection of an extract prepared from the devices (see Gad and Gad-McDonald, 2016), a specific set of vehicles (really a subject of those in the Appendix) is used for extraction and subsequently injected. These are summarized in Table 9.2, and are also addressed in the Appendix.

Excipients

There are several definitions for an excipient. In some cases, the definition can be simple, yet in other cases the definition is more encompassing and complex. For our purposes here, we will define excipients as substances that do not have as a primary purpose dissolving or "carrying" the therapeutic (and excipients) into the target biologic system. They may or may not be liquids, and serve a broad range of functional purposes.

TABLE 9.1

Nonactive Formulation Components in Oral Drug Candidate Delivery

Type	Purpose	Examples
Tablets, sachets, and capsules		
Binder	Facilitates agglomeration of powder into granules	Povidones, starches
Capsule shell	Contains powders or liquids	Gelatin, hypromellose
Coating agent	May mask unpleasant tastes or odors, improve ingestion or appearance, protect ingredients from the environment, or modify release of the active ingredient	Shellac, hypromellose
Colorant	Produces a distinctive appearance and may protect light-sensitive ingredients	FD&C colors, titanium dioxide
Disintegrant	Promotes rapid disintegration to allow a drug to dissolve faster	Sodium starch glycolate, crospovidones
Enteric coatings	Protects from dissolution in stomach ("gastro resistant")	HPMC, methyl acrylate copolymers
Filler or diluent	Increases volume or weight	Calcium phosphate, lactose
Glidant or anticaking agent	Promotes powder flow and reduces caking or clumping	Talc, colloidal silicon dioxide
Lubricant	Reduces friction between particles themselves and between particles and manufacturing equipment	Magnesium stearate, glycerides
Release modifier	Provides extended-release capability	Ethyl cellulose, guar gum
Solvent/vehicle	Improves stability and bioavailability	Water++ (see Gad et al., 2016)
Surfactant	Solubilizing agent	Irganox, SLS
Oral liquids		
Antimicrobial preservative	Prevents growth of bacteria, yeast, and mold	Glycerin, benzyl alcohol
Antioxidant	Reduces oxidative reactions that could alter ingredients	Ascorbic acid, butylated hydroxyanisole
Chelating or complexing agent	Stabilizes ions	Ethylenediaminetetraacetic acid salts, cyclodextrins
Liposome preparations	Improve bioavailability and protect from first-pass metabolism	Special vehicle with at least one lipid bilayer
pH modifier	Controls pH to improve drug stability or avoid irritation when consumed	Citric acid and its salts, salts of phosphoric acid
Surfactants and solubilizing agent	Promotes dissolution of insoluble ingredients	Sodium lauryl sulfate, polysorbates
Sweetening agent	Improves palatability	Sucrose, saccharin

Sources: Gad SC, *Pharmaceutical Manufacturing Handbook: Production and Processes*, Wiley, Hoboken, 2008; Gad SC, *Drug Safety Evaluation*, Wiley, Hoboken, 2016.

Note: Other routes may have different types of excipients and purposes for them.

TABLE 9.2

Vehicles Used for Medical Device Studies

Vehicle	Route
Normal (0.9%) saline	IV
Ethanol (5%)	IV
Polyethylene glycol (PEG) 400	IP
Cottonseed oil	IP

Abbreviations: IV, intravenous; IP, intraperitoneal.

Excipients may not be active drug or prodrug and must have been appropriately evaluated for safety. Their inclusion into a drug "product" is generally for any of the following purposes:

1. Improvement in drug delivery aspects (for nonclinical uses, generally bioavailability and tolerance).
2. Aid in processing of the system during manufacture.
3. Protection and/or enhancement of stability and bioavailability.
4. Enhancement of a specific attribute of the drug. Such attributes could include effectiveness of the drug during storage or use.
5. Aid in product identification.

Some of the functions that solid excipients can perform are fillers, diluents, binders, disintegrants, lubricants, antiadherents, glidants, wetting agents, surface-active agents, pigments, colors, flavors, sweeteners, and taste maskers.

Regarding topical and transdermal delivery systems, there are a variety of ointment bases. Hydrocarbon bases have the following properties: they are greasy, hydrophobic, not water-washable, and occlusive, and are emollients. An example of a hydrocarbon base would be petrolatum. Emulsion bases (oil/water) are water-washable, nonocclusive, and nongreasy, and can be diluted with water. An emulsion base would include any hydrophilic ointment. Water-soluble bases are lipid-free, nongreasy, water-soluble and washable, nonocclusive, and usually anhydrous. A water-soluble base would include a polyethylene glycol ointment. An absorption base is greasy, anhydrous, occlusive, and emollient and absorbs water. An absorption base would include hydrophilic petrolatum. Emulsion bases (water/oil) are greasy, emollient, and occlusive, contain water, and absorb additional water. An example of an emulsion base is lanolin.

When evaluating and assessing the exposure of an excipient, the following considerations should be included:

1. Identify the anticipated route of exposure.
2. Identify the intended dose (average daily dose, maximal daily dose, maximal single dose).

3. Identify the dosage form.
4. Clarify the intended duration of treatment.
5. Ascertain the bioavailability for the intended route of administration.
6. Identify any issues of hyper-susceptibility in the target patient population.
7. Determine concentration-dependent issues or concerns for route of administration.

The proper evaluation of the safety of excipients to which the public is exposed involves the strict adherence of sound scientific principles in the design, execution, and interpretation of any and all studies, both *in vitro* and *in vivo*. The principles to which one should adhere are as follows:

1. Studies should be designed to show a dose-response relationship.
2. Effects produced in studies need to be compound-related and specific and dose-dependent to be considered biologically significant.
3. Interspecies extrapolation of data should be performed with extreme caution. The mere use of safety factors should be avoided at all costs.
4. The test material must be completely and thoroughly characterized and include physical, chemical, and biological property evaluation.
5. Make sure that the test material is the exact same chemical moiety that will be used in the final preparation of the final product.
6. Experimental exposure to the test material must simulate human exposure as exactly as possible.
7. Keep a balance between the use of animals, while at the same time maintaining statistical power and relevance.
8. The selection of an animal species must be appropriate and defensible to the agency. Behavior should be as close as possible to the human.
9. Both biological and statistical significance must be completely and thoroughly referenced.
10. The degree and extent of testing is a function of the chemical and physical nature of the test material and the conditions of exposure. Such testing should include consideration of the extent of human exposure, frequency of exposure, and duration of exposure.

Specific impurities that can be found in excipients are inorganic impurities, which include heavy metals and sulfites. Organic impurities can include proteins. Other miscellaneous impurities can include various additives, degradation products, residual solvents, pesticide residues, mycotoxins, microbes, and residues from antimicrobial treatment.

Vehicles

Strictly speaking, vehicles are a subset of excipients. They are distinguished by being liquids, and their primary purpose is to dissolve or "carry" the active drug substance and solid-phase excipients. These solvents are characterized in two manners – either by solubility (Table 9.3) or their chemical nature (Table 9.4). The solubility and miscibility (ability to mix with other solvents) are other principle characteristics of a solvent (Yalkowsky, 1999) that can be directly influenced by temperature (increased temperature, up to the vaporization temperature of the liquid in question – increased atmospheric pressure increases solubility, and pressure increases the vaporization pressure of a liquid); the use of cosolvents, surfactants, and complexing agents; and changes in pH (if an aqueous system) to optimize solubility characteristics.

TABLE 9.3

Solubility Definitions

Descriptive Term	Parts of Solvent Required for 1 Part of Solute	Solubility Range (mg/mL)	Solubility Assigned (mg/mL)
Very soluble	<1	>1000	1000
Freely soluble	1–10	100–1000	100
Soluble	10–30	33–100	33
Sparingly soluble	30–100	10–33	10
Slightly soluble	100–1000	1–10	1
Very slightly soluble	1000–10,000	0.1–1	0.1
Practically insoluble or insoluble	≥10,000	<0.1	0.01

TABLE 9.4

Classification of Solvents

Bipolar Aprotic	Protic	Lewis Basic	Lewis Acidic	Aromatic	Nonpolar
DMF	Water	Acetone	Chloroform	Toluene	Heptane
DMSO	Ethanol	THF	Dicholoromethane	*p*-Xylene	Hexanes
N-Methyl-2-pyrolidinone	Methanol	Ethyl acetate	–	Pyridine	Cyclohexane
Acetonitrile	*n*-Butanol Acetic acid *n*-Propanol 2-Propanol	2-Pentanone Methyl-*t*-butyl ether Butyl acetate	–	Anisole Ethylbenzene	–

Abbreviations: DMF, dimethyl formamide; DMSO, dimethylsulfoxide; THF, tetrahydrofuran

For most vehicle systems, an optimum balance must be achieved between solubilizing the drug and achieving permeability across, through, and into lipid components. The balance of these two is sometimes expressed as the partitioning coefficient, and can be modeled in *in silico* and *in vitro* systems. Many may also be residual solvents in pharmaceutical active ingredients, which have the same purposes in chemical synthesis as vehicles have in facilitating delivery of the therapeutics into (and perhaps through) a biologic system. They are distinguished by having to be tolerated by the biologic systems they contact. The nature and size of solvent molecules also govern their solubility and physical characteristics.

Of our three objectives, vehicles primarily contribute in a positive manner to achieving systemic bioavailability, though they also significantly determine tolerance.

Aqueous vehicles are generally preferred, but the utilization of oily vehicles for lipid-soluble materials is acceptable. Food oils such as corn oil can be used, but mineral oils should never be used. Keep in mind that food oils add to the normal daily dietary caloric intake and furthermore can affect the absorption of fat-soluble nutrients and fat-soluble vitamins. Absorption levels can be quite different for a given material when using an oily vehicle versus an aqueous vehicle.

Solubility of the test article is chiefly considered as one of the most challenging physical properties of a moiety to predict. While drug developers seek a complete understanding of the connection between molecular mechanisms, expressions of in vivo toxicity, and adverse drug reactions, increasing focus is now being placed upon exercising a better understanding and control of basic molecular properties such as molecular weight, lipophilicity, and drug solubility. Ideally the ultimate goal which is sought after is the accurate prediction of these properties upstream in the path of drug development prior to synthesis, in order to optimize test article absorption, distribution, metabolism, and excretion. While there is a strong effort to accomplish the determination of test article solubility and the choice of a vehicle from mathematical models and computational methods, the most effective method for the choice and selection of a vehicle is by experience of the scientist, published sources, and trial and error.

Additional Readings

Ardeef A. (2012) *Absorption and Drug Development: Solubility, Permeability and Change State*, 2nd Ed. Hoboken, NJ: Wiley.

Blodinger J. (1983) *Formulation of Veterinary Dosage Forms, Drugs and the Pharmaceutical Sciences*, Vol. 17. New York, NY: Marcel Dekker.

Gibson M. (2009) *Pharmaceutical Preformulation and Formulation – A Practical Guide from Candidate Drug Selection to Commercial Dosage Form, 2nd Ed, Drugs and the Pharmaceutical Sciences*, Vol. 199. New York, NY: Informa Healthcare.

Hauss DJ. (2007) *Oral Lipid-Based Formulations – Enhancing the Bioavailability of Poorly Water-Soluble Drugs, Drugs and the Pharmaceutical Sciences,* Vol. 170. New York, NY: Informa Healthcare.

Kemsley J. (2014) Eyes on excipients. *C&EN,* January 3, 2014, pp. 9–12.

Thayer AM. (2015) Antibiotics: Will the bugs always win? *C&EN, Cover Story,* 94(35):36–43.

Weiner M, Bernstein IL. (1989) *Adverse Reactions to Drug Formulation Agents – A Handbook of Excipients.* New York, NY: Marcel Dekker.

Weiner ML, Kotkoskie LA. (2000) *Excipient Toxicity and Safety, Drugs and the Pharmaceutical Sciences,* Vol. 103. New York, NY: Marcel Dekker.

References

Gad SC. (2008) *Pharmaceutical Manufacturing Handbook: Production and Processes.* Hoboken, NJ: Wiley.

Gad SC. (2016) *Drug Safety Evaluation,* 3rd Ed. Hoboken, NJ: Wiley.

Gad SC, Chengelis CP. (1988) *Acute Toxicology Testing.* Caldwell, NJ: Telford Press.

Gad SC, Gad-McDonald SE. (2016) *Biomaterials, Medical Devices, and Combination Products.* Boca Raton, FL: CRC Press.

Gad SC, Spainhour CB, Shoemake C, Pallman DR, Stricker-Krongrad A, Downing PA, Seals RE, Eagle LA, Polhamus K, Daly J. (2016) Tolerable levels of nonclinical vehicles and formulations used in studies by multiple routes in multiple species with notes on methods to improve utility. *Int J Toxicol,* 35(2):95–178.

Yalkowsky SH. (1999) *Solubility and Solubilization in Aqueous Media.* New York, NY: Oxford University Press.

10

Practical Ways of Improving Regional and Selective Drug Delivery

Having now reviewed all of the contributing factors (save one) that can influence both the efficiency of delivery of a therapeutic to a region (the "bioavailability"), the specificity of delivery to either target tissues or receptors (the "specificity"), and the tolerance of animal models (and humans) to such delivery, we need to examine how the various factors can be used to optimize the balance of these three while meeting regulatory requirements for evaluating potential toxicity of the therapeutic (in its formulation) at the supraclinical levels required.

First, however, we should mention the factor not discussed in this book – choice of animal models. This is a critical but also quite extensive subject on its own, and has been previously covered in *Animal Models in Toxicology*, 3rd Ed. (Gad, 2015) and no attempt will be made to summarize the contents of that volume here. It should be pointed out that animal model selection is a critical aspect of properly assessing the potential safety of a new therapeutic, yet is often ignored in favor of simply defaulting to "standard species."

The three central objectives to be addressed in this volume are here to be each matched up with application, selection, and modification of the factors that have been presented in the preceding chapters. A final integrated approach which clarifies choices to achieve the optimal balance for use in therapeutic evaluation will then finish up our journey.

1. The first objective presented in the introduction is also the most commonly sought – how to improve the amount of a therapeutic delivered to a measurable target site – most frequently (but not always), the concentration in blood as expressed as an area under the curve (AUC – see Chapter 3). While I have already explored how this measure can be misleading as to what systemic exposure is, that said, we will look at the general case of achieving maximum (or at least sufficient supraclinical or highest feasible) levels to meet expectations of regulatory authorities. As will be shown, these approaches can also be applied to optimize clinical approaches (Touitou and Barry, 2007; Washington et al., 2001).

We will then proceed to examine what technique can be applied to improve/achieve delivery levels in certain specific tissues (such as the eye and brain).

2. Improving specificity of delivery of therapeutics to the desired target tissues and receptors.

Note that each of these first two (and, indeed, the next to be discussed) areas of optimizing tolerance by animal models (and clinically) requires consideration of both what the route of administration is and what the target tissue(s) are (or where desired target receptors are).

3. Optimizing animal model and (subsequently) human tolerance of administered agents. Particularly with the increased interest in pursuing alternate routes for use with already approved drugs, the assessment of tolerance with the new routes is a specific part of the regulatory program (FDA, 2015).

Optimization of Systemic or Target Organ/Tissue Delivery by the Principal Routes of Administration

Oral (PO)

For the oral route of administration, we here are going to focus on the traditionally most common route, where the mouth (oral cavity) is serving only as a gateway to the GI tract. The routes specific to the oral cavity (sublingual and buccal) are addressed separately. The GI tract, of course, can be further broken down into the esophageal tract, stomach, and small and large intestines.

Most absorption occurs in the intestinal tract. Improvement in systemic absorption is traditionally achieved by (1) avoiding conditions in the stomach which may degrade the drug (i.e., the acid environment in the stomach) and improving either nonpolar solubility and residence time in the GI region of the greatest absorption potential. These traditional approaches will not be addressed here.

Approaches to Improving Systemic Absorption

1. Dividing doses over the course of the day will both generally improve absorption (greater tissue to solubilized drug ratios, as well as avoiding absorption mechanism saturation) and tolerance. This also allows administration of greater volumes of formulation.
2. Use of better tolerated formulations. All too often the focus of oral formulation is to maximize the concentration of therapeutic in a solution for delivery to test animals. A more balanced focus on solubility and

tolerance (see Appendix) would allow use of cumulatively greater volumes over longer periods of time.

3. Improved residence time in the stomach (when that is a target for absorption) has been approached with a number of drug device combinations employing drug-impregnated bio-absorbed polymer or pasta matrices in capsules which expand to a size that cannot pass through the sphincter and leave the stomach.

Approaches to Improving Tolerance

Tolerance can be either meeting necessary systemic exposure levels while keeping the systemic C_{max} levels down to tolerated levels or achieving local tissue (stomach) tolerance levels.

Stomach tolerance is somewhat dependent on species – dogs are a prime example here, being both prone to emesis and very sensitive systemically to cationic surfactants.

Dividing oral doses by going to BID (twice a day) or even TID (three times a day) dosing. Not even is it thus possible to achieve higher total systemic doses while keeping C_{max} levels to tolerable levels, but also allows the use of less "rigorous" formulations.

Approaches to Improving Target Tissue Section

Unless the target tissue is a region of the GI tract, target tissue section is not a relevant aspect of oral administration. Various forms of encapsulation can effectively target different GI regions by remaining intact until that region is reached.

It is possible to produce and use capsules for rodents ("mini-encapsulation") to use this approach to bypassing the acid environment of the stomach.

A recent development is "smart pills" (really capsules) that can be programmed to deliver their contents only under specific conditions or in selected GI regions.

Intravenous (IV)

By current definition, direct injection of a therapeutic formulation defines the theoretical maximum; however, this does not mean that systemically administered doses by this route cannot be improved.

IV administration over a protracted period of time (as opposed to periodic bolus administration) will both increase the volume of formulation that can be administered per day, increase tolerance by test animals, and in most cases also increase NOAEL levels (by reducing administered drug C_{max} levels).

Dermal

Dermal administration may have one of two objectives: either to deliver the drug to a region of the skin (and as limited as possible to this region) or to deliver the therapeutic through the skin (generally to avoid first-pass metabolism). This directory of targets is shared with other topical or natural orifice targets (buccal, rectal, vaginal), but the first is more pronounced/frequent with the dermal route than with others.

Tools for enhancing delivery just to the skin itself are primarily limited to formulations or physical form, the administration of minimally soluble products via the hair follicles as particles too large to be carried further. Localized sound of the correct mechanical stimulation frequencies can enhance such uptake into the follicles.

The usual therapeutic objectives for delivery limited to the skin are skin (or primarily skin) limited diseases – acne, psoriasis, and atypical dermatitis, for example.

Systemic delivery through the skin is generally intended primarily to bypass first-pass metabolism and/or to achieve gradual delivery over a protracted period of time. Large molecules (proteins) traditionally have not been readily administered systemically because their large size prohibits passage through the skin, but this can be overcome by iontophoresis or selective administration through hair follicles.

Use of selective vehicles or formulations can also enhance the systemic absorption of therapeutics (particularly smaller molecules) across the skin.

The most common forms of dermal formulation are creams, lotions, and gels. Increasing the surface area of application of these formulations directly increases the potential for absorption, but this approach has (both nonclinically and clinically) practical limitations.

There are also penetration enhancers that can be used (such as dimethyl sulfoxide or DMSO) but some of these (while they can be used nonclinically) are not allowed for clinical use.

Physical approaches to enhancing skin penetration (and therefore systemic absorption) include the following:

Iontophoresis uses an electronic current applied to the skin to carry drugs through the epithelium, improving systemic delivery through the skin (Washington et al., 2001).

Electro perforation (also called electropermeabilization) utilizes short, high-voltage (200–1000 V) pulses of electricity to create aqueous pores in the lipid bilayers in membranes. Use is not limited to the dermal route.

Sonophoresis is the use of low-frequency ultrasound to improve passage through the skin of therapeutics, especially higher molecular weight molecules such as proteins (including insulin).

Transdermal patches have become a popular means of enhanced but also gradual delivery. The use of heat (warming the skin) leads to blood vessel dilation, and thus can improve transdermal absorption.

Except insofar as direct delivery to just the skin (when the skin is the target tissue) is a means of improving or optimizing target tissue delivery, approaches to improving target selectivity are not (currently) readily apparent.

Accompanying tolerance problems for dermal toxicity are irritation, sensitization, and (with the increased and protracted exposure to light) phototoxicity and photosensitivity.

Buccal/Sublingual

The buccal and sublingual routes have become increasingly popular for the administration of therapeutics. Advantages include bypassing first-pass metabolism in the liver, increased passage of therapeutics into the CNS, and improved systemic absorption of larger molecules, including proteins. These features are common to all the oral mucosal routes – the blood supply from these and the anal sphincter, unlike the rest of the gastrointestinal tract, do not proceed to the hepatic portal vein. Additionally, substances absorbed here are not subject to the range of acidity or to interaction with enzymes located in the main tract.

The potential region for administration by these oral mucosal routes is linked in humans and laboratory animals. The best animal model for buccal administration is the hamster, due to its comparatively large available oral mucosal surface, and also its tendency to physically retain materials in its pouch. The rat and dog (but not the hamster) can be used readily for sublingual administration.

Adherent buccal and sublingual films impregnated with the therapeutic (or formulated therapeutic) serve to be in direct contact with the mucosal surface and slowly release the desired moiety into the mucosal surface, while limiting loss into the flow of saliva and down into the stomach and rest of the GI tract. Rates of relevance can be "engineered" in the composition of the adherent film. Tissue and systemic tolerance to the therapeutic can also be improved by modifying concentrations at the tissue-contacting surface and likewise "spreading out" the systemic exposure levels. Systemic absorption can be improved overall and specifically for problematic molecules (particularly larger peptides and proteins). Additionally, greater specificity of delivery to the CNS and the local mucosal surface can be achieved. There are carrier-mediated transports for some amino acids and some vitamins.

Local tissue tolerance issues for the buccal route are uncommon, mostly due to either the presence of particles, presence of hygroscopic materials in a formation, or extremes of pH in a gel or film formulation.

Disadvantages (for which formulation steps can be used to modify) include limited active transport, and poor absorption of polar molecules. Keeping the dose form in place is an issue, unless a mucous gel or adherent film is used.

Inhalation

Historically, the inhalation route has been a limited one for therapeutics (Gad, 2014) – primarily used for drugs to treat respiratory disorders, such as asthma, respiratory infections, and cystic fibrosis. This has changed dramatically over the last thirty years with new formulations (insulin), biologics (peptides and proteins), and gene and cell therapy products, though respiratory route therapeutics still represent just over 1% of the market. Improving local or systemic delivery has similar requirements to what must be done for buccal and sublingual routes, but solutions are different – one cannot use a gel or film in the deep lungs to extend the available time for absorption across pulmonary membranes. One can, however, improve solubility to facilitate absorption. Particles are not your friend here – they are more readily cleared from the lung by the mucociliary "elevator" and can damage the cells they physically contact.

The route has its own set of delivery systems (metered dose inhalers, or MDIs), nebulizers, and respiratory devices which aerosolize or vaporize therapeutics and mix these forms with the stream of supplied atmosphere for patient breathing ("respiratory support devices").

A major driver in the increased interest in the inhalation route has been determining that proteins and peptides can be effectively delivered systemically through the deep lungs (Tamulinas and Leach, 2000; Gad, 2014).

Also called the pulmonary (to distinguish from the upper respiratory or nasal) route, this highly specialized route offers its own set of advantages and disadvantages, as well as challenges in assessing nonclinical safety and tolerance (Gad, 2014).

Nasal

Upper respiratory tract delivery of therapeutics have their own distinct set of delivery systems, as shown in Table 10.1 (Gad, 2014), and the route has been utilized for both legal and illicit drug delivery for years.

There is, however, increasing awareness that the nasal epithelial mucosal orifice may serve as an ideal port of entry for biopharmaceuticals. In addition to tissue-specific delivery of therapeutics, intranasal delivery can also serve as a portal of entry to the systemic circulation and the lymphatic system, and as a direct route to administering drugs to the brain. In part, this is due to the large mucosal surface area and six arterial branches that serve the nasal cavity. While anatomically relevant animal models are limited (rodents have complex turbinate structures in the nose that humans, dogs, swine, and primates do not), species differences in anatomy, cell types, and innervation, as well as functional differences (vascular changes, mucus secretion, airway humidification and filtration, irritant and olfactory responses) pertinent to

TABLE 10.1

Routes: Potential Actions to Improve Objectives

Route	Improve Systemic Exposure	Improve Tolerance[a]	Improve Target Specificity
Oral	• Formulation • Increase stomach retention delivery forms • Avoid stomach degradation • Dose with food to improve absorption and tolerance in stomach	• Divided doses • Well-tolerated formulation • Use of capsules in dogs and minipigs	• Smart pills
IV	• Slow infusion	• Infusion over time • Decrease rate of infusion in decreased systemic/local C_{max} for drug • Ensure sterility of formulation • For repeated injections, need to move site around	• Targeted delivery/ formulation
Dermal	• For animal, tape stripping of barrier layer (stratus corneum) • Occlusive wrapping • Warm application region • Iontophoresis • Electro perforation • Sonophoresis • Maximize administration surface area	• Non-irritating	• Particles sized to enter but not transit hair follicle
Buccal	• Embedded in films and gels	• Isotonic and neutral pHs • No surfactants	• Better adherence to buccal surfaces
Inhalation	• Better deep lung penetration • Hydrophilic formulations	• Avoid fibers • Physiologic osmolarity	• More complete deep lung penetration and retention
Nasal	• Prolong adherence tissue to epithelial tissue • Hydrophilic formulations	• Isotonic and neutral pHs • No surfactants	• If local tissue is target, long-term exposure at low rates of transfer

(*Continued*)

TABLE 10.1 (***Continued***)

Routes: Potential Actions to Improve Objectives

Route	Improve Systemic Exposure	Improve Tolerance[a]	Improve Target Specificity
Rectal	• Maximize exposure area and duration	• Physiologic osmolarity • No abrasive particulates • No high-potency vehicles	• Suppositories
Vaginal/ Uterine	• Maximize exposure area and duration	• No physiologic osmolarity • No abrasive particulates • No high-potency vehicles	• Suppositories and tampons
Ocular	• Drug-loaded stents • Enhanced corneal penetration • Nanoparticles in colloidal carriers	• pH in range of pH 5–10 (6–8 if internal) • Limited volume of administration • Sterility	• Limited delivery volumes to target tissue
SC	• Gels for sustained release • Can achieve administration of larger volume of dosage form in a short period	• Low volume spread over wide area or several injections • Nonpolar organic vehicles all defat tissues if injected	• May be effective means for lymphoid tissue administration • Use of pegylated liposomes in local region of neck can serve to improve specificity for head and neck cancer
IM	• Hydrophilic formulation	• Adjust injection volume to multiple sites	–
Implantation	–	–	–

[a] In general, any action which reduces systemic and local C_{max} levels improves tolerance.
A = Bypass first-pass metabolism
B = Invasive (I) or Noninvasive (N)

intranasal delivery, as well as the pathological response of the nose, will be briefly discussed.

Formulation for use in nasal therapeutic delivery has seen significant advances, and offers some unique advantages (Gad, 2014):

- Increased permeability of the nasal mucosa by interaction of the formulation components with the nasal membrane in a safe, effective, and reversible manner.

- Increase in drug solubility and protection against enzymatic degradation.
- Increase in the residence time of the drug in the nasal cavity. Commonly used formulation approaches include the following:

  - Liquid formulations
  - Aqueous solutions
  - Synthetic surfactants
  - Bile salts
  - Phospholipids
  - Cyclodextrins
  - Micelles
  - Liposomes
  - Emulsions
  - Polymeric microspheres

A wide range of excipients and excipient technologies have become available for use with nasally administered products since the 1980s. A broad overview of these follows.

Penetration enhancers act by decreasing the viscosity of the mucous and creating transient hydrophilic pores in the epithelial membrane. Examples include bile salts, EDTA fatty acid salts, and nonionic surfactants.

Surfactants. The effect of surfactants on drug absorption across nasal mucosa has been studied since the 1970s. In 1981, Hirai et al. compared nonionic, anionic, amphoteric synthetic surfactants and natural anionic surfactants for in vivo nasal absorption of insulin. The authors reported the following:

- A majority of surfactants enhanced insulin absorption relative to the extraction of membrane components.
- Enhancing effects correlate with the extraction efficiency of the membrane components by the surfactants.
- Alkyl glycosides constitute a novel class of sugar-derived surfactants used in cosmetics.
- Maltoside derivatives with an alkyl chain length between 12 and 14 enhance insulin absorption at low surfactant concentrations.
- The mechanism involved the loosening of tight junctions and increasing the paracellular transport.
- The enhancement effect produced by surfactant monomers is related to the ability of the surfactant molecules to penetrate and fluidize the lipid bilayers.

- Carboxymethyl cellulose (CMC) monomers aggregate into micelles, which can solubilize components, particularly cholesterol phospholipids.

Bile Salts

- Enhancement of insulin absorption but with milder effects on the biomembrane.
- Additionally, the absorption-promoting effects appear to arise from an inhibitory effect on the enzyme degradation of insulin.
- Bile salts appear to be the most promising and effective absorption enhancers of peptides and proteins and still remain widely used as a permeation enhancer.

Cyclodextrins

- Cyclodextrins are cyclic oligosaccharides containing a minimum of 6-D-glycopyranose units attached by an alpha 1,4 linkage.
- Cyclodextrins are produced by enzymatic conversion of prehydrolyzed starch.
- Natural cyclodextrins are designated by a Greek letter: α, β, and γ. The β is the most soluble of the three.
- The ring structure resembles a truncated cone with characteristic cavity volume.
- The internal surface of the cavity has slight hydrophobic properties, whereas the outer surface is hydrophilic.
- Cyclodextrins form inclusion complexes with lipophilic molecules.
- Generally appear to be less irritating to the nasal mucosa than bile salts and surfactants.
- Large increase in solubility through complex formation and protection from enzymatic degradation improves nasal absorption of lipophilic drugs.
- Very potent absorption enhancers for hydrophilic peptides that are not complexes.
- Shao and Mitra (1992) evaluated several cyclodextrins for their absorption-promoting effect on insulin.

Note that bioadhesive polymers and microspheres, as previously discussed under the buccal route, also present an opportunity here.

Viscosity Modifiers

- Increase the decreased mucociliary clearance from the nose, increasing residence time (and the opportunity to absorb through the epithelia).

The best results for absorption-promoting effect on insulin were obtained with dimethyl-β-cyclodextrin:

- Promoting order correlated well with the extent of nasal mucosal perturbation.

Mixed Micelles

- Nasal absorption of insulin in the presence of sodium glycocholate (NaGC) and linoleic acid increased relative to the increase by NaGC and linoleic acid alone.
- Mixed micelles of bile salts and fatty acid appear to have a synergistic effect on the absorption of peptides.
- Maximal nasal absorption enhancement of [D-Arg] kytorphin has been observed with mixed micelles of NaGC and linoleic acid; the effect is greater than that with glycocholate alone.

Formulation and Potential Mucosal Damage

- Improved absorption involves interactions with the mucosal membrane.
- Proposed enhancement mechanisms are as follows:
  - Extraction of membrane components.
  - Penetration and fluidization of membrane.
  - Loosening of tight junctions.
  - Perturbation of nasal mucociliary clearance system.
  - Simultaneous transport of environmental toxins.
  - Adverse effects have to be of short duration, mild, and rapidly reversible.
  - Kinetics of lipid and protein extraction from the membrane are measures of the extent of damage evaluated by measuring the activity of membrane marker enzymes.
  - Lactate dehydrogenase: cytosolic enzyme related to intracellular damage.

  - 5′-nucleotidase: membrane-bound enzyme, indicator of membrane perturbations.
  - Alkaline phosphatase: membrane-bound enzyme, related to membrane damage.

- Ideal characteristics of absorption enhancers include the following:
  - Pharmacological inertness.
  - Nonirritant, nontoxic, and nonallergenic.
  - Effect on nasal mucosa should be transient and completely reversible.
  - Potent in low concentrations.
  - Compatible with other adjuvants.
  - Possesses no offensive odor or taste.
  - Inexpensive and readily available.

- The factors influencing mucosal damage include the following:
  - Drug administration
  - Dose
  - Frequency
  - Interspecies difference
  - Sensitivity toward absorption enhancers

- Clinical signs of nasal irritation studies in rats include the following:
  - In stu®dies for less than 90 days:
    - Struggling, sneezing, salivation, head shaking, and nose rubbing.
  - In studies for more than 90 days:
    - Histological signs of nasal irritation, including inflammation of septal and turbinate mucosal surfaces, epithelial and submucosal infiltration of inflammatory cells, purulent exudates, and mucosal hyperplasia.

There are specific concerns for tolerance of formulation used nasally and tests for assessing them (see Gad, 2014; Zhang and Jung, 2001).

As made clear, a significant advantage for the nasal route is improved specificity of delivery to not only the local tissue but also the central nervous system (CNS). This specificity can be improved by the following:

1. Slow administration over a period of time. Preferential partition of absorbed to the CNS is greater at lower levels of therapeutic. Also, there is a lower chance for the therapeutic to pass through the nose and into either the GI tract or the lower respiratory tract.

2. Careful attention to formulation. Use of some of the penetration enhancers can help here, but consideration should be given to nanoparticles or carriers which improve the efficacy of desired transporter systems.

Note also that the nasopharyngeal region has an extensive lymphatic plexus draining into the cervical lymphatics. Accordingly, if these are (or serve) the therapeutic target, an opportunity presents for selective delivery to these sites.

Rectal

This route shares many characteristics with the other natural channel epithelial tissue routes – first-pass metabolism is bypassed, and absorption of nonpolar moieties (especially below 1KD in molecular weight) is generally good. However, local tissue tolerance of formulations is a significant concern, and, as with buccal and sublingual, retaining the dosage form in the delivery region (here, the rectum) requires special considerations (more so in nonclinical assessments than clinical use).

Rectal also offers a practical alternative to oral when nausea, vomiting (emesis), convulsions, and loss of consciousness may be issues (this last because of subsequent risk of choking). It has a long history of clinical (and nonclinical) use, but patient acceptance is poor and it is infrequently considered as a nonclinical alternative. An additional complication is defecation by the recipient.

The region is well served by both circulatory and lymphatic vasculature (Watanabe, 2007). When, in general toxicity studies, a second route with greater systemic absorption is required because the primary route (i.e., dermal) and the therapeutic are not highly polar, rectal is a viable alternative to oral or IV, with the additional attraction that first-pass metabolism is bypassed. There are route-specific absorption enhancers (surfactants, straight-chain fatty acids, cyclodextrins, and salicylates) which can improve absorption of hydrophilic compounds, though one must be careful to ensure minimal adverse effects on the rectal mucosa.

Nitrous oxide (NO) can also significantly improve rectal absorption while not damaging the mucosa (Utoguchi et al., 1998). More recently, Dodou (2012) has reported on the use of thermoreversible polymers (poloxamers in combination with mucoadhesive polymers such as cyclodextrins), strong chelating agents that act as liquid suppositories to increase rectal retention time and systemic absorption.

The traditional dosage forms are suppositories, drug-coated tampons (excellent for administration of polar therapeutics, for local tissue region targeting), and enemas, with which generally transitory absorption levels can approach or equal those achieved by intravenous administration, with the advantage of achieving desired plasma levels over longer periods of time with lower C_{max} levels (Prasanna et al., 2012).

Compared to the other natural channel routes, rectal provides a substantial absorptive surface, accommodating a substantial dosage form. Degradant enzymes are present at low concentrations in the rectum as opposed to the GI tract. Absorption is improved by spreading the dose over as large an area of the rectal lumen as possible.

Primary tolerance issues are local tissue tolerance concerns – particularly irritation/inflammation, but also occasional mucosal ulceration or (rarely) sensitization (Prasanna et al., 2012).

The route is generally well tolerated in juveniles (even neonates) and older patients or test animals (Washington et al., 2001).

Vaginal and Uterine

Most therapeutics administered by the vaginal route have functions associated with reproduction (preventing it, primarily) or regional issues such as menstrual cramps and vaginal infections. There are numerous marginally exploited therapeutic opportunities (such as vaccine and other biologics delivery, and chemotherapy). This route shares the potential advantages of no first-pass metabolism, a large (and well served by blood supply) absorptive tissue surface area, and ready absorption of hydrophilic substances. As with rectal, the most common animal model is the rabbit, followed by the dog. A rat model has been developed, but is extremely test-volume limited, and the space is too small to allow testing of tampons, suppositories, or devices such as IUDs (which, as is true with many of the newer candidates, have copper parts and are therefore regulated as drugs). Hwang et al. (1976, 1977), Ho et al. (1976), and Yotsuyanagi et al. (1975) have published a series of papers on determining the influence of factors on vaginal absorption in rabbits.

Considerations are different if local (rather than systemic) effects are desired. For this, instead of a semisolid gel or cream, an intravaginal ring (IVR) or a bioadhesive (and biodegradable) polymer form or a surface coating on a tampon would be optimal approaches.

Disadvantages are (1) presence in only one gender and (2) a steady vaginal flow which serves to usually limit residence time of a dosage form. The use of drug-implanted IVRs (in humans and nonrodent animal models) provides an approach to get around this limitation. The most significant challenges to increasing systemic absorption from the vaginal route are achieving systemic surface area contact and maximizing contact time.

A complication is the complex and somewhat delicate microbial flora of the vagina. One must always remember the occurrence of TSS (toxic shock syndrome) in the 1970s, associated with the introduction of "super absorbent" tampons. Formulations and dosage forms (IVRs, vaginal films, and tampons, for example) require evaluation of this potential before clinical testing or use.

Stability of a formulation and/or therapeutic in the vaginal environment must also be considered. Simulated vaginal fluids are available for doing so (Owen and Katz, 1999).

Ocular

The ocular route, whether topically (on the surface of the orb) or into the eye itself, is intended and utilized for therapy of that specific organ and tissue.

Sterility, lack of protein-disrupting potential, pH in the range of 5–10 (if topical, 6–8 if internal), and limited volume of administration are essential aspects for ocular formulations. If long-term receptor site/tissue activity is desired, then either nanoparticles in colloidal carriers for intravitreal injection or a drug-containing permeable polymer capillary "stent" that is manually inserted into the globe are the best current options for prolonged delivery of therapeutics (such as anti-inflammatory steroids). The surface properties of the nanoparticles (Thassu and Chader, 2013; Patel et al., 2013) can be adjusted to optimize migration of the particles from the vitreous human to retinal tissues if that is the target tissue.

Volumes of injection are limited, and local tissue tolerance is essential. Currently, the best screening model for local tissue tolerance is the bovine corneal opacity and permeability (BCOP) assay.

Subcutaneous (SC)

Long used in nonclinical safety testing and for administering vaccines and for "depot" parenteral administrations, here the therapeutic is administered under the dermis and epidermis, serving to enhance systemic exposure while still avoiding the first-pass effect associated with IV, IP or IM administration. SC is easier to perform than other parenteral routes, less painful and traumatizing, and, as long as proper aseptic materials and technique are employed, less likely to lead to injection site reactions (Dychter et al., 2012). Commonly used for insulin, morphine, and tuberculosis exposure testing, it is more immunogenic than the IV route (Ponce et al., 2009).

A separate set of vehicles can be and is utilized to formulate gels for sustained release of therapeutics. SC also allows ready access to the lymphatic system (McLennan et al., 2005). The route also allows for the administration of larger volumes of dosage form in a short period, while also achieving systemic absorption over a period of time, "flattening" the absorption curve. This route also tends to achieve longer systemic half-lives for drugs.

SC administration can be used to improve target tissue selectivity for the lymphatic system and organs. Use of pegylated liposomes as vehicles in the local region of the neck can also serve to improve target specificity for head and neck cancer and associated lymph node metastases (Harrington et al., 2000).

Intramuscular (IM)

Intramuscular administration has many of the advantages of achieving good systemic absorption (but with a more gradual buildup in systemic blood and therefore associated other tissues) while sharing the disadvantage of first-pass metabolism.

IM injections do act to create "depots" for larger systemic exposure for lipophilic substances and formulations. Injection leads to release of the therapeutic from the dose form into the intracellular fluid (ICF) followed by absorption into blood and lymphatics.

While not having hematologic or (usually) direct vascular damage as tolerance issues, muscle irritation and injection site tissue damage are possible, though these are most commonly an expression of formulation physicochemical factors. The advantage of achieving target levels and duration in the blood with reduced C_{max} levels is an attraction.

No effective route-related target tissue specificity (other than that of the specific region of administration) is offered by the IM route.

Conclusion

An initial evaluation shows that every route has available modifications for improving total ability to deliver a larger systemic (and also local) dose (that is, to improve desired "bioavailability"). Likewise, each route also has available means of improving tolerance (again, systemic and local). In other words, the means are available to achieve greater tolerated (at least in the short to mid-term) exposure, that is, to manipulate increases in what the NOAEL (and the other metrics) of a therapeutic are, both in nonclinical and clinical situations.

Available clear means of improving target tissue specificity are less universal or clear; they are available for most systemic access routes, though with use of more complicated combinations of tools. Clearly this stage of drug delivery devices, generally implants, are the most useful means of maximizing target tissue specificity (while also, at least in the short term, measuring performance for the first two objectives, though probably not as well for improving systemic delivery as other means).

References

Dodou K. (2012) Exploring the unconventional routes—Rectal and vaginal dosage formulations. *Pharm J*, (April):289:238.

Dychter SS, Gold DA, Haller MF. (2012) Subcutaneous drug delivery: A route to increased safety, patient satisfaction, and reduced costs. *J Infus Nurs*, 35(3):154–160.

FDA. (2015) *Nonclinical Safety Evaluation of Reformulated Drug Products and Products Intended for Administration by an Alternate Route.* Silver Spring, MD: Center for Drug Evaluation and Research (CDER).

Gad SC. (2014) Safety evaluation of therapeutic agents administered by the respiratory route. In: H Salem and SA Katz (Eds.) *Inhalation Toxicology*, 3rd Ed. Philadelphia, PA: Taylor & Francis.

Gad SC. (2015) *Animal Models in Toxicology*, 3rd Ed. Boca Raton, FL: CRC Press.

Harrington KJ, Rowlinson-Busza G, Syrigos KN, Uster PS, Vile RG, Stewart JSW. (2000) Pegylated liposomes have potential as vehicles for intratumoral and subcutaneous drug delivery. *Clin Cancer Res*, 6(6):2528–2537.

Hirai S, Yashiki T, Mima H. (1981) Mechanisms for the enhancement of the nasal absorption of insulin by surfactants. *Int J Pharm*, 9(2):173–184.

Ho NF, Suhardja L, Hwang S, Owada E, Molokhia A, Flynn GL, Higuchi WI, Park JK. (1976) Systems approach to vaginal delivery of drugs III: Simulation studies interfacing steroid release from silicone matrix and vaginal absorption in rabbits. *J Pharm Sci*, 65(11):1578–1585.

Hwang S, Owada E, Suhardja L, Ho JF, Flynn GL, Higuchi WI. (1977) Systems approach to vaginal delivery of drugs IV: Methodology for determination of membrane surface pH. *J Pharm Sci*, 66(6):778–781.

Hwang S, Owada E, Yotsuyanagi T, Suhardja L, Ho NF, Flynn GL, Higuchi WI. (1976) Systems approach to vaginal delivery of drugs II: In situ vaginal absorption of unbranched aliphatic alcohols. *J Pharm Sci*, 65(11):1574–1578.

McLennan DN, Porter CJH, Charman SA. (2005) Subcutaneous drug delivery and the role of the lymphatics. *Drug Discov Today Technol*, 2(1):89–96.

Owen DH, Katz DF. (1999) A vaginal fluid simulant. *Contraception*, 59(2):91–95.

Patel A, Cholkar K, Agrahari V, Mitra AK. (2013) Ocular drug delivery systems: An overview. *World J Pharmacol*, 2(2):47–64.

Ponce R, Abad L, Amaravadi L, Gelzleichter T, Gore E, Green J, Gupta S, Herzyk D, Hurst C, Ivens IA, Kawabata T, Maier C, Mounho B, Rup B, Shankar G, Smith H, Thomas P, Wierda D. (2009) Immunogenicity of biologically-derived therapeutics: Assessment and interpretation of nonclinical safety studies. *Regul Toxicol Pharmacol*, 54:164–182.

Prasanna JL, Deepthi B, Rao NR. (2012) Rectal drug delivery: A promising route for enhancing drug absorption. *Asian J Res Pharm Scii*, 2(4):143–149.

Shao Z, Mitra AK. (1992) Nasal membrane and intracellular protein and enzyme release by bile salts and bile salt-fatty acid mixed micelles: Correlation with facilitated drug transport. *Pharm Res*, 9(9):1184–1189.

Tamulinas CB, Leach CL. (2000) Routes of exposure: Inhalation and intranasal. In: ML Weiner and LA Kotkoskie (Eds.) *Excipient Toxicology and Safety*. New York, NY: Marcel Dekker, pp. 185–205.

Thassu D, Chader GJ. (2013) *Ocular Drug Delivery Systems: Barriers and Application of Nanoparticulate Systems.* Boca Raton, FL: CRC Press.

Touitou E, Barry BW. (2007) *Enhancement in Drug Delivery.* Boca Raton, FL: CRC Press.

Utoguchi N, Watanabe Y, Shida T, Matsumoto M. (1998) Nitric oxide donors enhance rectal absorption of macromolecules in rabbits. *Pharm Res*, 15(6):870–876.

Washington N, Washington C, Wilson CG. (2001) *Physiologic Pharmaceutics: Barriers to Drug Absorption.* Philadelphia, PA: Taylor & Francis.

Watanabe Y. (2007) Permeation pathways in rectal absorption. In: E Touitou and BW Barry (Eds.) *Enhancement in Drug Delivery.* Boca Raton, FL: CRC Press.

Yotsuyanagi T, Molokhia A, Hwang S, Ho NF, Flynn GL, Higuchi WI. (1975) Systems approach to vaginal delivery of drugs I: Development of in situ vaginal drug absorption procedure. *J Pharm Sci*, 64(1):71–76.

Zhang Y, Jiang X. (2001) Detoxification of nasal toxicity of nasal drug delivery system. *Zhongguo Yiyao Gongye Zazhi*, 32:323–327.

Appendix: Vehicles and Excipients

TABLE A.1

Index of Vehicles and Excipients – Codex and Details

Excipient/Vehicle	Data in Table #s (T#) Combination #s (C:#)	CAS#	Synonyms	Formula	Key Toxicity Review Articles / Sources	Animal Species Evaluated	Administration Routes Evaluated
Acacia	T2, T111 C:1–7	9000-01-5	*Acaciae gummi*	Natural product	Anderson (1986); Bachmann et al. (1978); TOXNET (2015a)	Nonhuman Primate, Rat	PO
Acetate Buffer	T111 C:8, 9, 119, 138–141	71-50-1	Acetate ion	C2-H3-O2	TOXNET (2015b)	Nonhuman Primate, Minipig, Mouse, Rat	IM, IV, PO
Acetic Acid	T3, T111 C:10, 12	64-19-7	Ethanolic acid	C2H4O2	Schonwald (2004a); Szilagyi (2012a); TOXNET (2015c)	Dog, Mouse, Rat	IV, PO
Acetone	T4, T111 C:11	67-64-1	2-Propanone	C3-H6-O	TOXNET (2015d)	Guinea Pig, Mouse, Rabbit, Rat	Dermal, PO
Acetonitrile	T111 C:12	75-05-8	–	C2-H3-N	TOXNET (2015e)	Dog	PO
Acetylated Lanolin Alcohol	T111 C:174	61788-49-6	–	N/A	TOXNET (2015f)	Minipig	Topical
Acetylmethylamide	T5 –	79-16-3	N-methyl-acetamide	C3H7NO	TOXNET(2015g)	Nonhuman Primate	PO
Alcohol Denatured SDA	T111 C:234	–	–	–	–	Minipig	Topical
Alginic Acid	T6 –	9005-32-7	Norgine	–	JECFA (1997); TOXNET (2015h)	Rat	IP
Anecortave Acetate	T7 –	7753-60-8	–	C23H30OS	Jockovich et al. (2006); Talsma (2004); TOXNET (2015i)	Rat	SC
Antifoam 1510-US™	T111 C:1, 100, 101	–	Silicone emulsion	–	Dow Corning (2015a)	Rat	PO

(*Continued*)

TABLE A.1 (*Continued*)

Index of Vehicles and Excipients – Codex and Details

Excipient/Vehicle	Data in Table #s (T#) Combination #s (C:#)	CAS#	Synonyms	Formula	Key Toxicity Review Articles / Sources	Animal Species Evaluated	Administration Routes Evaluated
Avicel CL-611™	T8 –	–	Microcrystalline cellulose and carboxymethylcellulose sodium, NF, Ph. Eur	–	TOXNET (2015j)	Dog	PO
Balanced Salt Saline	T9 –	–	–	–	–	Rabbit	Intravitreal
Basal Salt Solution	T10 –	–	–	–	–	Mouse	Subretinal Injection
Benzoic Acid	T11 –	65-85-0	Benzoic acid	C7H6O2	David et al. (2012); Nair (2001); TOXNET (2015k)	Rat	PO
Benzyl Alcohol	T111 C:8, 9, 13, 14, 31	100-51-6	–	C7-H8-O	Nair (2001); TOXNET (2015l)	Cat, Dog, Nonhuman Primate, Rat	IV, PO, Topical
Beta-Cyclodextrin	T12 –	7585-39-9	Beta-dextrin; Betadex	C42H70O35	Albers and Muller (1995); Challa et al. (2005); Marttin et al. (1998); Rajewski et al. (1995); TOXNET (2015m); Toyoda et al. (1997); Waner et al. (1995)	Dog, Mouse, Nonhuman Primate, Rat	IP, IV, PO
BHA	T111 C:73	25013-16-5	Butylated hydroxyanisole	C11-H16-O2	TOXNET (2015n)	Minipig	Topical

(*Continued*)

TABLE A.1 (*Continued*)

Index of Vehicles and Excipients – Codex and Details

Excipient/Vehicle	Data in Table #s (T#) Combination #s (C:#)	CAS#	Synonyms	Formula	Key Toxicity Review Articles / Sources	Animal Species Evaluated	Administration Routes Evaluated
BHT	T111 C:73, 127, 164, 165	128-37-0	Butylated hydroxytoluene	C15H24O	Briggs et al. (1989); Lanigan and Yamarik (2002a); Nakagawa et al. (1984); TOXNET (2015o)	Cat, Dog, Minipig, Mouse, Rat	IV, PO, Topical
Bicarbonate Buffer	T13 –	71-52-3	Bicarbonate ion	C-H-O3	TOXNET (2015p)	Mouse	PO
Calcium Chloride	T14 –	10035-04-8	–	Ca-Cl2.2H2-O	TOXNET (2015q)	Mouse	IV, SC
Canola Oil	T15 –	120962-03-0	Canbra oil	Natural product	Evangelista et al. (2004); TOXNET (2015r)	Dog	PO
Capmul MCM™	T111 C:15	26402-22-2,26402-26-6	Medium-chain mono- and diglycerides	N/A	Susananta et al. (1995)	Nonhuman Primate	PO
Capmul MCM NF™	T111 C:16, 255	91744-32-0,26402-22-2,26402-26-6	Glyceryl caprylate/ Caprate	N/A	–	Dog, Rat	PO
Capmul PG 8™	T111 C:242, 243	–	–	–	–	Dog, Rat	PO
Capryol 90™	T16 –	31565-12-5	Propylene glycol monocaprylate (type II) NF; Capmul PG-8™	C11H22O3	Li et al. (2005); Cho and Gwak (2004)	Dog, Rabbit, Rat	Dermal, Ocular, PO
Captisol™ (See also: Sulfobutylether-β-cyclodextrin (SBECD))	T17, T111 C:17, 191	182410-00-0	Beta-cyclodextrin sulfobutyl ether, sodium salt (CDSBE); Sulfobutylether-β-cyclodextrin (SBECD)	C42H70-nO35 * (C4H8SO3Na) n	Albers and Muller (1995)[9]; Challa et al. (2005); Marttin et al. (1998); TOXNET (2015s)	Dog, Mouse, Nonhuman Primate, Rat,	IV, PO, SC

(Continued)

TABLE A.1 (*Continued*)

Index of Vehicles and Excipients – Codex and Details

Excipient/Vehicle	Data in Table #s (T#) Combination #s (C:#)	CAS#	Synonyms	Formula	Key Toxicity Review Articles / Sources	Animal Species Evaluated	Administration Routes Evaluated
Carbomer 974P™	T111 C:234, 235, 272	151687-96-6	Carbomer homopolymer type B (allyl pentaerythritol crosslinked)	N/A	TOXNET (2015t)	Minipig, Mouse	Topical
Carbopol Ultrez 10™	T111 C:61	195739-91-4	–	N/A	TOXNET (2015u)	Minipig	Topical
Carboxymethy-lcellulose	T18, T111 C:20-30, 301	9000-11-7	CMC; Acetic acid; 2,3,4,5,6-penta-hydroxy-hexanal	N/A	Gupta et al. (1996); Mehlman (2012a); TOXNET (2015v)	Dog, Minipig, Mouse, Nonhuman Primate, Rabbit, Rat	IA, PO, SC
Carboxymethyl-cellulose Calcium	T19 –	9050_04_8	Calcium CMC; Carmellose calcium	N/A	TOXNET (2015w)	Dog	PO
Carboxymethyl-cellulose Sodium	T20, T111 C:269, 270	9004-32-4	Sodium CMC; Carmellose sodium	N/A	Bachmann et al. (1978); Bar et al. (1995); Cavender (2012d); Freeman et al. (2003); TOXNET (2015x)	Mouse, Rabbit, Rat	PO
Cavisol™	T111 C:185, 186	–	–	–	–	Dog, Rat	PO
Cetostearyl Alcohol	T111 C:288	67762-27-0	–	N/A	TOXNET (2015y)	Minipig	Topical
Cetyl Acetate	T111 C:174	629-70-9	–	C18-H36-O2	TOXNET (2015z)	Minipig	Topical
Cetyl Alcohol	T21, T111 C:163	36653-82-4	Hexadecan-1-ol; 1-Hexadecanol	C16H34O1	Bevan (2012c); TOXNET (2015aa)	Minipig, Mouse	IP, Topical

(*Continued*)

TABLE A.1 (*Continued*)

Index of Vehicles and Excipients – Codex and Details

Excipient/Vehicle	Data in Table #s (T#) Combination #s (C:#)	CAS#	Synonyms	Formula	Key Toxicity Review Articles / Sources	Animal Species Evaluated	Administration Routes Evaluated
Citrate Buffer	T22, T111 C:18, 19, 126, 147, 229	77-92-9	Sodium citrate-citric acid buffer	C6-H8-O7	Schonwald (2004b)	Dog, Nonhuman Primate, Rat	IV, PO, SC
Citric Acid Buffer	T23, T111 C:13, 38, 58, 95, 302	77-92-9	–	C6H807*H20	Szilagyi (2012b)	Cat, Mouse, Nonhuman Primate, Rat	PO, Topical
Coconut Oil	T111 C:288	8001-31-8	–	Natural product	National Toxicology Program (2001); Shadnia et al. (2005); TOXNET (2015ab)	Minipig	Topical
Collagen Matrix	T24 –	9007-34-5	Collagen human	Natural product	McCarthy et al. (2002); Clark et al. (1989)	Nonhuman Primate, Rabbit	Implantation in humerus bone, Implantation
Corn Oil	T25, T111 C:31-33	8001-30-7	Corn germ oil, glyceridic	Natural product	DeWitt et al. (2005); Dupont et al. (1990); TOXNET (2015ac); Wu et al. (2004)	Chicken Embryo, Dog, Mouse, Nonhuman Primate, Rabbit, Rat	Injection into egg, PO
Cottonseed Oil	T26 –	8001-29-4	–	Natural product	TOXNET (2015ad)	Dog	SC
Cyclodextrin	T111 C:34–37	12619-70-4	–	–	TOXNET (2015ae)	Rat	PO
Cyclohexane	T27, T111 C:11	110-82-7	Hexahydro-benzene; Hexamethy-lene; Hexanaphthene	C6H12	Gad (2014); Kreckmann et al. (2000); Malley et al. (2000); TOXNET (2015af)	Rabbit, Rat	PO, Dermal
Cyclomethicone NF	T111 C:174, 288	69430-24-6	–	N/A	TOXNET (2015ag)	Minipig	Topical

(*Continued*)

TABLE A.1 (*Continued*)

Index of Vehicles and Excipients – Codex and Details

Excipient/Vehicle	Data in Table #s (T#) Combination #s (C:#)	CAS#	Synonyms	Formula	Key Toxicity Review Articles / Sources	Animal Species Evaluated	Administration Routes Evaluated
DAM	T111 C:179–182	57-71-6	Diacetyl-monoxime; 2,3-Butane-dione 2-oxime	C4-H7-N-O2	TOXNET (2015ah)	Dog, Rat	IV
Dextrose	T28, T111 C:17, 39, 40, 108–111, 184, 277	50-99-7	Glucose; D-glucose, anhydrous; Dextrosol	C6H12O6	Buard et al. (2003); Robertson et al. (2003); TOXNET (2015ai)	Cat, Dog, Minipig, Nonhuman Primate, Rabbit, Rat	IV, IV/PO, Perivascular, PO, SC
Dichlorvos	T29	62-73-7	–	C4H7Cl2O4P	TOXNET (2015aj)	Nonhuman Primate	IV
Diethylacetamide	T30, T111 C:41	685-91-6	N,N-Diethyl-acetamide	C6-H13-N-O	Budden et al. (1978); Caujolle et al. (1970); Diethylacetamide (2015) TOXNET (2015ak)	Cat, Chicken, Dog, Mouse, Rabbit, Rat	IP, IV
Diethyleneglycol-monoethylether	T31 –	111-90-0	2-(2-Ethoxyethoxy) ethanol; DEGEE; Carbitol	C6-H14-O3	Hardin (1983); Hardin et al. (1984); TOXNET (2015al)	Nonhuman Primate	IV
Dimethicone	T111 C:301	9006-65-9	–	N/A	TOXNET (2015am)	Rabbit	PO
Dimethiconol Blend 20	T111 C:38	70131-67-8 and 63148-62-9	Dimethicone and Dimethiconol	–	Dow Corning (2015b)	Minipig	Topical

(*Continued*)

TABLE A.1 (*Continued*)

Index of Vehicles and Excipients – Codex and Details

Excipient/Vehicle	Data in Table #s (T#) Combination #s (C:#)	CAS#	Synonyms	Formula	Key Toxicity Review Articles / Sources	Animal Species Evaluated	Administration Routes Evaluated
Dimethyl Acetamide	T32, T111 C:187–188	127-19-5	DMA; N,N-Dimethylacetamide; Acetdimethylamide	C4-H9-N-O	TOXNET (2015an)	Chicken, Dog, Mouse, Rabbit, Rat	Dermal, Inhalation, IP, IV, PO
Dimethylsulfoxide	T33, T111 C:2–5, 43–57, 91, 94, 157, 158, 193–197	67-68-5	DMSO	C2H6OS	Ali (2001); Augustine et al. (2000); Bartsch et al. (1976); Leadscope (2008a); Pestel et al. (2006); Pharmpk (2002a); RTECS (2015); Ruble et al. (2006); Schonwald (2004a); Sodicoff et al. (1990); TOXNET (2015ao); White et al. (1984); Wood et al. (1971)	Dog, Guinea Pig, Minipig, Mouse, Nonhuman Primate, Rabbit, Rat	IP, IV, PO, SC
Di-sodium hydrogen phosphate dihydrate	T111 C:42	10140-65-5	Sodium phosphate, dibasic	H3-O4-P.x-H2-O.2Na	TOXNET (2015ap)	Nonhuman Primate	SC
Docosanol	T111 C:288	661-19-8	1-Docosanol	C22-H46-O	TOXNET (2015aq)	Minipig	Topical
Dulbecco's modified PBS	T34 –	–	Dulbecco's modified phosphate buffered saline solution	–	–	Rat	IV (tail vein)

(*Continued*)

TABLE A.1 (*Continued*)

Index of Vehicles and Excipients – Codex and Details

Excipient/Vehicle	Data in Table #s (T#) Combination #s (C:#)	CAS#	Synonyms	Formula	Key Toxicity Review Articles / Sources	Animal Species Evaluated	Administration Routes Evaluated
Dulbecco's PBS	T35 –	–	Dulbecco's phosphate buffered saline solution	–	–	Rat	PO
EDTA	T111 C:58–61	60-00-4	Ethylenediamin-eetraacetic acid; Edetic acid	C_{10}-H_{16}-N_2-O_8	Cavender (2012e); Heimbach et al. (2000); Lanigan and Yamarik (2002c); TOXNET (2015ar)	Minipig, Nonhuman Primate, Rat	Dermal, IV, Topical
Ethanol	T36, T111 C:32, 33, 38, 40, 61–73, 98, 99, 108–111, 124, 148–156, 160, 161, 164, 165, 175, 184, 191, 196–205, 211, 212, 223, 224, 239–250, 253, 254, 259, 260, 283, 285, 286, 290	64-17/5	Ethyl alcohol	C_2H_6O	Bartsch et al. (1976); Bevan (2012a); Church and Witting (1997); Fort et al. (1984); Moorman et al. (2001); Rowe et al. (2012); Ruble et al. (2006); Sivilotti (2004); TOXNET (2015as)	Dog, Guinea Pig, Minipig, Mouse, Nonhuman Primate, Rat	Dermal, IP, IV, PO, SC, Topical
Fumaric Acid	T111 C:90	110-17-8	2-Butenedioic acid	C_4-H_4-O_4	TOXNET (2015at)	Dog	PO
Gelatin	T37, T111 C:74, 120	9000-70-8	–	–	TOXNET (2015au)	Dog, Nonhuman Primate	PO
Gelatin Phosphate Buffer	T28	–	–	–	–	Minipig	Topical
Gelucire™	T111 C:75	–	–	–	–	Minipig	PO

(*Continued*)

TABLE A.1 (*Continued*)

Index of Vehicles and Excipients – Codex and Details

Excipient/Vehicle	Data in Table #s (T#) Combination #s (C:#)	CAS#	Synonyms	Formula	Key Toxicity Review Articles / Sources	Animal Species Evaluated	Administration Routes Evaluated
Gelucire 44/14™	T39, T111 C:76, 77	121548-04-7	PEG-32 glyceryl laurate	–	Cavender (2012b); Dordunoos et al. (1996); Kawakami et al. (2004); Ratsimbazafy et al. (1999); TOXNET (2015av); Working et al. (1997)	Dog, Minipig, Mouse Rabbit, Rat	Dermal, Ocular, PO
Gelucire 50/13™	T40 –	121548-05-8	G-50-13	–	Fini et al. (2005); Passerini et al. (2002); Ratsimbazafy et al. (1999); Gattefosse (2014); Sharma et al. (2006); TOXNET (2015aw)	Rat	PO
Gluconic Acid	T41 –	133-42-6	–	C6-H12-O7	TOXNET (2015ax)	Dog, Rat	PO
Glycerol	T42, T111 C:61, 90, 239, 240	56-81-5	Glycerine; Glycerin	C3H8O3	Anderson et al. (1950); Bartsch et al. (1976); Cosmetic Ingredient Review (2004); TOXNET (2015ay)	Dog, Guinea Pig, Minipig, Mouse, Nonhuman Primate, Rabbit, Rat	IP, IV, PO, SC, Topical
Glyceryl Stearate SE	T111 C:163	11099-07-3	–	C18-H36-O2.x-C3-H8-O3	Cosmetic Ingredient Review (2004a)	Minipig	Topical
Glycofurol	T43 –	31692-85-0	–	(C2-H4-O) mult-C5-H10-O2	Ruble et al. (2006); TOXNET (2015az)	Dog	IV

(Continued)

TABLE A.1 (*Continued*)

Index of Vehicles and Excipients – Codex and Details

Excipient/Vehicle	Data in Table #s (T#) Combination #s (C:#)	CAS#	Synonyms	Formula	Key Toxicity Review Articles / Sources	Animal Species Evaluated	Administration Routes Evaluated
Glycol Dimethacrylate Cross Polymer	T111 C:127	–	–			Rat	Topical
Gum Tragacanth	T44 –	9000-65-1	–	Natural product	Anderson DM (1989); Bachmann et al. (1978); Hagiwara et al. (1992); TOXNET (2015ba)	Mouse	PO
Gum Xanthane	T45, T111 C:163, 305	11138-66-2	Keltrol	$(C_{35}H_{49}O_{29})_n$	TOXNET (2015bb)	Minipig, Rabbit, Rat	PO, Topical
Hexylene Glycol	T111 C:38	107-41-5	–	C6-H14-O2		Minipig	Topical
Histidine	T111 C:78–80, 287	71-00-1	–	C6-H9-N3-O2	TOXNET (2015bc)	Dog, Rat	IV, SC
Hydrochloric Acid	T46 –	7647-01-0	Hydrogen chloride; Muriatic acid; Chlorohydric acid; Chlorane	HCI	TOXNET (2015bd)	Dog, Rat	PO
Hydrogenated Castor Oil	T111 C:288	8001-78-3	–	–	TOXNET (2015be)	Minipig	Topical
Hydroxyethyl-cellulose	T47, T111 C:100, 101	9004-62-0	Natrosol; 2-Hydroxyethyl ether cellulose	–	Pestel et al. (2006); TOXNET (2015bf)	Rat	PO

(*Continued*)

TABLE A.1 (*Continued*)
Index of Vehicles and Excipients – Codex and Details

Excipient/Vehicle	Data in Table #s (T#) Combination #s (C:#)	CAS#	Synonyms	Formula	Key Toxicity Review Articles / Sources	Animal Species Evaluated	Administration Routes Evaluated
Hydroxypropyl Beta-Cyclodextrin	T48, T111 C:91–99	128446-35-5	2-hydroxypropyl Beta-cyclodextrin; HPβCD; Hydroxypropyl betadex; Cavasol W7 HP™	C54H102O39	Aerpio Technical Document (2015); Albers and Muller (1995); Challa et al. (2005); Coussement et al. (1990); Gerloczy et al. (1994); Gould and Scott (2005); Leadscope (2008b); Marttin et.al (1998); Pestel et al. (2006); Ruble et al. (2006); Stella and He (2008); Thackaberry et al. (2009); TOXNET (2015bg)	Dog, Mouse, Nonhuman Primate, Rabbit, Rat,	Intranasal, IP, IV, PO, SC
Hydroxypropyl Cellulose	T49, T111 C:38	9004-64-2	Hyprolose (INN), 2-Hydroxy-propyl cellulose	N/A	Cavender (2012c); TOXNET (2015bh)	Minipig, Rat	PO, Topical
Hydroxypropyl Methylcellulose	T50, T111 C:81–87, 90	9004-65-3	Benecel MHPC, Hypromellose, HPMC; Methocel E50 Premium LV™	N/A	Feitoza et al. (2003); Geerling et al. (2001); Maki et al. (2000); Mehlman (2012b); Obara et al. (1992); Rosen et al. (1987); Thackaberry et al. (2009); TOXNET (2015bi)	Dog, Minipig, Mouse, Nonhuman Primate, Rat	IP, PO

(*Continued*)

TABLE A.1 (*Continued*)

Index of Vehicles and Excipients – Codex and Details

Excipient/Vehicle	Data in Table #s (T#) Combination #s (C:#)	CAS#	Synonyms	Formula	Key Toxicity Review Articles / Sources	Animal Species Evaluated	Administration Routes Evaluated
Hydroxypropyl Methylcellulose Acetate Succinate	T111 C:88, 89, 291–294	71138-97-1	Hypromellose acetate succinate; HPMC AS	N/A	TOXNET (2015bj)	Dog, Minipig, Mouse, Rabbit, Rat	PO
Hymetellose	T111 C:102	9032-42-2	Methyl hydroxy-ethylcellulose; Tylose MH50	N/A	TOXNET (2015bk)	Rabbit	PO
Hypotonic PBS	T51 –	–	–	–		Dog, Rat	IV
Imwitor 742™	T111 C:103–105	–	Caprylic/Capric glycerides; Glyceryl monocapry-locaprate, Type l	–	Susananta et al. (1995)	Hamster, Nonhuman Primate, Rat	PO
Isopropyl Alcohol	T52, T111 C:14	67-63-0	2-propanol; Isopropanol; Sec-propyl alcohol	C3H8O	Allen et al. (1998); Bevan C (2012b); Burleigh-Flayer et al. (1998); Church and Witting (1997); Sivilotti (2004); TOXNET (2015bl); Tyl et.al (1994)	Dog, Rabbit	Dermal, Topical
Isopropyl Myristate	T53, T111 C:163, 234, 235	110-27-0	1-Methylethyl tetradecanoate; Crodamol IPM	C17H34O2	Campbell and Bruce (1981); Komatsu et al. (1979); TOXNET (2015bm)	Minipig, Mouse, Rabbit	Dermal, Topical
Isotonic Saline	T111 C:117	–		–	–	Rat	PO

(*Continued*)

TABLE A.1 (*Continued*)

Index of Vehicles and Excipients – Codex and Details

Excipient/Vehicle	Data in Table #s (T#) Combination #s (C:#)	CAS#	Synonyms	Formula	Key Toxicity Review Articles / Sources	Animal Species Evaluated	Administration Routes Evaluated
Kolliphor™	T111 C:43, 48	–	Cremophor; Polyoxyn castor oil	–	–	Nonhuman Primate	IV, PO
Kolliphor EL™	T54, T111 C:15, 16, 62, 63, 106, 108–111, 170–173, 209	61791-12-6	Cremophor EL™; Polyoxyl castor oil; Polyoxyl 35 castor oil	N/A	Gelderblom et al. (2001); Gupta et al. (1996); Lorenz et. al (1977); PharmPK (2002a); pharmpk (2002b); Ramadan et al. (2001); Stokes et al. (2013); TOXNET (2015bn)	Dog, Mouse, Nonhuman Primate, Rat	IV, PO
Kolliphor ELP™	T55	61791-12-6	Cremophor EL™; Polyoxyl castor oil; Polyoxyl 35 castor oil	N/A	TOXNET (2015bn); Xcess (2015)	Dog	–
Kolliphor HS15™ (See also: Solutol HS15)	T111 C:113	–	–	–	–	Minipig	PO

(Continued)

TABLE A.1 (*Continued*)

Index of Vehicles and Excipients – Codex and Details

Excipient/Vehicle	Data in Table #s (T#) Combination #s (C:#)	CAS#	Synonyms	Formula	Key Toxicity Review Articles / Sources	Animal Species Evaluated	Administration Routes Evaluated
Kolliphor RH40™	T56, T111 C:107, 207–209, 231	61788-85-0	Cremophor RH40™; PEG-40 hydrogenated castor oil; Polyoxyl 40 Hydrogenated Castor Oil; Macrogol-glycerol Hydroxy-stearate	N/A	Gupta et al. (1996); Stokes et al. (2013); TOXNET (2015bo)	Dog, Rat	IV, PO
Labrafil M1944™	T57, T111 C:112, 114	62563-68-2	Labrafil™	N/A	Beckwith-Hall et al. (2002); Gattefosse (2003a); TOXNET (2015bp)	Dog, Rabbit, Rat	Dermal, PO
Labrasol™	T58, T111 C:113–115, 209	85536-07-8	Polyglycolyzed glycerides	N/A	Hu et al. (2001); Hu et al. (2002)	Dog, Minipig, Rabbit, Rat	Dermal, IV, Ocular, PO
Lactated Ringer's	T111 C:116	8026-79-7	Sodium chloride, sodium lactate, potassium chloride and calcium chloride; Compound sodium lactate injection	C3-H6-O3. Ca-Cl2.Cl-K. Na	TOXNET (2015bq)	Nonhuman Primate	IV

(*Continued*)

TABLE A.1 (*Continued*)

Index of Vehicles and Excipients – Codex and Details

Excipient/Vehicle	Data in Table #s (T#) Combination #s (C:#)	CAS#	Synonyms	Formula	Key Toxicity Review Articles / Sources	Animal Species Evaluated	Administration Routes Evaluated
Lactic Acid	T111 C:296–298	50-21-5	2-Hydroxy-propanoic acid	C3-H6-O3	TOXNET (2015br)	Dog, Rat	IA, IV
Lactose	T59, T111 C:261–264, 278–281	63-42-(anhydrous)	O-β-D-alactopy-ranosyl-(1->4)-α-Dgluco-pyranose	C12H22O11 (anhydrous)	Ahmad et al. (2004); Baldrick and Bamford (1997); TOXNET (2015bs)	Dog, Nonhuman Primate, Rat	Inhalation, IV, SC
Lanolin	T60 –	8006-54-0	Wool wax	N/A	Kligman (1983); TOXNET (2015bt)	Minipig, Rabbit	Dermal, Topical
Lanolin Alcohol NF	T111 C:174	8027-33-6	Eucerin	Natural product	TOXNET (2015bu)	Minipig	Topical
L-Arginine HCl USP	T111 C:163, 289	15595-35-4	L-Arginine hydrochloride	C6-H14-N4-O2.x-Cl-H	TOXNET (2015bv)	Minipig, Nonhuman Primate	SC, Topical
L-Ascorbic Acid	T111 C:117	50-81-7	Cevatine, Cevex, Cevital	C6H8O6	Bendich and Cohen (1990); Dykes and Meier (1975); Temple (2004) TOXNET (2015bw)	Rat	PO
Lauroglycol 90™	T61 –	27194-74-7	Propylene glycol monolaurate; Lauric acid, monoester with propane-1,2-diol	C15-H30-O3	Bartsch et al. (1976); Liu et al. (2006a); TOXNET (2015bx)	Rabbit, Rat	Dermal, Ocular, PO

(Continued)

TABLE A.1 (*Continued*)
Index of Vehicles and Excipients – Codex and Details

Excipient/Vehicle	Data in Table #s (T#) Combination #s (C:#)	CAS#	Synonyms	Formula	Key Toxicity Review Articles / Sources	Animal Species Evaluated	Administration Routes Evaluated
Maltitol Solution	T62 –	9053-46-7	Liquid Maltitol; Lycasin	C12H24O11 + C6H14O6	Modderman (1993); Walker and El Harith (1978); TOXNET (2015by)	Rat	IP
Maltol	T63 –	118-71-8	2-Methyl pyromeconic acid; 2-Methyl-3-hydroxy-4-pyrone	C6H6O3	Hironishi et al. (1996); Murakami et al. (2006); TOXNET (2015bz)	Guinea Pig, Rabbit	PO
Mannitol	T64, T111 C:43, 48, 74, 118–123, 159, 233, 274, 275, 282	69-65-8	D-Mannitol	C6H14O6	Horvath et al. (1982); Lina et al. (1996); TOXNET (2015ca)	Dog, Minipig, Mouse, Nonhuman Primate, Rabbit, Rat	IM, IV, PO, SC
Methane Sulfonic Acid	T111 C:124	75-75-2	Methylsulfonic acid	CH4O3S	Shertzer (2012); TOXNET (2015cb)	Rat	PO
Methocel™	T111 C:125	–	–	–	–	Nonhuman Primate; Rat	PO
Methyl Methacrylate	T111 C:127	80-62-6	–	C5-H8-O2	TOXNET (2015cc)	Rat	Topical
Methylcellulose	T65, T111 C:88, 89, 126, 129–146, 178, 183, 295	9004-67-5	Cellulose methyl ether; Methocel A4M Premium	N/A	Bachmann et al. (1978); Gupta et al. (1996); Mehlman (2012c); Sellers et al. (2005); TOXNET (2015cd)	Dog, Guinea Pig, Mouse, Nonhuman Primate, Rabbit, Rat	IV, PO, Topical

(Continued)

TABLE A.1 (*Continued*)

Index of Vehicles and Excipients – Codex and Details

Excipient/Vehicle	Data in Table #s (T#) Combination #s (C:#)	CAS#	Synonyms	Formula	Key Toxicity Review Articles / Sources	Animal Species Evaluated	Administration Routes Evaluated
Methylparaben	T111 C:128, 147, 174	99-76-3	4-Hydroxy-benzoic acid, methyl ester	C8-H8-O3	TOXNET (2015ce)	Minipig	SC, Topical
Methylpyrrolidone	T66, T111 C:75–77	872-50-4	N-methyl-2-pyrrolidone, 1-methyl-2-pyrrolidone, Pharmasolv, NMP	C5H9NO	Bartsch et al. (1976); Kennedy (2012); Lee et al. (1987); Ruble et al. (2006); Solomon et al. (1996); TOXNET (2015cf)	Dog, Minipig, Mouse	IV, PO
Mineral Oil	T67, T111 C:288	8012-95-1	Liquid paraffin	Natural product	Carlton et al. (2001); Dalbey and Biles (2003); Nash et al. (1996); TOXNET (2015cg); Trimmer et al. (2004)	Cat, Dog, Guinea Pig, Minipig, Mouse, Rat	PO, Topical
Myristyl Alcohol	T111 C:288	112-72-1	1-Tetradecanol	C14-H30-O	TOXNET (2015ch)	Minipig	Topical
Neobee 1053 Oil™	T111 C:164, 165	73398-61-5	Medium-chain triglycerides	N/A	Bellantone et al. (1999); Traul et al. (2000); Susananta et al. (1995); Wieland et al. (1993)	Mouse, Rat	IV, PO
Octoxynol-40	T111 C:167–169	9002-93-1	–	(C2-H4-O) mult-. C14-H22-O (C2-H4-O) mult-C14-H22-O C34-H62-O11	TOXNET (2015ci)	Dog, Rabbit	Ocular

(*Continued*)

TABLE A.1 (*Continued*)

Index of Vehicles and Excipients – Codex and Details

Excipient/Vehicle	Data in Table #s (T#) Combination #s (C:#)	CAS#	Synonyms	Formula	Key Toxicity Review Articles / Sources	Animal Species Evaluated	Administration Routes Evaluated
Oleic Acid NF	T111 C:163, 170–174	112-80-1	9-Octadecenoic acid	C18-H34-O2	TOXNET (2015cj)	Minipig	Topical
Oleyl Alcohol NF	T111 C:174	143-28-2	–	C18-H36-O	TOXNET (2015ck)	Minipig	Topical
Olive Oil	T68, T111 C:174	8001-25-0	–	Natural product	Evangelista et al. (2004); TOXNET (2015cl)	Minipig, Rat	PO, Topical
OraPlus™ suspension	T111 C:34–37	–	Purified water, microcrystalline cellulose, carboxymethylcellulose sodium, xanthan gum, carrageenan, calcium sulfate, trisodium phosphate, citric acid and sodium phosphate as buffers, dimethicone antifoam emulsion. Preserved with methylparaben and potassium sorbate.	–	Paddock Laboratories (2015)	Rat	PO
Panthenol	T111 C:61	16485-10-2	Dexpanthenol	C9-H19-N-O4	TOXNET (2015cm)	Minipig	Topical

(*Continued*)

TABLE A.1 (*Continued*)

Index of Vehicles and Excipients – Codex and Details

Excipient/Vehicle	Data in Table #s (T#) Combination #s (C:#)	CAS#	Synonyms	Formula	Key Toxicity Review Articles / Sources	Animal Species Evaluated	Administration Routes Evaluated
Peanut Oil	T69, T111 C:175	8002-03-7	Arachis oil; Fletcher's	Natural product	Cosmetic Ingredient Review (2001); Patel et al. (2005); TOXNET (2015cn)	Dog, Rat	PO, SC
Peceol	T111 C:176, 177	25496-72-4	Glyceryl monooleate NF; Monoolein	C21-H40-O4	TOXNET (2015co)	Dog, Rat	PO
PEG 200	T70, T111 C:131, 183, 184	25322-68-3	Polyethylene glycol 200	(C2-H4-O) mult-H2-O	Cavender (2012b); Dordunoos et al. (1996); Smyth et al. (1945); Smyth et al. (1950); Smyth et al. (1955); Quadbeck (1950); TOXNET (2015cp); Working et al. (1997)	Minipig, Nonhuman Primate, Rabbit, Rat	IP, IV, PO
PEG 300	T71, T111 C:106, 185–190, 290	25322-68-3	Polyethylene glycol 300	(C2-H4-O) mult-H2-O	Carpenter and Shaffer (1952); Cavender (2012b); Dordunoos et al. (1996); Patel et al. (2005); Rowe and Wolf (1982); Smyth et al. (1945); Smyth et al. (1950); Smyth et al. (1955); TOXNET (2015cp); Working et al. (1997)	Cat, Dog, Guinea Pig, Mouse, Rabbit, Rat	IP, IV, PO, PO mucosa

(Continued)

TABLE A.1 (*Continued*)

Index of Vehicles and Excipients – Codex and Details

Excipient/Vehicle	Data in Table #s (T#) Combination #s (C:#)	CAS#	Synonyms	Formula	Key Toxicity Review Articles / Sources	Animal Species Evaluated	Administration Routes Evaluated
PEG 400	T72, T111 C:44–47, 50–52, 75–77, 107, 115, 151, 152, 170–173, 176, 177, 191–231, 254, 276	25322-68-3	Polyethylene glycol 400	(C2-H4-O) mult-H2-O	Bartsch et al. (1976); Cavender (2012b); Dordunoos et al. (1996); Fort et al. (1984); Gupta et al. (1996); Gutiérrez-Cabano (2000); Hermansky et al. (1995); Li et al. (2011); Patel et al. (2005); Rowe and Wolf (1982); Ruble et al. 2006); Shideman and Procita (1951); Smyth et al. (1945); Smyth et al. (1950); Smyth et al. (1955); Stokes et al. (2013); Strickley (2008); Thackaberry (2014); TOXNET (2015cp); Working et al. (1997)	Dog, Guinea Pig, Minipig, Mouse, Nonhuman Primate, Rabbit, Rat	Dermal, IP, IV, PO, Topical
PEG 600	T73 –	25322-68-3	Polyethylene glycol 600	(C2-H4-O) mult-H2-O	Pfordte (1971); Rowe and Wolf (1982); Smyth et al. (1955); TOXNET (2015cp)	Rat	IP, IV, PO
PEG 810	T74 –	25322-68-3	Polyethylene glycol 810	(C2-H4-O) mult-H2-O	Kärber (1951); TOXNET (2015cp)	Rat	IV, SC

(*Continued*)

TABLE A.1 (*Continued*)

Index of Vehicles and Excipients – Codex and Details

Excipient/Vehicle	Data in Table #s (T#) Combination #s (C:#)	CAS#	Synonyms	Formula	Key Toxicity Review Articles / Sources	Animal Species Evaluated	Administration Routes Evaluated
PEG 1000	T75 –	25322-68-3	Polyethylene glycol 1000	(C2-H4-O) mult-H2-O	Shideman and Procita (1951); Smyth et al. (1950); TOXNET (2015cp)	Mouse, Rabbit, Rat	IP, IV, PO
PEG 1500	T76 –	25322-68-3	Polyethylene glycol 1500	(C2-H4-O) mult-H2-O	Rowe and Wolf (1982); Smyth et al. (1947); Smyth et al. (1950); Smyth et al. (1942); Smyth et al. (1955); TOXNET (2015cp)	Rat	IP, IV, PO
PEG 1540	T77 –	25322-68-3	Polyethylene glycol 1540	(C2-H4-O) mult-H2-O	Smyth et al. (1950); Smyth et al. (1955); TOXNET (2015cp)	Dog, Rabbit, Rat	IP, IV, PO
PEG 4000	T78, T111 C:53, 54	25322-68-3	Polyethylene glycol 4000	(C2-H4-O) mult-H2-O	Rowe and Wolf (1982); Shideman and Procita (1951); Smyth et al. (1942); Smyth et al. (1947); Smyth et al. (1950); Smyth et al. (1955); TOXNET (2015cp)	Dog, Mouse, Rabbit, Rat	IP, IV, PO
PEG 6000	T79 –	25322-68-3	Polyethylene glycol 6000	(C2-H4-O) mult-H2-O	Smyth et al. (1950); Smyth et al. (1955); TOXNET (2015cp)	Rabbit, Rat	IP, IV, PO
PEG 10000	T80 –	25322-68-3	Polyethylene glycol 10000	(C2-H4-O) mult-H2-O	Smyth et al. (1950); TOXNET (2015cp)	Rat	IP, PO

(*Continued*)

TABLE A.1 (*Continued*)

Index of Vehicles and Excipients – Codex and Details

Excipient/Vehicle	Data in Table #s (T#) Combination #s (C:#)	CAS#	Synonyms	Formula	Key Toxicity Review Articles / Sources	Animal Species Evaluated	Administration Routes Evaluated
PEG 400000	T81 –	25322-68-3	Polyethylene glycol 400000	(C2-H4-O) mult-H2-O	Smyth et al. (1950); Smyth et al. (1970); TOXNET (2015cp)	Rat	IV, PO
Petrolatum	T82 –	8009-03-8	Yellow soft paraffin; Petroleum jelly	N/A	TOXNET (2015cq)	Rabbit	Dermal
Phenoxyethanol	T111 C:234, 235, 272	122-99-6	2-Phenoxy-ethanol	C8-H10-O2	TOXNET (2015cr)	Minipig, Mouse	Topical
Phosal 53 MCT™	T111 C:231	–	Lecithin in caprylic/capric triglycerides, alcohol, glyceryl stearate, oleic acidand ascorbyl palmitate	–	American Lecithin (2007); Susananta et al. (1995)	Rat	PO
Phosphate	T111 C:59, 60, 232	14265-44-2	Phosphate ion	O4-P	TOXNET (2015cs)	Nonhuman Primate	IV
Phosphate Buffer	T111 C:233	–	–	–	–	Rat	SC
Phosphate Buffered Saline	T83, T111 C:28–30	–	PBS	–	–	Dog, Minipig, Mouse, Nonhuman Primate, Rabbit, Rat	IA, IV, PO, SC, Topical
Polawax	T111 C:234, 235	–	Emulsifying wax	–	Carlton et al. (2001)	Minipig, Mouse	Topical

(Continued)

TABLE A.1 (*Continued*)

Index of Vehicles and Excipients – Codex and Details

Excipient/Vehicle	Data in Table #s (T#) Combination #s (C:#)	CAS#	Synonyms	Formula	Key Toxicity Review Articles / Sources	Animal Species Evaluated	Administration Routes Evaluated
Polaxamer 124™	T111 C:225–228, 231	–	–	–	–	Dog, Mouse, Rat	PO
Poloxamer 188™	T84, T111 C:42, 83–85, 102, 236–238	9003-11-6	Poloxalene	(C3-H6-O. C2-H4-O)x-	Benita (1999) Curry et al. (2004); Frim et al. (2004); Grindel et al. (2002); Lemieux et al. (2000); Serbest et al. (2005); TOXNET (2015ct)	Dog, Minipig, Mouse, Nonhuman Primate, Rabbit, Rat	PO, SC
Polyethylene terephthalate	T111 C:162	–	PET	–	–	Rat	IV
Poly(glycolide-co-dl-lactide) Microspheres	T85 –	26780-50-7	–	–	–	Dog	Into periodontal pockets
Polyglyceryl Oleate	T86 –	9007-48-1	1,2,3-propanetriol, homopolymer, (Z)-9-octadecenoate; Decaglyceryl monooleate	C18-H34-O2.x-(C3H8-O3)x-	TOXNET (2015cu); Gattefosse (2003b)	Rabbit, Rat	Dermal, Ocular, PO
Polyvinyl-pyrrolidone	T87, T111 C:216, 217, 256–258, 291–293	9003-39-8	Povidone; PVP; PVP K30	(C6-H9-N-O)x-	Beji et al. (2006); PharmPK (2002a); pharmpk (2002b); TOXNET (2015cv)	Dog, Nonhuman Primate, Rabbit, Rat	IM, PO
Potassium Chloride	T111 C:163	7447-40-7	KCl	KCl	TOXNET (2015cw)	Minipig	Topical

(*Continued*)

TABLE A.1 (*Continued*)

Index of Vehicles and Excipients – Codex and Details

Excipient/Vehicle	Data in Table #s (T#) Combination #s (C:#)	CAS#	Synonyms	Formula	Key Toxicity Review Articles / Sources	Animal Species Evaluated	Administration Routes Evaluated
Propylene Glycol	T88, T111 C:16, 61, 65–69, 73, 98, 99, 107, 148, 153–156, 163, 190, 212–215, 234, 235, 239–255	57-55-6	1,2-Dihydroxypropane	C3H8O2	Cavender (2012a); Fort et al. (1984); Ruble et al. (2006); Thackaberry et al. (2009); TOXNET (2015bx)	Dog, Minipig, Guinea Pig, Nonhuman Primate, Mouse, Rabbit, Rat	Dermal, IP, IV, PO, SC, Topical
Propylene Glycol Dicaprylate/ Dicaprate	T111 C:127	68583-51-7	Caprylic, capric acid, propylene glycol diester	C10-H20-O2. C8-H16-O2. C3-H8-O2	Cosmetic Ingredient Review (1999); TOXNET (2015cx)	Rat	Topical
Propylparaben	T111 C:128, 147, 174	94-13-3	4-Hydroxybenzoic acid, propyl ester	C10-H12-O3	TOXNET (2015cy)	Minipig	SC, Topical
PVP VA 64	T111 C:223, 224, 294	25086-89-9	Vinylpyrrolidone-vinyl acetate copolymer; Copovidone; PVP VA	(C6-H9-N-O. C4-H6-O2)x-	TOXNET (2015cz)	Dog, Rat	PO
RAMEB	T89 –		Randomly Methylated-beta-cylodextrins		Challa et al. (2005)	Nonhuman Primate	Intranasal
Safflower Oil	T90 –	8001-23-8	Carthamus tinctorius oil	Natural product	TOXNET (2015da)	Dog	SC
Salicylic Acid	T111 C:61	69-72-7	Benzoic acid, 2-hydroxy-	C7-H6-O3	TOXNET (2015db)	Minipig	Topical
Saline (pH adjusted, pH 4.5)	T91 –	–	–	–	–	Mouse	IM

(*Continued*)

TABLE A.1 (*Continued*)

Index of Vehicles and Excipients – Codex and Details

Excipient/Vehicle	Data in Table #s (T#) Combination #s (C:#)	CAS#	Synonyms	Formula	Key Toxicity Review Articles / Sources	Animal Species Evaluated	Administration Routes Evaluated
Sesame Oil	T92, T111 C:259, 260	8008-74-0	Gingilli oil; Gingelly oil; Til oil	Natural product	Farber et al. (1976); Genovese et al. (1999); Prasamthi et al. (2005); TOXNET (2015dc)	Dog, Mouse, Rabbit, Rat	PO
Shea Butter	T111 C:174	194043-92-0	–	Natural product	TOXNET (2015dd)	Minipig	Topical
Simethicone	T111 C:126, 143	8050-81-5	Silicone antifoam agent S 184; Gas-x	N/A	TOXNET (2015de)	Rat	PO
Sodium Acetate	T111 C:92, 93, 261–265, 299, 300	127-09-3	Acetic acid sodium salt	C2H3NaO2	TOXNET (2015df)	Dog, Mouse, Nonhuman Primate, Rat	IV, PO, SC
Sodium Acetate Trihydrate Buffer	T93, T111 C:266, 267	6131-90-4	–	–	–	Mouse, Nonhuman Primate	IV, PO
Sodium Chloride	T94, T111 C:41, 49, 58–60, 64, 80, 96, 97, 117, 121, 148–163, 166, 189, 210, 211, 232, 236, 237, 251–253, 265, 268, 271, 273, 278–281, 285, 289	7467-14-5	Salt; Saline; Halite	NaCl	Barrie et al. (2012); Caraccio et al. (2004); Meneely et al. (1953); Meneely and Ball (1958); Moore et al. (1988)	Cat, Dog, Minipig, Mouse, Nonhuman Primate, Rabbit, Rat	ID, IP, IM, IV, Ocular, Perivascular, PO, SC, Topical
Sodium Citrate	T111 C:251–258, 268, 294, 303	994-36-5	–	C6-H8-O7.x-Na	TOXNET (2015dg)	Nonhuman Primate, Rat	IV, PO

(Continued)

TABLE A.1 (*Continued*)
Index of Vehicles and Excipients – Codex and Details

Excipient/Vehicle	Data in Table #s (T#) Combination #s (C:#)	CAS#	Synonyms	Formula	Key Toxicity Review Articles / Sources	Animal Species Evaluated	Administration Routes Evaluated
Sodium Dihydrogen Phosphate Dihydrate	T95, T111 C:42	13472-35-0	Sodium phosphate, monobasic, dihydrate; SDPD	H2-O4-P. Na.2H2-O	TOXNET (2015dh)	Mouse, Nonhuman Primate	PO, SC
Sodium Hydroxide	T111 C:147, 163, 234, 235, 271, 272, 277, 304	1310-73-2	Caustic soda	NaOH	TOXNET (2015hi)	Minipig, Mouse, Nonhuman Primate, Rat	IV, SC, Topical
Sodium Lauryl Sulfate	T111 C:142–146	151-21-3	Sodium dodecyl sulfate	C12-H26-O4-S. Na	TOXNET (2015dj)	Dog, Rat	PO
Sodium Metabisulfite	T96 –	7681-57-4	–	–	TOXNET (2015dk)	Mouse, Nonhuman Primate, Rat	PO
Sodium Methylparaben	T111 C:269, 270	5026-62-0	Benzoic acid, 4-hydroxy-, methyl ester, sodium salt	C8-H7-Na-O3	TOXNET (2015dl)	Minipig, Mouse, Rat	PO, Topical
Sodium Phosphate Buffer	T97, T111 C:121, 273–277, 289	7558-80-7	–	–	Jefferson (2012); Moore et al. (1988); TOXNET (2015dm)	Dog, Mouse, Nonhuman Primate, Rat	IV, PO
Sodium Propylparaben	T111 C:269, 270	35285-69-9	4-Hydroxy-benzoic acid, propyl ester, sodium salt	C10-H11-Na-O3	TOXNET (2015dn)	Mouse, Rat	PO

(*Continued*)

TABLE A.1 (*Continued*)

Index of Vehicles and Excipients – Codex and Details

Excipient/Vehicle	Data in Table #s (T#) Combination #s (C:#)	CAS#	Synonyms	Formula	Key Toxicity Review Articles / Sources	Animal Species Evaluated	Administration Routes Evaluated
Sodium Succinate	T111 C:118, 122, 123, 278–282	150-90-3	Succinic acid sodium salt; Succinic acid; Disodium butanedioate	C4H4Na2O4	Szilagyi (2012c); TOXNET (2015do)	Dog, Mouse, Rat	IV, SC
Sodium Sulfite	T98	7757-83-7	Sulfurous acid, disodium salt	H2-O3-S.2NaO3-S.2Na	TOXNET (2015dp)	Rabbit	Ocular (Topical)
Solutol HS15™ (See also: Kolliphor HS15™)	T99, T111 C:55, 70, 71, 218–222, 283-286	61909-81-7	Polyethylene glycol-15-hydroxy-stearate; Polyethylene glycol 660 hydroxy stearate	(C2-H4-O) mult-. C18-H36-O3	Cavender (2012b); Coon et al. (1991); Dordunoos et al. (1996); Ruchatz (1998); Stokes et al. (2013); TOXNET (2015dq)	Dog, Mouse, Nonhuman Primate, Rat	IV, IP, PO, Any
Sorbitan Tristearate	T111 C:174	26658-19-5	Sorbitan, triocta-decanoate	C60-H114-O8	Lanigan and Yamarik (2002b) TOXNET (2015dr)	Minipig	Topical
Sorbitol	T111 C:287, 299, 300	50-70-4	D-Sorbitol	C6-H14-O6	TOXNET (2015ds)	Dog, Nonhuman Primate, Rat	IV
Soybean Oil	T 100, T111 C:288	8001-22-7	–	Natural product	Cambridge MedChem (2012); Earl et al. (2002); Farber et al. (1976); Kawashima et al. (2009); TOXNET (2015dt)	Dog, Minipig, Rat	PO, Topical

(*Continued*)

TABLE A.1 (*Continued*)

Index of Vehicles and Excipients – Codex and Details

Excipient/Vehicle	Data in Table #s (T#) Combination #s (C:#)	CAS#	Synonyms	Formula	Key Toxicity Review Articles / Sources	Animal Species Evaluated	Administration Routes Evaluated
Squalene NF	T111 C:163	111-02-4	–	C30-H50	TOXNET (2015du)	Minipig	Topical
Stearic Acid	T111 C:288	57-11-4	n-Octadecanoic acid	C18-H36-O2	TOXNET (2015dv)	Minipig	Topical
Stearyl Alcohol	T111 C:288	112-92-5	1-Octadecanol	C18-H38-O	TOXNET (2015dw)	Minipig	Topical
Sucrose	T111 C:42, 78, 79, 274, 275, 289	57-50-1	Sugar	C12-H22-O11	TOXNET (2015dx)	Cat, Dog, Nonhuman Primate, Rabbit, Rat	IV, PO mucosa, PO, SC
Sucrose Acetate Isobutyrate	T111 C:290	27216-37-1	SAIB	C40-H62-O19	TOXNET (2015dy)	Cat	PO mucosa
Sulfobutylether β-cyclodextrin (See also: Captisol™)	T101 –	182410-00-0	SBECD; Captisol™	N/A	Albers and Muller (1995); Challa et al. (2005); Kim et al. (1998); Marttin et al. (1998); TOXNET (2015dz); Ueda et al. (1998)	Mouse	PO
Tartaric Acid	T102 –	87-69-4	D-tartaric acid; 2,3-dihydroxy-butanedioic acid	C4-H6-O6	Sourkes and Koppanyi (1950); Szilagyi (2012d); TOXNET (2015ea)	Rabbit, Rat	PO
Terbinafine HCL™ Placebo Nail Lacquer	T103 –	78628-80-5	Terbinafine hydrochloride	C21-H25-N. Cl-H	TOXNET (2015eb)	Pig	Dermal
Tetraglycol	T111 C:56, 157, 158	15826-19-4	–	–	TOXNET (2015ec)	Minipig, Rat	IV

(*Continued*)

TABLE A.1 (*Continued*)

Index of Vehicles and Excipients – Codex and Details

Excipient/Vehicle	Data in Table #s (T#) Combination #s (C:#)	CAS#	Synonyms	Formula	Key Toxicity Review Articles / Sources	Animal Species Evaluated	Administration Routes Evaluated
Transcutol™	T104, T111 C:75–77, 113, 114	111-90-0	Diethylene glycol monoethyl ether; DEGEE; 2-(2-Ethoxyethoxy) ethanol	$C_6H_{14}O_3$	Liu et al. (2006b); Sullivan et al. (2014); TOXNET (2015al)	Cat, Dog, Guinea Pig, Minipig, Mouse, Rabbit, Rat	Dermal, IM, IP, IV, Inhalation, Ocular, PO, SC
Trehalose	T111 C:39, 296–298	99-20-7	Alpha-D-Trehalose	C12-H22-O11	TOXNET (2015ed)	Dog, Rat	IA, IV
Tris Buffer	T111 C:52	77-86-1	Tromethamine	C4-H11-N-O3	TOXNET (2015ee)	Minipig, Nonhuman Primate	IV, Topical
Trisodium Citrate Dihydrate	T105 –	6132-04-3	Trisodium citrate; Sodium citrate	C6H5Na3O7·2H2O	TOXNET (2015dg)	Dog, Hamster, Mouse, Rat	PO
Tween 20™	T106, T111 C:18, 19, 61, 163, 225–228, 273, 299, 300	9005-64-5	Polysorbate 20 NF	N/A	Bartsch et al. (1976); TOXNET (2015ef)	Dog, Minipig, Mouse, Nonhuman Primate, Rat	IV/SC/IP, IV, PO, SC, Topical

(Continued)

TABLE A.1 (*Continued*)

Index of Vehicles and Excipients – Codex and Details

Excipient/Vehicle	Data in Table #s (T#) Combination #s (C:#)	CAS#	Synonyms	Formula	Key Toxicity Review Articles / Sources	Animal Species Evaluated	Administration Routes Evaluated
Tween 80™	T107, T111 C:6, 7, 20–30, 57, 81, 82, 86, 90, 100, 101, 103–105, 107, 112, 120, 125, 129–141, 159, 174, 176, 175, 213–215, 229, 230, 232, 266, 267, 287, 301, 302, 305	9005-65-6	Polysorbate 80; Armotan pmo-20, Tween 80™; Polyoxy-ethylene (20) sorbitan monooleate	N/A	Daher et al. (2003); Fisherman and Cohen (1974); Gelperina et al. (2002); National Toxicology Program (1992); O'Sullivan et al. (2004); Oz et al. (2004); Sellers et al. (2005); Thackaberry et al. (2009); TOXNET (2015eg)	Dog, Guinea Pig, Hamster, Minipig, Mouse, Nonhuman Primate, Rabbit, Rat	IA, IV/SC/IP, Intranasal, IV, IP, PO, SC, Topical
Vitamin E	T111 C:167–169, 303	59-02-9	Alpha-tocopherol	C29-H50-O2	TOXNET (2015eh)	Dog, Rabbit, Rat	Ocular, PO
Vitamin E TPGS	T108, T111 C:107, 176, 177, 216, 217, 223, 224, 227, 228, 255, 291–295	9002-96-4	Tocophersolan (USAN)	(C2-H4-O) mult-C33-H54-O5 (C2-H4-O) n-C33-H54-O5	TOXNET (2015ei)	Dog, Mouse, Rabbit, Rat	PO

(*Continued*)

TABLE A.1 (*Continued*)

Index of Vehicles and Excipients – Codex and Details

Excipient/Vehicle	Data in Table #s (T#) Combination #s (C:#)	CAS#	Synonyms	Formula	Key Toxicity Review Articles / Sources	Animal Species Evaluated	Administration Routes Evaluated
Water (NOTE: Only combination formulations expressly listing water in their ingredients are listed here.)	T109, T111 C:1, 17, 23–27, 42, 53–55, 58, 61–63, 65–72, 74, 80, 86, 87, 90–94, 98–101, 108–111, 118, 121, 122–125, 138–141, 143–146, 163, 174, 190, 191, 200–205, 212–215, 234, 235, 241–250, 261–264, 273–275, 278–283, 286, 287, 289, 291–295, 301, 304, 305	7732-18-5	–	H2O	–	Dog, Guinea Pig, Minipig, Mouse, Nonhuman Primate, Pig, Rabbit, Rat	Dermal, IA/SC, IP, IV, PO, SC, Topical
White Wax	T111 C:288	8012-89-3	Beeswax	Natural product	Carlton et al. (2001); TOXNET (2015ej)	Minipig	Topical
Xylitol	T110	87-99-0	Xylite	C5H12O5	Takahashi et al. (1999); TOXNET (2015ek)	Nonhuman Primate	Intranasal

TABLE A.2

Acacia

Species	Route	Duration	Dose	Adverse Reactions/Toxicity	Notes
Nonhuman Primate	PO	Efficacy	100 mg/kg/day	Well tolerated; some weight loss, reduction in food intake	Arabic gum; 3% solution in water

(*Continued*)

TABLE A.2 (*Continued*)

Acacia

Species	Route	Duration	Dose	Adverse Reactions/Toxicity	Notes
Rat	PO	1 month	500 mg/kg	Well tolerated	–
	PO	90 days	10 mL/kg	Well tolerated	20% solution

TABLE A.3

Acetic Acid

Species	Route	Duration	Dose	Adverse Reactions/Toxicity	Notes
Mouse	PO	1 month	5 mL/kg	Well tolerated	15% solution
	PO (gavage)	90 days	5 mL/kg	Well tolerated	3% solution
Rat	PO	Acute	5 mL/kg	Well tolerated	10% solution
	PO	1 month	10 mL/kg	Well tolerated	20% solution
	PO (gavage)	90 days	5 mL/kg/day	Well tolerated	3% in purified water (92/8)

TABLE A.4

Acetone

Species	Route	Duration	Dose	Adverse Reactions/Toxicity	Notes
Guinea Pig	Dermal	1 month	1 mL	Well tolerated	–
Mouse	Dermal	2 year	0.5 mL	Well tolerated	–
	PO	2 weeks	3 mL/kg	Higher doses cause acidosis; transient neurobehavioral effects at this dose	–

(*Continued*)

TABLE A.4 (*Continued*)

Acetone

Species	Route	Duration	Dose	Adverse Reactions/Toxicity	Notes
Rabbit	Dermal	90 days	1 mL	Defatting of application site	–
Rat	Dermal	30 days	5 mL/kg	Well tolerated	–
	Dermal	90 days	1.5 mL/kg 6 hours daily, 5 d/wk	Well tolerated	Sham treatment group included, vehicle similar to sham treatment; 100% acetone; age 60 days; ♂/♀
	PO	2 weeks	5 mL/kg	Higher doses cause acidosis; transient neurobehavioral effects at this dose	–

TABLE A.5

Acetylmethylamide

Species	Route	Duration	Dose	Adverse Reactions/Toxicity	Notes
Nonhuman Primate	PO	1 month (ADME)		Well tolerated	In water

TABLE A.6

Alginic Acid

Species	Route	Duration	Dose	Adverse Reactions/Toxicity	Notes
Rat	IP	1 month	100 mg/kg	Well tolerated	–

TABLE A.7

Anecortave Acetate

Species	Route	Duration	Dose	Adverse Reactions/Toxicity	Notes
Rat	SC (bolus)	4 doses	2 mL/kg	Well tolerated	–

TABLE A.8

Avicel CL-611

Species	Route	Duration	Dose	Adverse Reactions/Toxicity	Notes
Dog	PO (gavage)	Single dose	1 mL/kg	Soft feces	2.4% in sterile water; age 5 months; ♂/♀

TABLE A.9

Balanced Salt Saline

Species	Route	Duration	Dose	Adverse Reactions/Toxicity	Notes
Rabbit	Intravitreal	43 days	50 µL /eyeq 14d	None	Non-GLP; age 5 months; 2♂/2♀

TABLE A.10

Basal Salt Solution

Species	Route	Duration	Dose	Adverse Reactions/Toxicity	Notes
Mouse	Subretinal Injection	SD for 9 months	2.0 µL	None	Non-GLP; age 5–7 weeks; 44♂

TABLE A.11
Benzoic Acid

Species	Route	Duration	Dose	Adverse Reactions/Toxicity	Notes
Rat	PO	Acute	100 mg/kg	Well tolerated	–

TABLE A.12
Beta-cyclodextrin

Species	Route	Duration	Dose	Adverse Reactions/Toxicity	Notes
Dog	IP	1 month	50 mg/kg	Well tolerated	–
	IV	1 month	100 mg/kg	Well tolerated	–
	PO	1 month	200 mg/kg	Well tolerated	–
Mouse	IP	1 month	10 mg/kg	Well tolerated	–
Nonhuman Primate	PO	12 months	–	Tubular hypertrophy at doses above 100 mg/kg/day at 3 months or longer	–
Rat	IV	≥ 3 months	–	Tubular hypertrophy at doses above 100 mg/kg/day at 3 months or longer	–
	PO	12 month	500 g/kg	Hepatitis, nephrosis, acute tubular necrosis at dose levels above 20 g/kg	–

TABLE A.13

Bicarbonate Buffer, pH 9.5

Species	Route	Duration	Dose	Adverse Reactions/Toxicity	Notes
Mouse	PO (gavage)	QD for 2 months	10 mL/kg	None	Age 8–10 weeks; ♂/♀

TABLE A.14

Calcium Chloride

Species	Route	Duration	Dose	Adverse Reactions/Toxicity	Notes
Mouse	IV	1x/wk for 4 weeks	1.61 mL/kg	None	0.5M; age 5 weeks; ♂/♀
	SC	1x/wk for 4 weeks	1.61 mL/kg	None	0.5M; age 5 weeks; ♂/♀

TABLE A.15

Canola Oil

Species	Route	Duration	Dose	Adverse Reactions/Toxicity	Notes
Dog	PO	1 month	2 mL/kg	Well tolerated	–

TABLE A.16

Capryol 90™

Species	Route	Duration	Dose	Adverse Reactions/Toxicity	Notes
Dog	PO	28 days	1000 mg/kg	Nontoxic	–
	PO	28 days	2500 mg/kg	Nontoxic	–
Rabbit	Dermal	Acute	No dilution	Mildly irritant	–
	Ocular	Acute	No dilution	Moderately irritant	–
Rat	PO	Acute	–	Nontoxic LD_{50}> 5 g/kg	–
	PO	7 days	300, 1000, 2500 mg/kg	Well tolerated	–
	PO	28 days	500, 1500, 2500 mg/kg	NOAEL of 2500 mg/kg	–

TABLE A.17

Captisol™ (See also Table 101: Sulfobutylether-β-cyclodextrin (SBECD))

Species	Route	Duration	Dose	Adverse Reactions/Toxicity	Notes
Dog	PO (gavage)	2x/wk for 28 days	5 mL/kg	None	15% in DI water; age 5–6 months; ♂/♀
	IV (bolus)	–	1 mL/kg	Well tolerated	12% solution in water pH 3–11
	IV (infusion)	–	2 mL/kg	Well tolerated	12% solution in water pH 3–11
Mouse	PO	1 month	500 mg/kg	Well tolerated	10% solution
	SC	90 days	1200 mg/kg	NOEL	–
	SC	6 months	1200 mg/kg	NOAEL	–

(Continued)

TABLE 17 (*Continued*)

Captisol™ (See also Table 101: Sulfobutylether-β-cyclodextrin (SBECD))

Species	Route	Duration	Dose	Adverse Reactions/Toxicity	Notes
Nonhuman Primate	PO	9 months	1 g/kg	Well tolerated	10% solution
	SC	3x/wk for 12 months	120 mg/kg	Well tolerated	–
Rat	IV	1 month	4 mL/kg	Well tolerated	12% in water
	IV (bolus)	–	2 mL/kg	Well tolerated	12% solution in water pH 3–11
	IV (infusion)	–	5 mL/kg	Well tolerated	12% solution in water pH 3–11
	PO	1 month	10 mL/kg	Well tolerated	12% in water
	PO (gavage)	2x/wk for 28 days	10 mL/kg	None	15% in DI water; age 7–8 weeks; ♂/♀

TABLE A.18

Carboxymethylcellulose (CMC)

Species	Route	Duration	Dose	Adverse Reactions/Toxicity	Notes
Minipig	PO	14 days	8.0 mL/kg SD	None	0.5% CMC in water; GLP; age 3 months; 1♂/1♀
	PO	28 days	8.0 mL/kg QD	None	0.5% CMC in water; GLP; age 3–4 months; 4♂/4♀
Nonhuman Primate	PO	30 days		Well tolerated	5% in water
	SC	Acute	10 mL/kg	Well tolerated	–

(*Continued*)

TABLE A.18 (*Continued*)

Carboxymethylcellulose (CMC)

Species	Route	Duration	Dose	Adverse Reactions/Toxicity	Notes
Rat	PO		20 mg/kg	NOEL	5 % in water
	PO	14 days	8.0 mL/kg SD	None	0.5% CMC in water; GLP; age 8 weeks; 5♂/5♀
	PO	28 days	8.0 mL/kg QD	None	0.5% CMC in water; GLP; age 8 weeks; 10♂/10♀
	PO (gavage)	93 weeks	10 mL/kg QD	None	1% CMC (medium viscosity) in DI water; age 6 weeks; ♂/♀

TABLE A.19

Carboxymethylcellulose Calcium (Calcium CMC)

Species	Route	Duration	Dose	Adverse Reactions/Toxicity	Notes
Dog	PO	90 days	1 mL/kg	Well tolerated	1% solution

TABLE A.20

Carboxymethylcellulose Sodium (Sodium CMC)

Species	Route	Duration	Dose	Adverse Reactions/Toxicity	Notes
Rabbit	PO	1 month	0.5 mL/kg	Well tolerated	1% solution

TABLE A.21

Cetyl Alcohol

Species	Route	Duration	Dose	Adverse Reactions/Toxicity	Notes
Mouse	IP	1 month	100 mg/kg	Well tolerated	–

TABLE A.22

Citrate Buffer

Species	Route	Duration	Dose	Adverse Reactions/Toxicity	Notes
Dog	IV infusion	8 doses	30 mL/kg/day	Well tolerated	0.1 M, Aqueous
	SC	30 days	10 mL/kg QD	Well tolerated	–
Rat	IV	4 weeks	nd	Hypoactivity, pain at injection site	100 mM at pH 5
	PO	2 weeks	10 mL/kg	Well tolerated	50 mM
	PO	2 weeks	15 mL/kg	Well tolerated	50 mM

TABLE A.23

Citric Acid Buffer

Species	Route	Duration	Dose	Adverse Reactions/Toxicity	Notes
Mouse	PO (gavage)	182 days	2 mL/kg QD	None	0.015 M at pH 4.50; age 6–7 weeks; ♂/♀
Nonhuman Primate	PO (gavage)	39 weeks	7.5 mL/kg QD	None	10 mM; age 3–3.5 years; ♂/♀

(Continued)

TABLE A.23 (*Continued*)

Citric Acid Buffer

Species	Route	Duration	Dose	Adverse Reactions/Toxicity	Notes
Rat	PO	2 weeks	10 mL/kg QD	Well tolerated	50 mM
	PO	2 weeks	15 mL/kg QD	Well tolerated	50 mM

TABLE A.24

Collagen Matrix

Species	Route	Duration	Dose	Adverse Reactions/Toxicity	Notes
Nonhuman Primate	Implantation in humerus bone	6 months	10 mL/kg/day	Well tolerated	Bovine type I and hydroxyapatite
Rabbit	Implantation	6 months	Single application, 5 mL/kg	Well tolerated	–

TABLE A.25

Corn Oil

Species	Route	Duration	Dose	Adverse Reactions/Toxicity	Notes
Chicken Embryo	Injection into egg	Single dose	0.1 µL/g	Less mortality than 1.0 µL/g egg	–
	Injection into egg	Single dose	1.0 µL/g	Increased mortality, decreased activity during righting reflex, running time, visual discrimination, olfactory aversion test	–

(*Continued*)

TABLE 25 (*Continued*)

Corn Oil

Species	Route	Duration	Dose	Adverse Reactions/Toxicity	Notes
Dog	PO	1 month	3.0 mL/kg	Well tolerated	–
Mouse	PO	1 month	2.5 mL/kg	Well tolerated	–
Nonhuman Primate	PO	1 month	1 mL/kg	Well tolerated	–
Rabbit	PO	1 month	1 mL/kg	Well tolerated	–
Rat	PO (gavage)	Single dose	10 mL/kg	Well tolerated	–
	PO (gavage)	20 doses	5 mL/kg	Well tolerated	–
	PO (gavage)	90 days	5 mL/kg QD	None	Age 6 weeks; ♂/♀

TABLE A.26

Cottonseed Oil

Species	Route	Duration	Dose	Adverse Reactions/Toxicity	Notes
Dog	SC	Single dose	1 mL	Well tolerated; no evidence of irritation macroscopically or histologically	100% solution

TABLE A.27
Cyclohexane

Species	Route	Duration	Dose	Adverse Reactions/Toxicity	Notes
Rabbit	PO	30 days	0.5 mL/kg/day	Well tolerated	–
Rat	Dermal	30 days	1 mL/kg/day	Well tolerated	–
	PO (gavage)	4 weeks	5 mL/kg/day	Intermittent convulsive after dosing, piloerection, round back, emaciated appearance	–

TABLE A.28
Dextrose

Species	Route	Duration	Dose	Adverse Reactions/Toxicity	Notes
Cat	Oral mucosa	24 hours	0.6 mL SD	None	5%; non-GLP; age >6 months; 3♂/3♀
Dog	IV	Single dose	150 mL/hr	Well tolerated	5%, USP
	IV/ PO	ADME	2/10 mL/kg/day	Well tolerated	5% solution
Nonhuman Primate	PO (gavage)	13 weeks	0.78–9.3 mL/kg/day	Well tolerated	10% solution (w/w)
	PO (gavage)	ADME	5 mL/kg/day	Well tolerated	5% solution
	PO (gavage)	Card. vas.	5 mL/kg/day	Well tolerated	5% solution
Rabbit	IV (slow bolus)	12 doses		Well tolerated	5%, USP

(Continued)

TABLE A.28 (*Continued*)

Dextrose

Species	Route	Duration	Dose	Adverse Reactions/Toxicity	Notes
Rat	IV	Single dose	1.4 mL/animal	Well tolerated	5%, USP
	IV	7 days	5 mL/kg SD	None	5%; non-GLP; age 7–10 weeks; 2♂/2♀
	PO (gavage)	26 weeks	0.71–8.6 mL/kg/day	Well tolerated	10% solution (w/w)
	PO (gavage)	Prelim	5 mL/kg/day	Well tolerated	5% solution
	SC	2 weeks	0.75 mL/kg/day	Well tolerated	5% solution

TABLE A.29

Dichlorvos

Species	Route	Duration	Dose	Adverse Reactions/Toxicity	Notes
Nonhuman Primate	IV Infusion	2 weeks	2 mL/kg 10 Minutes 3×/week	Well tolerated	10 mg/mL dichlorvos; age 3–6.5 years; ♀

TABLE A.30

Diethylacetamide

Species	Route	Duration	Dose	Adverse Reactions/Toxicity	Notes
Cat	IV	Single dose	1 g/kg (1000 mg/kg)	LD_{LO}	Behavioral: altered sleep time (including change in righting reflex)
Chicken	IV	Single dose	3900 mg/kg	LD_{LO}	–
Dog	IV	Single dose	1 gm/kg (1000 mg/kg)	LD_{LO}	Behavioral: altered sleep time (including change in righting reflex)

(*Continued*)

TABLE A.30 (*Continued*)

Diethylacetamide

Species	Route	Duration	Dose	Adverse Reactions/Toxicity	Notes
Mouse	IP	Single dose	1600 mg/kg	LD_{50}	Sense organs and special senses: mydriasis (pupillary dilation)
	IV	Range finding	MTD: 1.4 g/kg; NOEL: 468 mg/kg	–	Published LD_{50} = 2.3–3.2 g/kg
Rabbit	IV	Single dose	1920 mg/kg	LD_{LO}	–
Rat	IP	Single dose	1840 mg/kg	LD_{50}	–
	IV	Single dose	1 g/kg (1000 mg/kg)	LD_{50}	Behavioral: altered sleep time (including change in righting reflex)

TABLE A.31

Diethyleneglycol-monoethylether (DEGEE)

Species	Route	Duration	Dose	Adverse Reactions/Toxicity	Notes
Nonhuman Primate	IV	1 month (ADME)	.355 mL/kg single injection	Well tolerated	Into saphenous vein

TABLE A.32

Dimethylacetamide (DMA)

Species	Route	Duration	Dose	Adverse Reactions/Toxicity	Notes
Chicken	IV	Acute	12000 mg/kg	LD_{Lo}	Lowest published lethal dose

(Continued)

TABLE A.32 (*Continued*)

Dimethylacetamide (DMA)

Species	Route	Duration	Dose	Adverse Reactions/Toxicity	Notes
Dog	Dermal	6 weeks	2690.476 mg/kg QD	TD_{Lo}; fatty liver degeneration; chronic death (nutritional and gross metabolic); reproductive	–
Mouse	IP	Acute	2800 mg/kg	LD_{50}	–
	IV	Acute	3020 mg/kg	LD_{50}	–
	IV (into tail vein)	Acute	469 mg/kg (dose vol 5 mL/kg)	NOEL	10% solution
	IV (into tail vein)	Acute	1405 mg/kg (dose vol 5 mL/kg)	MTD	30% solution; mild to moderate hypoactivity for up to 6 minutes is typical
	PO	Acute	4620 mg/kg	LD_{50}	Intragastric feeding or introduction with drinking water
Rabbit	Dermal	Acute	2240 mg/kg	LD_{50}	Application directly onto the skin, either intact or abraded
	IV	Acute	8340 mg/kg	LD_{Lo}	Lowest published lethal dose
	PO	13 days	3900 mg/kg	TD_{Lo}; specific developmental abnormalities (eye, ear, craniofacial including nose and tongue, musculoskeletal system); post-implantation mortality; fetotoxicity (except death, e.g., stunted fetus)	Intragastric feeding or introduction with drinking water; 6–18 days of pregnancy
Rat	Inhalation	6 hours/day for 10 days	281 ppm	TD_{Lo}	Inhalation in chamber by cannulation or through mask; 6–15 days of pregnancy

(Continued)

TABLE A.32 (*Continued*)

Dimethylacetamide (DMA)

Species	Route	Duration	Dose	Adverse Reactions/Toxicity	Notes
	IP	–	2 mg/kg	TD_{Lo}; post-implantation mortality; fetotoxicity (except death, e.g., stunted fetus)	1 day pregnant
	IP	Acute	2750 mg/kg	LD_{50}	–
	IV	Acute	2640 mg/kg	LD_{50}	–
	PO	14 days	5600 mg/kg	TD_{Lo}; effects on fertility, post-implantation mortality; fetal death; specific developmental abnormalities (craniofacial, including nose and tongue, musculoskeletal system, cardiovascular system, homeostasis); fetotoxicity (except death, e.g., stunted fetus)	Intragastric feeding or introduction with drinking water; 6–19 days of pregnancy
	PO	14 days	2240 mg/kg	TD_{Lo}	Intragastric feeding or introduction with drinking water; 6–19 days of pregnancy
	PO	Acute	4300 mg/kg	LD_{50}	Intragastric feeding or introduction with drinking water
	PO	10 days	1500 mg/kg QD	TD_{Lo}; findings in digestive system and liver; Chronic	Intragastric feeding or introduction with drinking water; lowest published toxic dose
	PO	26 weeks	2 mg/kg QD	TD_{Lo}; Enzyme inhibition, induction, or change in blood or tissue levels; Liver	Intragastric feeding or introduction with drinking water; lowest published toxic dose

(*Continued*)

TABLE A.32 (*Continued*)

Dimethylacetamide (DMA)

Species	Route	Duration	Dose	Adverse Reactions/Toxicity	Notes
	PO	90 days	50 mg/kg QD	TD_{Lo}; changes in erythrocyte (RBC) and leucocyte (WBC) counts	Intragastric feeding or introduction with drinking water; lowest published toxic dose

TABLE A.33

Dimethylsulfoxide (DMSO)

Species	Route	Duration	Dose	Adverse Reactions/Toxicity	Notes
Dog	IV	Single dose @ 0.2mL/minute for 18.5 minutes	0.43–0.46 mL/kg	During dose administration excessive salivation and labored respiration were noted at 1 and 4 hours post dose	MUST use IV catheter; 100% solution
	IV	–	2 mg/kg	–	–
	IV	–	2.5 g/kg	LD_{50}	MUST use IV catheter; 100% solution
	IV	Inter-mittent for 4 weeks	57,600 mg/kg	LD_{50}; lowest published toxic dose, hematuria, normocytic anemia, death	MUST use IV catheter
	IV	–	2,500 mg/kg	LD_{50}; cardiac changes, hematuria	–
		Single dose	1 mL/kg	Well tolerated	10% solution
	IV	–	0.1 mL/kg	–	MUST use IV catheter; 100% solution
	IV	1 month	1.25 mL/(0.112) × (BW)	Well tolerated	–

(*Continued*)

TABLE A.33 (*Continued*)

Dimethylsulfoxide (DMSO)

Species	Route	Duration	Dose	Adverse Reactions/Toxicity	Notes
Guinea Pig	IV	1 month	0.1 mL/kg	Well tolerated	–
	IP	Acute	6.5 g/kg	LD_{50}	100% solution
Mouse	IP	1 month	2.5 g/kg	LD_{50}	100% solution
	IP	–	3.82–10.73 g/kg	LD_{50}	100% solution
	IP	–	8.2 g/kg	LD_{50}	100% solution
	IP	–	20.06 g/kg	LD_{50}	100% solution
	IP	1 month	100 mg/kg	Well tolerated	–
	IP	3 days	10 mL/kg	Well tolerated	15% solution
	IV (into tail vein)	Acute	1650 mg/kg (dose vol 5 mL/kg)	NOEL	30% solution
	IV (into tail vein)	Acute	2200 mg/kg (dose vol 5 mL/kg)	MTD; rapid breathing, ataxia and muscle contractions, with full recovery by 1 min is typical	40% solution
	IV	Range finding	MTD: 2.2 g/kg; NOEL: 1.6 g/kg	–	Published LD_{50} = 3.8–7.6 g/kg
	IV	–	3,100 mg/kg	LD_{50}; eye hemorrhage, conjunctiva irritation	–
	IV	–	240 gm/kg	Lowest published toxic dose; post-implantation mortality	Age day 1–20 presumed pregnant
	PO (gavage)	–	15.0–22 g/kg	LD_{50}	100% solution
	PO (gavage)	–	7.9 g/kg	LD_{50}	100% solution
	PO	–	5 mL/kg	–	–

(*Continued*)

TABLE A.33 (*Continued*)
Dimethylsulfoxide (DMSO)

Species	Route	Duration	Dose	Adverse Reactions/Toxicity	Notes
	SC	–	13.9–20.5 g/kg	LD_{50}	100% solution
Nonhuman Primate	PO (gavage)	Efficacy	3 mL/kg/day	Well tolerated	–
Rabbit	SC	1 month	1 mL/kg	Erythema, inflammation	–
Rat	IV	–	4–5 mg/kg	–	–
	IV	–	5.25–5.36 g/kg	LD_{50}	100% solution
	IV	–	5.3 g/kg	LD_{50}	100% solution
	IV	Single dose	200 mg/kg	In serum, slightly and transiently changed metabolic parameters including glucose, lactate, triglycerides, free fatty acids, or creatinine as well as electrolytes (Na, Cl, Mg) and osmolality, increased ALT, impeded clinical chemistry measurements of various parameters at 4 hr post dose, kidney function – induced loss of protein and albumin	2% solution
	IV	–	5,360 mg/kg	LD_{50}; tremors, muscle weakness, dyspnea	–
	IV	1 month	200 mg/kg	Well tolerated	–
	SC	–	12 g/kg	LD_{50}	100% solution
	PO (gavage)	7 days	5 mL/kg/day	Well tolerated	–

(*Continued*)

TABLE A.33 (*Continued*)

Dimethylsulfoxide (DMSO)

Species	Route	Duration	Dose	Adverse Reactions/Toxicity	Notes
	PO (gavage)	4 weeks	5 mL/kg/day	Well tolerated	–
	PO (gavage)	–	16.0-28.3 g/kg	LD_{50}	100% solution
	PO (gavage)	–	14.5 g/kg	LD_{50}	100% solution
	PO (gavage)	Single dose	200 mg/kg	Did not affect stomach emptying, did not reduce intestinal transit time	2% solution
	PO (gavage)	Single dose	1000 mg/kg	Did not affect stomach emptying, did not reduce intestinal transit time	10% solution
	IP	28 doses	5 mL/kg	Well tolerated	15% solution
	IP	–	6.5–13.621 g/kg	LD_{50}	100% solution
	IP	–	8.2 g/kg	LD_{50}	100% solution

TABLE A.34

Dulbecco's Modified Phosphate Buffered Saline (PBS)

Species	Route	Duration	Dose	Adverse Reactions/Toxicity	Notes
Rat	IV (into tail vein)	1 month	1 mL/kg/day	Well tolerated	–

TABLE A.35

Dulbecco's Phosphate Buffered Saline (PBS)

Species	Route	Duration	Dose	Adverse Reactions/Toxicity	Notes
Rat	PO (gavage)	1 month	0.1, 0.8 and 1.2 mg/kg/day	Well tolerated	–

TABLE A.36

Ethanol (EtOH)

Species	Route	Duration	Dose	Adverse Reactions/Toxicity	Notes
Dog	IV	–	5 mL/kg at a rate of 0.3 mL/kg	Hemolysis *in vitro* dog blood	30% solution
	IV	–	5 mL/kg at a rate of 0.3 mL/kg	Hemolysis *in vitro* dog blood	40% solution
	IV	Single dose	1 mL/kg	Well tolerated	10% solution
	IV	Single dose	1 mL/kg	CNS depression, ataxia	30% solution
	IV	5 days	1 mL/kat 2 mL/minute	Excessive salivation	30% solution; MUST use winged infusion set
	IV	At 3, 7, 14, and 24 hours after ingestion of ethylene glycol	1584 mg/kg	Remained recumbent or severely ataxic for 36, depressed for 72	20% solution

(*Continued*)

TABLE A.36 (*Continued*)

Ethanol (EtOH)

Species	Route	Duration	Dose	Adverse Reactions/Toxicity	Notes
	IV	Every 4 hours for 5 treatments then every 6 hours for 4 treatments	5.5 mL/kg	–	20% solution
	PO	1 month	5 mL/kg	Well tolerated	7.5% solution
	PO	6 months	400 mL/kg	Hepatotoxicity, myopathy; CNS changes	–
	PO	90 days	5 mL/kg	Well tolerated	5% solution
Minipig	Dermal	ADME	nd (5 mg/cm2)	Well tolerated	60/40: purified water/ethyl alcohol absolute: v/v
Mouse	Dermal	13 weeks	100 µL/animal/day	Well tolerated	70% (62% m/m)
	Dermal	7 days	0.5 mL (fixed vol) QD	Well tolerated	80% solution; age 6 weeks; ♂/♀
	IP	Acute	5 mL/kg	Well tolerated	5% solution
	IV	Range finding	MTD: 986 mg/kg; NOEL: 197 mg/kg	–	Published LD_{50} = 1.6–4.3 g/kg
	IV (into tail vein)	Acute	197 mg/kg (dose vol 5 mL/kg)	NOEL	5% solution

(*Continued*)

TABLE A.36 (*Continued*)

Ethanol (EtOH)

Species	Route	Duration	Dose	Adverse Reactions/Toxicity	Notes
	IV (into tail vein)	Acute	986 mg/kg (dose vol 5 mL/kg)	MTD: ventral recumbency and "swimming" behavior immediately post-dose, ataxia for up to 6 min is typical	25% solution
	PO	1 month	2.5 mL/kg	Well tolerated	5% solution
	PO	6 months	2500 gm/kg	Well tolerated	–
Nonhuman Primate	PO	9 months	250 gm/kg	Behavioral changes	–
Rat	Dermal	ADME	nd (5 mg/cm2)	Well tolerated	60/40: purified water/ethyl alcohol absolute: v/v
	Dermal	91 days	0.5 mL (fixed vol) QD	Well tolerated	80% solution; age 6 weeks; ♂/♀
	IP	Acute	3.75 g/kg	LD_{50}	–
	IV	Acute	1.44 g/kg	LD_{50}	–
	IV	Acute	5 mL/kg at a rate of 0.3 mL/kg	Hematuria	30% solution
	IV	9 days	5 mL/kg at 2 mL/minute	Ataxia, respiratory depression, and death if dosed faster than 2mL/minute	30% solution
	IV	12 months	250 g/kg	Nephrosis, ATN, bladder changes, weight loss	–
	PO	Acute	7.06 g/kg	LD_{50}	–
	PO (gavage)	7 days	0.8, 2 and 5 mL/kg/day	Well tolerated	70% (62% m/m)
	PO	7 days	10 mL/kg	Well tolerated	10% solution

(*Continued*)

TABLE A.36 (*Continued*)
Ethanol (EtOH)

Species	Route	Duration	Dose	Adverse Reactions/Toxicity	Notes
	PO	Acute	5 mL/kg	Depression	–
	PO (gavage)	4 weeks	2 mL/kg QD	Hypokinesia, dyspnea regurgitation, distended lungs/ ileum and swollen abdomen	70% (62% m/m)
	PO	28 doses	175 g/kg	Depression, decreased RBC	–
	PO	90 days	8 mL/kg	Well tolerated	10% solution
	PO	12 months	1000 mg/kg	Fatty liver	–

TABLE A.37
Gelatin Capsules

Species	Route	Duration	Dose	Adverse Reactions/Toxicity	Notes
Dog	PO	5 days	QD	Well tolerated	–
	PO	6 days	QD	Well tolerated	–
	PO	8 days	QD	Well tolerated	–
	PO	14 days	QD	Well tolerated	–
	PO	16 days	QD	Well tolerated	–

TABLE A.38
Gelatin Phosphate Buffer

Species	Route	Duration	Dose	Adverse Reactions/Toxicity	Notes
Minipig	Topical	28 days	10 mL q14d × 2 doses	None	GLP; age 4–6 months; 5♂/5♀

TABLE A.39
Gelucire 44/14™

Species	Route	Duration	Dose	Adverse Reactions/Toxicity	Notes
Dog	PO	3 months	400, 1000, 2500 mg/kg/day	NOAEL: >2500 mg/kg/day	–
	PO	14 days	400, 1000, 2500 mg/kg/day	NOAEL: >2500 mg/kg/day	–
Rabbit	Dermal	Acute	0.5 mL	Not irritant	–
	Ocular	Acute	0.1 mL	Slight irritant	–
Rat	PO	28 days	600, 1500, 2400 mg/kg/day	NOEL: 2400 mg/kg/day	–
	PO	7 days	600, 1500, 2400 mg/kg/day	NOEL: 2400 mg/kg/day	–
	PO	Acute	No dilution	LD_{50}: >2004 mg/kg/day	–

TABLE A.40
Gelucire 50/13™

Species	Route	Duration	Dose	Adverse Reactions/Toxicity	Notes
Rat	PO	Acute	–	Well tolerated	No dilution; $LD_0 \geq$ 20,000 mg/kg/day

TABLE A.41

Gluconic Acid

Species	Route	Duration	Dose	Adverse Reactions/Toxicity	Notes
Dog	PO (gavage)	14 days	2 mL/kg QD	Well tolerated	0.3M gluconic acid pH 3.0; age 11 months; ♂/♀
Rat	PO (gavage)	14 days	10 mL/kg QD	Well tolerated	0.3M gluconic acid pH 3.0; age 10 weeks; ♂/♀

TABLE A.42

Glycerol

Species	Route	Duration	Dose	Adverse Reactions/Toxicity	Notes
Dog	SC	28 days	20 mL/day (fixed volume)	None	2% solution in sterile water; age 5–6 months; ♂/♀
Guinea Pig	PO	1 month	500 mg/kg	Well tolerated	–
Mouse	IV	1 month	100 mg/kg	Well tolerated	–
	IP	1 month	250 mg/kg	Well tolerated	–
	PO	90 days	500 mg/kg	Depression and reduced respiration	–
	SC	Acute	10 mg/kg	Well tolerated	–
Rabbit	IV	Acute	10 mg/kg	Well tolerated	–
Rat	PO	Acute	1000 mg/kg	Well tolerated	–
	PO	1 month	15 g/kg	Reduced adrenal weights	–
	PO	1 month	1000 mg/kg	Well tolerated	–
	SC	Acute	10 mg/kg	Well tolerated	–

TABLE A.43
Glycofurol

Species	Route	Duration	Dose	Adverse Reactions/Toxicity	Notes
Dog	IV	Single dose	1 mL/kg	Well tolerated	50% solution

TABLE A.44
Gum Tragacanth

Species	Route	Duration	Dose	Adverse Reactions/Toxicity	Notes
Mouse	PO (gavage)	2 weeks	10 mL/kg QD	Well tolerated	In distilled water, 0.5%

TABLE A.45
Gum Xanthane

Species	Route	Duration	Dose	Adverse Reactions/Toxicity	Notes
Rabbit	PO (gavage)	Tolerance	3 mL/kg/day	Well tolerated	0.4% aqueous solution
	PO (gavage)	Segmt. II	3 mL/kg/day	Well tolerated	0.4% aqueous solution

TABLE A.46

Hydrochloric Acid (HCl)

Species	Route	Duration	Dose	Adverse Reactions/Toxicity	Notes
Dog	PO (gavage)	4 weeks	4 mL/kg QD	None	0.1 N; age 5–6 months; ♂/♀
	PO (gavage)	Daily	10 mL/kg	None	0.1%–10% in water; beagle age 5 months; ♂/♀
Rat	PO (gavage)	26 weeks	10 mL/kg QD	None	0.05 M HCl; Age 6 weeks; ♂/♀
	PO (gavage)	Daily	10 mL/kg	None	0.1%–10% in water; Age 6–8 weeks; ♂/♀

TABLE A.47

Hydroxyethyl Cellulose

Species	Route	Duration	Dose	Adverse Reactions/Toxicity	Notes
Rat	PO	Single dose	20 mg/kg	Intestinal transit was slightly enhanced, not dose dependent	0.5%
	PO	Single dose	100 mg/kg	Intestinal transit was slightly enhanced, not dose dependent	1%
	PO	28 days	50 mg/kg	Easiest and most tolerable formulation for PO administration	–

TABLE A.48

Hydroxypropyl-β-Cyclodextrin (HPβCD)

FDA has informed sponsors that this excipient is no longer permitted for use in subcutaneous clinical formulations due to perceived risk of carcinogenicity[354] and has advised that it not be so used in repeat dose nonclinical safety studies.

Species	Route	Duration	Dose	Adverse Reactions/Toxicity	Notes
Dog	IV	1 dose	1 mL/kg	Well tolerated	40% solution
	IV (slow bolus)	1 dose	1.2 mL/kg	Well tolerated	6% solution
	IV (2-hr infusion)	1 month	10 mL/kg	Well tolerated	40% solution
	Intranasal	14 days	1 mL/nostril TID 2 hours apart	Well tolerated	45%; beagle dogs (Marshall) age ~ 6 months; ♂/♀
	PO (gavage)	1 dose	10 mL/kg	Well tolerated	6% solution
	PO (gavage)	90 days	Dose vol 5 mL/kg; daily dose 500, 1000 mg/kg	Loose/soft stools in high dose group	Dose concentration 100, 200 mg/mL (respectively)
	PO (gelatin capsules)	28 doses	–	Emesis, fecal changes	6% solution
	SC	91 days	0.86 mL/kg QD	None	10% solution in sterile water; Age 5.5–6 months; ♂/♀
Mouse	IV	Acute	5000 mg/kg	LD	Administration directly into the vein by hypodermic needle; >5 gm/kg
	PO (gavage)	90 days	Dose vol 10 mL/kg; daily dose 500, 1000 mg/kg	Produced elevated transaminase (aspartate and alanine aminotransferase) levels; use with caution	Dose concentrations 50, 100 mg/mL (respectively)

(Continued)

TABLE A.48 (*Continued*)

Hydroxypropyl-β-Cyclodextrin (HPβCD)

FDA has informed sponsors that this excipient is no longer permitted for use in subcutaneous clinical formulations due to perceived risk of carcinogenicity[354] and has advised that it not be so used in repeat dose nonclinical safety studies.

Species	Route	Duration	Dose	Adverse Reactions/Toxicity	Notes
	PO (gavage)	104 weeks	500 mg/kg/day	Well tolerated	–
Nonhuman Primate	PO (gavage)	13 doses	5 mL/kg	None	11% solution
	PO (gavage)	90 days	Dose vol 5 mL/kg; Daily dose 500, 1000 mg/kg	Loose/soft stools in high dose group	Dose concentration 100, 200 mg/mL (respectively)
	IP	Acute	10,000 mg/kg	Well tolerated, LD > 10,000 mg/kg	–
Rabbit	PO (gavage)	12 doses	2 mL/kg	Well tolerated	11% solution
Rat	Intranasal	14 days	50 mL/nostril TID 2 hours apart	Well tolerated	45%; Sprague Dawley (Harlan) age ~8–10 weeks at initiation; ♂/♀; histopathology limited to purulent exudates (minimal to mild) involving the nasal turbinates
	IP injection	Single dose	1000 mg/kg	Increased glucose levels at 4 hr, minor transient changes for triglycerides and BUN, no functional changes were observed, only slight enhancement of ALT and AST	10% solution
	IV	Single dose	1 mL/kg	Well tolerated	20% solution

(*Continued*)

TABLE A.48 (*Continued*)

Hydroxypropyl-β-Cyclodextrin (HPβCD)

FDA has informed sponsors that this excipient is no longer permitted for use in subcutaneous clinical formulations due to perceived risk of carcinogenicity[354] and has advised that it not be so used in repeat dose nonclinical safety studies.

Species	Route	Duration	Dose	Adverse Reactions/Toxicity	Notes
	IV	Single dose	10 mL/kg	Death occurred within a few minute of receiving a bolus dose of the vehicle. The rate of administration was slowed to ~2 min, which was tolerated and clinical observations limited to red urine	45%; Sprague Dawley (Harlan); age ~8 weeks; ♂/♀
	IV (slow bolus)	10 doses	4 mL/kg	Well tolerated	12.5% solution
	IV (1-hr infusion)	1 month	10 mL/kg	Well tolerated	40% solution
	IV (slow bolus)	–	2 mL/kg	Well tolerated	12.5% solution
	PO	Acute	5,000 mg/kg	LD_{50} > 5,000 mg/kg	–
	PO	Single dose	up to 2000 mg/kg	No effect on gastric emptying, modestly inhibited intestinal transit	20% solution
	PO (gavage)	Single dose	10 mL/kg	Well tolerated	6% solution
	PO (gavage)	10 doses	10 mL/kg	Well tolerated	11% solution
	PO (gavage)	4 weeks	10 mL/kg QD	None	20%; Age 6 weeks; ♂/♀

TABLE A.48 (*Continued*)

Hydroxypropyl-β-Cyclodextrin (HPβCD)

FDA has informed sponsors that this excipient is no longer permitted for use in subcutaneous clinical formulations due to perceived risk of carcinogenicity[354] and has advised that it not be so used in repeat dose nonclinical safety studies.

Species	Route	Duration	Dose	Adverse Reactions/Toxicity	Notes
	PO (gavage)	90 days	Dose vol 5 mL/kg; daily dose 500, 1000 mg/kg	Produced elevated transaminase (aspartate and alanine aminotransferase) levels; use with caution	Dose concentrations 100, 200 mg/mL (respectively)
	PO (gavage)	2 years	500 mg/kg	No effects	–
	SC	28 days	QD	Erosion/ulceration and/or necrosis generally observed. At end of 28-day recovery period, fibrosis and subacute/chronic inflammation at the injection sites persisted but were resolving; injection site necrosis not observed	10% HPβCD
	SC	91 days	QD	At end of 28-day recovery period, fibrosis and subacute/chronic inflammation at the injection sites persisted but were resolving; injection site necrosis not observed	10% HPβCD
	SC	91 days	1.14 mL/kg QD	None	10% solution in sterile water; age 7 weeks; ♂/♀

(*Continued*)

TABLE A.49
Hydroxypropyl Cellulose

Species	Route	Duration	Dose	Adverse Reactions/Toxicity	Notes
Rat	PO	90 day	1000 gm/kg	Well tolerated	–

TABLE A.50
Hydroxypropyl Methylcellulose (HPMC)

Species	Routes	Duration	Dose	Adverse Reactions/Toxicity	Notes
Dog	IP	Acute	200 mg/kg	Well tolerated	
	PO (gavage)	–	5 mL/kg	None	1% (methocel E5 premium LV, 5cp) in DI water; age 6.5–7 months; ♂/♀
	PO (gavage)	90 days	Dose vol 5 mL/kg, daily dose 20 mg/kg (0.4% w/v)	Well tolerated	
Minipig	PO (gavage)	7 days	5 mL/kg QD	None	0.5% in distilled water: age 6.5–11.5 months; ♂/♀
Mouse	IP	Single dose	5 mL/kg	Well tolerated	0.5%
	IP	Acute	50 mg/kg	Well tolerated	–
	PO	5 days	20 mL/kg BID	None	0.5%; Non-GLP; age 9 weeks; 6♀
	PO (gavage)	Single dose	10 mL/kg	Well tolerated	0.5%
	PO (gavage)	10 doses	10 mL/kg	Well tolerated	0.2%

TABLE 50 (*Continued*)

Hydroxypropyl Methylcellulose (HPMC)

Species	Routes	Duration	Dose	Adverse Reactions/Toxicity	Notes
	PO (gavage)	90 days	Dose vol 10 mL/kg, daily dose 20 mg/kg (0.4% w/v)	Well tolerated	–
Nonhuman Primate	PO (gavage)	28 days	2 mL/kg QD	None	0.5% in distilled water; age 2–3 years; ♂/♀
	PO (gavage)	28 days	5 mL/kg QD	Soft feces (non-adverse)	1% (Methocel E5 premium LV, 5cp) in DI water; age 2.5–3.5 years; ♂/♀
	PO (gavage)	90 days	Dose vol 5 mL/kg, Daily dose 20 mg/kg (0.4% w/v)	Well tolerated	–
	PO (gavage)	91 days	5 mL/kg QD	None	0.2% in distilled water; cynomolgulous monkeys age 2 years; ♂/♀
Rat	IP	Single dose	5 mL/kg	Well tolerated	0.5%
	PO (gavage)	Single dose	10 mL/kg	Well tolerated	0.2%
	PO (gavage)	Single dose	10 mL/kg	Well tolerated	0.5%
	PO (gavage)	90 days	Dose vol 5 mL/kg, Daily dose 20 mg/kg (0.4% w/v)	Well tolerated	–
	PO (gavage)	91 days	5 mL/kg QD	None	1% (Methocel E5 premium LV, 5cp) in DI water; age 6 weeks; ♂/♀
	PO (gavage)	Up to 104 weeks	10 mL/kg QD	None	0.2% in distilled water; age 6–11 weeks; ♂/♀
	PO (gavage)	182 days	10 mL/kg QD	None	0.5% in distilled water; age 6 weeks; ♂/♀

(*Continued*)

TABLE A.51

Hypotonic Phosphate Buffered Saline (PBS)

Species	Route	Duration	Dose	Adverse Reactions/Toxicity	Notes
Dog	IV	2 days	2 mL/kg SD	None	GLP; age 6 months; 5♂/5♀
Rat	IV	2 days	2 mL/kg SD	None	GLP; age ≥8 weeks; 10♂/10♀

TABLE A.52

Isopropyl Alcohol

Species	Route	Duration	Dose	Adverse Reactions/Toxicity	Notes
Rabbit	Dermal	1 month	1000 gm/kg	Well tolerated	–

TABLE A.53

Isopropyl Myristate

Species	Route	Duration	Dose	Adverse Reactions/Toxicity	Notes
Rabbit	Dermal	1 month	500 mg/kg	Well tolerated	–

TABLE A.54

Kolliphor EL™

Species	Route	Duration	Dose	Adverse Reactions/Toxicity	Notes
Dog	IV	1 month	2 mL/kg	Well tolerated	–
Rat	PO	1 month	100 mg/kg	Well tolerated	–

TABLE A.55
Kolliphor ELP™

Species	Route	Duration	Dose	Adverse Reactions/Toxicity	Notes
Dog	–	Acute	–	Anaphylaxis	Up to 50%; never use for dog studies!

TABLE A.56
Kolliphor RH40™

Species	Route	Duration	Dose	Adverse Reactions/Toxicity	Notes
Dog	IV	1 month	2 mL/kg	Well tolerated	–

TABLE A.57
Labrafil M1944™

Species	Route	Duration	Dose	Adverse Reactions/Toxicity	Notes
Dog	PO	1 month	2 mg/kg	Well tolerated	–
Rabbit	Dermal	Acute/PDI		Non-irritant	No dilution; 0.38 dermal irritation index
Rat	PO	Acute	20 g/kg	Well tolerated	No dilution; LD_0 ≥20,000 mg/kg/day

TABLE A.58

Labrasol™

Species	Route	Duration	Dose	Adverse Reactions/Toxicity	Notes
Dog	PO	14 day	100, 300, 1000 and 3000 mg/kg/day	In high dose group, moderate suppurative inflammation of the lungs. No adverse effects on survival and clinical observations	–
	PO	13 week	0, 300, 1000 and 3000 mg/kg/day	NOEL: 1000 mg/kg/day; NOAEL: 3000 mg/kg/day	–
Rabbit	Dermal	Patch test	0.5 mL	Well tolerated	–
	Ocular	Acute	0.1 mL	Slight irritant	–
Rat	Dermal	Acute		Very well tolerated	–
	Dermal	Patch test	0.02 mL/animal	Well tolerated	–
	IV	ADME	10 mg/kg/day		–
	Ocular			Slight irritant	–
	PO	Acute	20, 22.4, 25.1, 28.21 and 31.60 g/kg	$LD_{50} > 22$ g/kg; well tolerated	–
	PO	ADME	10, 150 mg/kg/day	Well tolerated	–
	PO	Segment II: Embryofetal development	1000, 2000 or 3000 mg/kg/day	NOEL: 3000 mg/kg/day with no indication of a teratogenicity	–
	PO	14 day	100, 300, 1000, 3000 mg/kg/day	NOAEL: 3000 mg/kg/day	–
	PO	6 months	300, 1000 and 3000 mg/kg/day	NOEL: 300 mg/kg/day; NOAEL: 3000 mg/kg/day	–

TABLE A.59

Lactose

Species	Route	Duration	Dose	Adverse Reactions/Toxicity	Notes
Nonhuman Primate	Inhalation	2 weeks	1 L/min/animal	Well tolerated	Lactose 200 μM; anhydrous

TABLE A.60

Lanolin

Species	Route	Duration	Dose	Adverse Reactions/Toxicity	Notes
Rabbit	Dermal	90 days	1000 mg/kg	Well tolerated	–

TABLE A.61

Lauroglycol 90™

Species	Route	Duration	Dose	Adverse Reactions/Toxicity	Notes
Rabbit	Dermal	Acute	No dilution	Moderately irritant	–
	Ocular	Acute	No dilution	Slightly irritant	–
Rat	PO	Acute	–	LD_{50}: >2003 mg/kg/day	–

TABLE A.62

Maltitol Solution

Species	Route	Duration	Dose	Adverse Reactions/Toxicity	Notes
Rat	IP	1 month	500 mg/kg	Well tolerated	–

TABLE A.63
Maltol

Species	Route	Duration	Dose	Adverse Reactions/Toxicity	Notes
Guinea Pig	PO	1 month	75 mg/kg	Well tolerated	–
Rabbit	PO	1 month	100 mg/kg	Well tolerated	–

TABLE A.64
Mannitol

Species	Route	Duration	Dose	Adverse Reactions/Toxicity	Notes
Minipig	SC	Tolerance	0.2 mL/kg/day	Well tolerated	5% solution
Nonhuman Primate	PO (gavage)	2 sem	10 mL/kg/day	Well tolerated	–
Rabbit	IV	ADME	0.8 mL/kg	Well tolerated	–
	PO	ADME	1.6 mL/kg	Well tolerated	–

TABLE A.65
Methylcellulose

Species	Route	Duration	Dose	Adverse Reactions/Toxicity	Notes
Dog	IV		40 mL	Anemia, decreased WBC, increased sedimentation rate in 24 hrs	0.7–2.8% solution
	PO	14 days	5.0 mL/kg QD	None	GLP; age 6 months; 6♂/6♀
	PO (dietary)	90 days	6%	Well tolerated	–

(Continud)

TABLE A.65 (*Continued*)

Methylcellulose

Species	Route	Duration	Dose	Adverse Reactions/Toxicity	Notes
	PO (gavage)	14 doses	10 mL/kg/dose	Well tolerated	0.5%
	PO (gavage)	39 weeks	5 mL/kg QD	Soft/mucoid feces	0.5% solution (400 cps) in DI water; age 6–10.5 months; ♂/♀
Guinea Pig	PO	12 doses	4 mL/kg	Well tolerated	0.5%
	Topical	3 weeks	0.4 mL Once/week	None	0.5% solution (400 cps) in DI water; age 2–3 months; ♂/♀
Mouse	PO	90 days	10 mL/kg	Well tolerated	0.5%
Nonhuman Primate	IV	30-minute infusion in a single dose	1 mL/kg	None	0.5% solution (400 cps) in DI water; ♂/♀
	PO (gavage)	14 doses	5 mL/kg	Well tolerated	0.5%
	PO	1 month	10 mL/kg/dose	Well tolerated	0.5%
	PO (gavage)	28 doses	5 mL/kg/dose	Well tolerated	1%
	PO (gavage)	28 doses	10 mL/kg	Well tolerated	0.1%
Rabbit	PO (gavage)	12 doses	4 mL/kg	Well tolerated	0.5%
Rat	PO (dietary)	90 days	10%	Well tolerated	–
	PO (gavage)		1020 mg/kg	NOAEL	–
	PO (gavage)	Single dose	10 mL/kg	None	1%
	PO (gavage)	Single dose	10 mL/kg	None	0.5%
	PO	Single dose	10 mL/kg	Well tolerated	2%
	PO	3 days	10 mL/kg QD	None	1% solution in water; non-GLP; age 11 weeks; 5♀
	PO (gavage)	5 doses	10 mL/kg	None	0.5%

(*Continud*)

TABLE A.65 (*Continued*)

Methylcellulose

Species	Route	Duration	Dose	Adverse Reactions/Toxicity	Notes
	PO (gavage)	9 doses	10 mL/kg/day	None	0.5%
	PO (gavage)	14 days	10 mL/kg QD	Well tolerated	1%
	PO	1 month	10 mL/kg	Well tolerated	0.5%
	PO	1 month	5 mL/kg	Well tolerated	0.5%
	PO (gavage)	28 doses	5 mL/kg/dose	None	1%
		2 year	120 mg/kg	Well tolerated	1%
	PO (gavage)	182 days	10 mL/kg BID	None	0.5% solution (400 cps) in DI water; age 6 weeks; ♂/♀

TABLE A.66

Methylpyrrolidone

Species	Route	Duration	Dose	Adverse Reactions/Toxicity	Notes
Dog	IV	Single dose	0.25 mL/kg	Well tolerated	50% solution
Mouse	IV	Range finding	MTD: 1.3 g/kg; NOEL: 257 mg/kg	–	Published LD_{50} = 54–36000 mg/kg
	IV (into tail vein)	Acute	257 mg/kg (dose vol 5 mL/kg)	NOEL	5% solution
	IV (into tail vein)	Acute	1285 mg/kg (dose vol 5 mL/kg)	MTD; struggling and vocalization at dosing, rapid breathing, stiff tail and splayed limbs immediately post dose. Hypoactivity for up to 15 mins is typical.	25% solution

TABLE A.67

Mineral Oil

Species	Route	Duration	Dose	Adverse Reactions/Toxicity	Notes
Cat	Topical	2x/wk × 2 doses for 28 days	1.15 mL total: 0.35 mL, then 0.4 mL every 60 min after initial application for 2 doses	None	GLP; age 54–57 days; 6♂/6♀
	Topical	q14d for 56 days	2.1 mL total: 0.5 mL, then 0.4 mL every 30 min after initial application for four applications	None	GLP; age 9 weeks; 6♂/6♀
Dog	PO	1 month	2.5 mL/kg	Well tolerated	–
Guinea Pig	Topical	28 days	0.4 mL q7d	None	GLP; age 6 weeks; 11♀
Mouse	PO	1 month	250 mg/kg	Well tolerated	–
Rat	PO	1 month	5 mL/kg	Well tolerated	–

TABLE A.68

Olive Oil

Species	Route	Duration	Dose	Adverse Reactions/Toxicity	Notes
Rat	PO (gavage)	28 days	10 mL/kg QD	Well tolerated	Age 6 weeks; ♂/♀

TABLE A.69
Peanut Oil

Species	Route	Duration	Dose	Adverse Reactions/Toxicity	Notes
Rat	PO	1 month	10 g/kg	Well tolerated	–
	PO	12 month	10 g/kg	Well tolerated	–
	PO	90 day	10 g/kg	Well tolerated	–
	PO (gavage)	–	5 mL/kg/day	Well tolerated	–
	SC	–	2 mL/kg/day	Well tolerated	–

TABLE A.70
PEG 200

Species	Route	Duration	Dose	Adverse Reactions/Toxicity	Notes
Nonhuman Primate	PO (gavage)	14 days	5 mL/kg QD	Soft/watery feces (non-adverse)	Age 2–3.5 years; ♂/♀
Rabbit	IV	Acute	>10 g/kg bw	LD_{50}	–
Rat	IP	–	8–9 g/kg bw	LD_{50}	–
	PO (dietary)	90 days	20, 40, 80, 160, 240 g/kg diet QD (1.5, 3. 6. 12. 18 g/kg bw QD)	NOAEL = 6 g/kg bw QD	At 12, 18 g/kg bw liver and liver/kidney weights increased (respectively); 5♂/5♀
	PO (drinking water)	90 days	4.8 g/kg bw QD	NOAEL	5♂

(*Continued*)

TABLE A.70 (*Continued*)

PEG 200

Species	Route	Duration	Dose	Adverse Reactions/Toxicity	Notes
	PO (drinking water)	90 days	10.9 g/kg bw QD	66% mortality; decreased body weight gain	5♂

TABLE A.71

PEG 300

Species	Route	Duration	Dose	Adverse Reactions/Toxicity	Notes
Dog	PO (gavage)	28 days	1 mL/kg BID	None	Age 5.5–6 months; ♂/♀
Guinea Pig	IV	1 month	1 mL/kg	Well tolerated	–
Mouse	PO (gavage)	ADME	10 mL/kg/day	Well tolerated	–
Rabbit	PO	1 month	500 gm/kg	Well tolerated	–
	IV	Acute	>10 g/kg bw	LD_{50}	–
Rat	IP	Acute	16–18 g/kg bw	LD_{50}	–
	IP	Acute	17 g/kg bw	LD_{50}	–
	IV	Acute	8 g/kg bw	LD_{50}	–
	IV	Acute	7.1 g/kg bw	LD_{50}	–

(Continued)

TABLE A.71 (*Continued*)

PEG 300

Species	Route	Duration	Dose	Adverse Reactions/Toxicity	Notes
	PO (dietary)	90 days	20, 40, 80, 160, 240 g/kg diet QD (1.5, 3, 6, 12, 18 g/kg bw QD)	NOAEL = 3 g/kg bw QD	At 6, 12, 18 g/kg bw decreased body weight gain, liver and kidney weight increase, and both decreased body weight gain and increased liver weight (respectively); 5♂/5♀
	PO (drinking water)	90 days	5.4 g/kg bw QD	NOAEL	5♂
	PO (drinking water)	90 days	20.5 g/kg bw QD	66% mortality; decreased body weight gain; liver and kidney changes	5♂
	PO (gavage)	2 weeks	7.5 mL/kg once, or 2.5 mL/kg TID	Well tolerated	50%
	PO (gavage)	4 weeks	5 mL/kg/day	Well tolerated	50%
	PO (gavage)	28 days	2 mL/kg BID	None	Age 8 weeks; ♂/♀

TABLE A.72
PEG 400

Species	Route	Duration	Dose	Adverse Reactions/Toxicity	Notes
Dog	IV	Single dose	Total Dose <2g/kg	–	100%
	IV	28 days	2–3 g/kg	Decreased blood pressure and reversible depression in respiration. These symptoms increased at doses of 3 g/kg or greater and eventually resulted in complete respiratory arrest. At necropsy dogs were found to have pulmonary edema and small infarcts in the lungs, but no changes in the heart or kidneys.	100%
	IV	–	–	–	NS solution; hemolysis occurs > 33% (v/v) concentration
	IV	Single dose	1 mL/kg	Well tolerated	30%
	IV	30 days, + 21 day recovery	8.45 g/kg QD	Dry mouth and dry nasal mucous membrane; histopathological change in kidney; reversible cloudy swelling of kidney cells and increased glomerular volume	NS solution 25% (v/v)
	IV	30 days, + 21 day recovery	5 mL/kg/day	Dry mouth and dry nasal mucous membrane	NS solution 25% (v/v)
	IV	30 days, + 21 day recovery	4.23 g/kg QD	Increased values of electrolytes (Na+ and Cl–)	NS solution 25% (v/v)

(Continued)

TABLE A.72 (*Continued*)

PEG 400

Species	Route	Duration	Dose	Adverse Reactions/Toxicity	Notes
	IV (bolus)	–	1 mL/kg	Well tolerated	60% solution in water pH 3–11
	IV (infusion)	–	2 mL/kg	Well tolerated	60% solution in water pH 3–11
	IV (infusion)	–	0.5 mL/kg	Well tolerated	80% solution in water pH 3–11
	PO (dietary)	1 year	20 g/kg diet QD (0.5 g/kg bw QD)	NOAEL = 0.5 g/kg bw QD	3♂/1♀
	PO (gavage)	28 days	0.25 mL/kg QD	Soft/watery feces (non-adverse)	Age 5–5.5 months; ♂/♀
Guinea Pig	PO	1 month	1000 mg/kg	Well tolerated	–
Minipig	Dermal	2 weeks	2.5 mL/kg	Well tolerated	–
	Topical	90 days	2 mL/kg QD	Mild dose-site inflammation after >30 days of administration	GLP; age 4–6 months; 6♂/6♀
Mouse	IP	3 days	10 mL/kg	Well tolerated	35%
	IP	1 month	2.5 mL/kg	Well tolerated	40% solution
	IP	28 doses	500 mg/kg	Well tolerated	–
	IP	–	14.5 g/kg bw	LD_{50}	–
	IP	–	9.2 g/kg bw	LD_{50}	–
	IV	–	8.6 g/kg bw	LD_{50}	–
	IV	Range finding	MTD: 4.5 g/kg; NOEL: 1.7 g/kg	–	Published LD_{50} = 8.6–9.7 g/kg
	IV (into tail vein)	Acute	1692 mg/kg (dose vol 5 mL/kg)	NOEL	30% solution

(*Continued*)

TABLE A.72 (*Continued*)

PEG 400

Species	Route	Duration	Dose	Adverse Reactions/Toxicity	Notes
	IV (into tail vein)	Acute	4512 mg/kg (dose vol 5 mL/kg)	MTD; tremors, ventral recumbency and splayed limbs shortly after dosing, hypoactivity for up to 12 min is typical	80% solution
	PO (gavage)	4 weeks	10 mL/kg/day	Well tolerated	–
	PO	13 weeks	5 mL/kg BID	Well tolerated	100%; CD-1 (Harlan); age ~7 weeks at study initiation; ♂/♀
Nonhuman Primate	PO (gavage)	28 days	1 mL/kg QD	Soft/watery feces (non-adverse)	In DI water; age 1–2 years; ♂/♀
	PO (gavage)	28 days	5 mL/kg QD	Soft/watery feces (non-adverse)	In DI water; age 1–2 years; ♂/♀
Rabbit	IV	–	>10 g/kg bw	LD_{50}	–
Rat	Dermal	13 weeks	2.5 mL/kg/day	Well tolerated	–
	Dermal	104 weeks	2.5 mL/kg/day	Well tolerated	–
	IP	–	14.7 g/kg bw	LD_{50}	–
	IP	–	12.3 g/kg bw	LD_{50}	–
	IP	1 month	5 mL/kg	Well tolerated	35%
	IV	–	4.7 g/kg bw	LD_{50}	–
	IV	Single dose	0.5 mL/kg	Well tolerated	–
	IV (bolus)	–	2 mL/kg	Well tolerated	60% solution in water pH 3–11
	IV (infusion)	–	5 mL/kg	Well tolerated	60% solution in water pH 3–11

(Continued)

TABLE A.72 (*Continued*)

PEG 400

Species	Route	Duration	Dose	Adverse Reactions/Toxicity	Notes
	PO (gavage)	–	2 mL/kg	Well tolerated	–
	PO (gavage)	Single dose	5 mL/kg	Well tolerated	–
	PO (gavage)	10 doses	1.67 mg/kg	Well tolerated	–
	PO	1 month	5 mL/kg	Well tolerated	–
	PO (gavage)	4 weeks	5 mL/kg/day	Well tolerated	–
	PO (gavage)	28 days	3 mL/kg QD	None	Age 6 weeks; ♂/♀
	PO (gavage)	28 days	10 mL/kg QD	None	In DI water; age 6 weeks; ♂/♀
	PO	13 weeks	10 mL/kg/day	Loose feces and decreased food consumption; increased water consumption; increases in relative kidney weights; reversible renal toxicity	–
	PO	26 weeks	5 mL/kg BID	Well tolerated	100%; Sprague Dawley (Harlan) age ~8 weeks at initiation; ♂/♀
	PO (dietary)	90 days	20, 40, 80, 160, 240 g/kg diet QD (1.5, 3, 6, 12, 18 g/kg bw QD)	NOAEL = 6 g/kg bw QD	At 12, 18 g/kg bw decreased body weight gain and liver and kidney weight increase (respectively); 5♂/5♀

(*Continued*)

TABLE A.72 (*Continued*)

PEG 400

Species	Route	Duration	Dose	Adverse Reactions/Toxicity	Notes
	PO (dietary)	2 years	10, 20, 40, 80 g/kg diet QD (0.75, 1.5, 3, 6 g/kg bw QD)	NOAEL = 1.5 g/kg bw QD	From 3 g/kg bw decreased body weight gain (male); 20♂/20♀
	PO (drinking water)	90 days	4.8 g/kg bw QD	NOAEL	5♂
	PO (drinking water)	90 days	16.4 g/kg bw QD	66% mortality, decreased body weight gain	5♂

TABLE A.73

PEG 600

Species	Route	Duration	Dose	Adverse Reactions/Toxicity	Notes
Rat	IP	–	14.1 g/kg bw	LD_{50}	–
	IV	–	7.7 g/kg bw	LD_{50}	–
	PO (dietary)	90 days	20, 40, 80, 160, 240 g/kg diet QD (1.5, 3, 6, 12, 18 g/kg bw QD)	NOAEL = 6 g/kg bw QD	From 12 g/kg bw decreased body weight gain and increased kidney weights; 5♂/5♀

TABLE A.74
PEG 810

Species	Route	Duration	Dose	Adverse Reactions/Toxicity	Notes
Rat	IV	–	13 g/kg bw	LD_{50}	–
	SC	–	16 g/kg bw	LD_{50}	–

TABLE A.75
PEG 1000

Species	Route	Duration	Dose	Adverse Reactions/Toxicity	Notes
Mouse	IP	–	2 g/kg bw	LD_{50}	–
Rabbit	IV	Acute	>10 g/kg bw	LD_{50}	–
Rat	IP	Acute	15.6 g/kg bw	LD_{50}	–
	PO (dietary)	90 days	20, 40, 80, 160, 240 g/kg diet QD (1.5, 3, 6, 12, 18 g/kg bw QD)	NOAEL = 6 g/kg bw QD	From 12 g/kg bw decreased body weight gain; 5♂/5♀

TABLE A.76
PEG 1500

Species	Route	Duration	Dose	Adverse Reactions/Toxicity	Notes
Rat	IP	–	17.7 g/kg bw	LD_{50}	–
	IV	–	8.5 g/kg bw	LD_{50}	–
	PO (dietary)	90 days	0.88, 4.05, 8.1, 22.9 g/kg bw QD	NOAEL = 2 g/kg bw QD	From 4.05 g/kg bw kidney damage
	PO (drinking water)	90 days	20, 40, 80, 160, 240 g/kg diet QD (1.5, 3, 6, 12, 18 g/kg bw QD)	NOAEL = 3 g/kg bw QD	From 6 g/kg decreased body weight gain; at 18 g/kg bw increased kidney weights; 5♂/5♀
	PO (drinking water)	2 years	0.2, 0.8, 4, 20 g/l QD (0.015, 0.059, 0.27, 1.69 g/kg bw QD)	1.69 g/kg bw QD: no effects on fertility, survival, hematology or histopathology	8♂/8♀

TABLE A.77
PEG 1540

Species	Route	Duration	Dose	Adverse Reactions/Toxicity	Notes
Dog	PO (dietary)	1 year	20 g/kg diet QD (0.5 g/kg bw QD)	NOAEL = 0.5 g/kg bw QD	–
Rabbit	IV	Acute	>10 g/kg bw	LD_{50}	–
Rat	IP	Acute	15.4 g/kg bw	LD_{50}	–

(*Continued*)

TABLE A.77 (*Continued*)

PEG 1540

Species	Route	Duration	Dose	Adverse Reactions/Toxicity	Notes
	PO (dietary)	90 days	20, 40, 80, 160, 240 g/kg diet QD (1.5, 3, 6, 12, 18 g/kg bw QD)	NOAEL = 3 g/kg bw QD	From 6 g/kg bw decreased body weight gain; from 18 g/kg bw increased kidney weights; 5♂/5♀
	PO (dietary)	2 years	0.2, 0.8, 4, 20, 40, 80 g/kg diet QD (0.015, 0.06, 0.3, 1.5, 3, 6 g/kg bw QD)	NOAEL = 3 g/kg bw QD	From 6 g/kg QD cloudy swelling in the liver; 35♂/35♀

TABLE A.78

PEG 4000

Species	Route	Duration	Dose	Adverse Reactions/Toxicity	Notes
Dog	PO (dietary)	1 year	20 g/kg diet QD (0.5 g/kg bw QD)	NOAEL = 0.5 g/kg bw QD	–
Mouse	IP		8.0 g/kg bw	LD_{50}	–
Rabbit	IV	Acute	>10 g/kg bw	LD_{50}	–
	PO (gavage)	5 weeks (6 days/wk)	5, 10, 20 g/kg bw QD	From 5 g/kg bw decreased body weight gain, decreased glycogen storage; from 20 g/kg bw decreased body weights	–

(*Continued*)

TABLE A.78 (*Continued*)

PEG 4000

Species	Route	Duration	Dose	Adverse Reactions/Toxicity	Notes
Rat	IP	Acute	11.6–13 g/kg bw	LD_{50}	–
	IP	–	9.7 g/kg bw	LD_{50}	–
	IV	–	7.5 g/kg bw	LD_{50}	–
	PO (dietary)	90 days	20, 40, 80, 160, 240 g/kg diet QD (1.5, 3, 6, 12, 18 g/kg bw QD)	NOAEL = 3 g/kg bw QD	From 6 g/kg bw decreased body weight gain; from 12 g/kg bw increased kidney weights; 5♂/5♀
	PO (dietary)	90 days	1.6 g/kg bw QD	NOAEL = 1.6 g/kg bw QD	–
	PO (dietary)	2 years	0.375, 0.75, 1.5, 3, 6 g/kg bw QD	NOAEL = 3 g/kg bw QD	At 6 g/kg bw decreased body weight gain; 20♂/20♀
	PO (drinking water)	90 days	0.04-19 g/kg bw QD	NOAEL = 0.8 g/kg bw QD	At 0.23 g/kg bw degeneration of the testis tubules, degenerated sperm; from 7 g/kg bw decreased body weight gain; at 19 g/kg bw kidney damage; 5♂
	PO (drinking water)	2 years	0.00085, 0.0036, 0.017, 0.062 g/kg bw QD	NOAEL = 0.062 g/kg bw QD	8♂/8♀

TABLE A.79

PEG 6000

Species	Route	Duration	Dose	Adverse Reactions/Toxicity	Notes
Rabbit	IV	Acute	>10 g/kg bw	LD_{50}	–
Rat	IP	Acute	6.8 g/kg bw	LD_{50}	–
	PO (dietary)	90 days	20, 40, 80, 160, 240 g/kg diet QD (1.5, 3, 6, 12, 18 g/kg bw QD)	NOAEL = 12 g/kg bw QD	At 18 g/kg bw kidney weights increased, decreased body weight gain; 5♂/5♀

TABLE A.80

PEG 10000

Species	Route	Duration	Dose	Adverse Reactions/Toxicity	Notes
Rat	IP	Acute	12.6 g/kg bw	LD_{50}	–
	PO (dietary)	90 days	1.6 g/kg bw QD	NOAEL	–

TABLE A.81

PEG 4000000

Species	Route	Duration	Dose	Adverse Reactions/Toxicity	Notes
Rat	IV	Acute	>10 g/kg bw	LD_{50}	–
	PO (dietary)	90 days	8.0, 18.4 g/kg bw QD	From 8 g/kg bw cloudy swelling in the renal tubules; at 18.4 g/kg bw decreased body weight gain (males), decreased relative liver weights	10♂/10♀
	PO (dietary)	2 years	Up to 2.76 g/kg bw QD	NOAEL = 2.76 g/kg bw QD	36♂/36♀
	PO (dietary)	2 years	Up to 0.56 g/kg bw QD	NOAEL = 0.56 g/kg bw QD	4♂/2♀

TABLE A.82

Petrolatum

Species	Route	Duration	Dose	Adverse Reactions/Toxicity	Notes
Rabbit	Dermal	1 month	1 gm/kg	Well tolerated	–

TABLE A.83

Phosphate Buffered Saline (PBS)

Species	Route	Duration	Dose	Adverse Reactions/Toxicity	Notes
Dog	IV (slow bolus)	28 days	2.5 mL/kg/dose (5 mL/kg/day) BID	Well tolerated	pH 7.2; age 5–6 months; ♂/♀
	1-hr infusion	1-hr infusion single dose	10 mL/kg	Well tolerated	0.01 M PBS (powder) in sterile water for injection; beagle dog; age 5 months; ♂/♀
	PO (gavage)	28 days	5 mL/kg QD	Well tolerated	pH 7.2; age 5–6 months; ♂/♀
	Topical	q7d × 2 doses for 7 weeks	1.4 mL/kg divided into 3 doses 60 minutes apart	Well tolerated	Non-GLP; age 7 weeks; 2♂/2♀
Minipig	IV	14 days	5 mL/kg QD	Well tolerated	GLP; pH 6.0; age 17–21 days; 5♂/5♀
	IV	14 days	1 mL/kg QD	Well tolerated	GLP; pH 6.0; age 5–8 months; 5♂/5♀
	PO	7 days	5 mL/kg BID	Well tolerated	GLP; age 4 days; 4♂/4♀
	PO	28 days	5 mL/kg BID	Well tolerated	GLP; pH 6.0; age 4–5 days; 10♂/10♀
Mouse	SC (infusion)	Continuous (24 hrs/day) infusion for 7 days	1.0 mL/hr (+0.15 mL/hr)	Well tolerated	pH 7.2; Age 10 weeks; ♂/♀

(Continued)

TABLE A.83 (*Continued*)

Phosphate Buffered Saline (PBS)

Species	Route	Duration	Dose	Adverse Reactions/Toxicity	Notes
	SC (bolus)	Every 2 days for 1 week then weekly for 26 weeks	10–11.83 mL/kg	Well tolerated	pH 7.2; age 6 weeks; ♂/♀
	SC	6 months	10 mL/kg	Well tolerated	–
Nonhuman Primate	IV (slow bolus)	28 days	10 mL/kg QD	Well tolerated	pH 7.2; age 2–3 years; ♂/♀
	PO (gavage)	2 weeks	10 mL/kg/dose	Well tolerated	–
	PO (gavage)	2 weeks	1.6 mL/kg	Well tolerated	–
	SC	1 week	0.2 mL/kg	Well tolerated	–
	SC	9 months	1 mL/kg	Well tolerated	–
Rabbit	IV	12 weeks	1.0 mL/kgq7d	Well tolerated	Non-GLP; pH 6.5; age 7–8 months; 9♂
Rat	IV	Single dose	1 mL/kg	Well tolerated	
	IV (slow bolus)	28 days	5 mL/kg Once weekly	Well tolerated	pH 7.2; age 6 weeks; ♂/♀
	PO (gavage)	5 doses	10 mL/kg	Well tolerated	–
	PO	28 doses	10 mL/kg	Well tolerated	–
	PO (gavage)	28 days	10 mL/kg/dose QD	Well tolerated	pH 7.2; age 6 weeks; ♂/♀

(*Continued*)

TABLE A.83 (*Continued*)

Phosphate Buffered Saline (PBS)

Species	Route	Duration	Dose	Adverse Reactions/Toxicity	Notes
	SC	1 month	1 mL/kg QD	Well tolerated	–
	Slow bolus injection	11 doses	1 mL/kg	Well tolerated	–

TABLE A.84

Poloxamer 188™

Species	Route	Duration	Dose	Adverse Reactions/Toxicity	Notes
Dog	SC	4 weeks	5 mL/kg QD	Dog vehicle changed following a single administration due to animal distress in response to the injections. The distress was attributed to the vehicle.	2% in DI water; pH 5 ±0.2; age 5–6 months; ♂/♀
Mouse	PO	1 month	10 mL/kg	Well tolerated	5% solution
Rat	PO	1 month	10 mL/kg	Well tolerated	7.5% solution
	SC	4 weeks	5 mL/kg QD	None	2% in DI water; pH 5 +0.2; age 6 weeks; ♂/♀

TABLE A.85

Poly(glycolide-co-dl-lactide) microspheres

Species	Route	Duration	Dose	Adverse Reactions/Toxicity	Notes
Dog	Into peri-odontal pockets	28 days	3.5 mg/pocket SD	Well tolerated	GLP; age 6–10 years

TABLE A.86

Polyglyceryl Oleate

Species	Route	Duration	Dose	Adverse Reactions/Toxicity	Notes
Rabbit	Dermal	Acute	–	Moderately irritant (irritation and corrosion test)	No dilution
	Ocular	Acute	–	Slightly irritant (irritation and corrosion test)	No dilution
Rat	PO	Acute	–	LD_{50} ≥2005 mg/kg/day	No dilution

TABLE A.87

Polyvinylpyrrolidone (PVP)

Species	Route	Duration	Dose	Adverse Reactions/Toxicity	Notes
Dog	PO	Acute	–	Causes histamine release in dogs. The reaction is highly variable ranging from no discernible effect, to reddening of extremities, to total collapse.	–
Rat	IM	Single dose	1 mL	Well tolerated	1%

TABLE A.88
Propylene Glycol

Species	Route	Duration	Dose	Adverse Reactions/Toxicity	Notes
Dog	IV	Single dose	1 mL/kg	NOAEL; some emesis & diarrhea	30%
	IV	14 days	5 mL/kg at a rate of 0.3 mL/kg	Hemolysis *in vitro* dog blood	60% solution in water
	IV (bolus)	–	1 mL/kg	Well tolerated	40% solution in water pH 3–11; 50% PG causes hemolysis
	IV (infusion)	–	2 mL/kg	Well tolerated	40% solution in water pH 3–11; 50% PG causes hemolysis
	PO	28 days	1.5 mL/kg	Well tolerated	100%
	PO	1 month	2.5 mL/kg	Well tolerated	–
	PO (gavage)	Up to 7 days	2 mL/kg/day	Well tolerated	–
	PO (gavage)	90 days	Dose vol 5 mL/kg, Daily dose 1000 mg/kg (20% w/v)	Well tolerated	Dose concentration 200 mg/mL
Minipig	Dermal	26 weeks	2.5 mL/kg	Well tolerated	–
Nonhuman Primate	PO (gavage)	90 days	Dose vol 5 mL/kg, Daily dose 1000 mg/kg (20% w/v)	Well tolerated	Dose concentration 200 mg/mL

(*Continued*)

TABLE A.88 (*Continued*)
Propylene Glycol

Species	Route	Duration	Dose	Adverse Reactions/Toxicity	Notes
Mouse	IP	1 month	2.5 mL/kg	Well tolerated	40% solution
	IV	Range finding	MTD: 1.5 g/kg; NOEL: 1 g/kg	–	Published LD_{50} = 5.0–8.6 g/kg
	IV (into tail vein)	Acute	1036 mg/kg (dose vol 5 mL/kg)	NOEL	20% solution
	IV (into tail vein)	Acute	1554 mg/kg (dose vol 5 mL/kg)	MTD; tremors and hind limb ataxia, full recovery by 1 min is typical	30% solution
	PO	1 month	10 mL/kg	Well tolerated	50% solution
	PO (gavage)	90 days	Dose vol 10 mL/kg, daily dose 1000 mg/kg (20% w/v)	Well tolerated	Dose concentration 100 mg/mL

(*Continued*)

TABLE A.88 (*Continued*)
Propylene Glycol

Species	Route	Duration	Dose	Adverse Reactions/Toxicity	Notes
Rat	IV (bolus)	–	2 mL/kg	Well tolerated	40% solution in water pH 3–11; 50% PG causes hemolysis
	IV (infusion)	–	5 mL/kg	Well tolerated	40% solution in water pH 3–11; 50% PG causes hemolysis
	PO	1 month	2.5 mL/kg	Well tolerated	–
	PO (gavage)	Prelim./ Segment II	5 mL/kg/day	Well tolerated	60/40: purified water/propylene glycol: W/W
	PO (gavage)	2 weeks	5 mL/kg/day	Well tolerated	60/40: purified water/propylene glycol: W/W
	PO (gavage)	Segment II	5 mL/kg/day	Well tolerated	60/40: purified water/propylene glycol: W/W
	PO (gavage)	2 weeks	2 mL/kg/day	Well tolerated	–
	PO (gavage)	90 days	Dose vol 5 mL/kg, daily dose 1000 mg/kg (20% w/v)	Well tolerated	Dose concentration 200 mg/mL
	SC	4 weeks	2.5 mL/kg/day	Well tolerated	–

TABLE A.89

Randomly Methylated Beta-Cyclodextrins (RAMEB)

Species	Route	Duration	Dose	Adverse Reactions/Toxicity	Notes
Nonhuman Primate	Intranasal	1 month	82.8 mg/mL (with treatment) 74.7 mg/mL (placebo) TID	Well tolerated	7.5%

TABLE A.90

Safflower Oil

Species	Route	Duration	Dose	Adverse Reactions/Toxicity	Notes
Dog	SC	Single dose	1 mL	Well tolerated, no evidence of irritation macroscopically or histologically	100%

TABLE A.91

Saline (pH Adj.)

Species	Route	Duration	Dose	Adverse Reactions/Toxicity	Notes
Mouse	IM	14 days	4 mL/kg SD	None	(pH 4.5); non-GLP; age 6 weeks; 8♂/8♀

TABLE A.92

Sesame Oil

Species	Route	Duration	Dose	Adverse Reactions/Toxicity	Notes
Dog	PO	1 month	5 mL/kg	Well tolerated	
	PO (gavage)	28 days	5 mL/kg QD	None	Age 7–8 months; ♂/♀
	PO (gavage)	9 months	1 mL/kg QD	None	Age 7–8 months; ♂/♀
Mouse	PO	1 month	0.25 mL/kg	Well tolerated	–
Rabbit	PO	1 month	0.5 mL/kg	Well tolerated	–
Rat	PO	1 month	1 mL/kg	Well tolerated	–
	PO (gavage)	26 weeks	1 mL/kg QD	None	Age 6 weeks; ♂/♀

TABLE A.93

Sodium Acetate Trihydrate Buffer

Species	Route	Duration	Dose	Adverse Reactions/Toxicity	Notes
Nonhuman Primate	IV	–	1 mL/kg	Well tolerated	–

TABLE A.94

Sodium Chloride

Species	Route	Duration	Dose	Adverse Reactions/Toxicity	Notes
Cat	SC	9 days	0.1 mL/kg QD	None	0.9% saline; non-GLP; age 4 months; 2♂/2♀
	SC	9 days	0.1 mL/kg QD	None	Non-GLP; 2♂/2♀
	IV	8 days	1.9 mL/kg SD	None	0.9% saline; non-GLP; age 8–9 months; 2♂/2♀
	IV	19 days	1.32 mL/kg 3x/wk × 2wk	None	0.9% saline; non-GLP; age 8–9 months; 2♂/2♀
Dog	IV	Single dose	10 mL/kg	Well tolerated	0.9%
	IV	Single dose	2 mL/kg	Well tolerated	0.9%
	IV	2 weeks	5 mL/kg/day	Well tolerated	0.9%
	IV (bolus)	Single dose	0.3 mL/kg	Well tolerated	0.9%
	IV	14 days	2.5 mL/kg QD	None	0.9% saline and water to make 0.8% saline; GLP; age 8 weeks; 10♂/10♀
	IV	14 days	2 mL/kg QD	None	0.9% saline and water to make 0.8% saline; GLP; age 10–11 months; 12♂/12♀
	IV (infusion)	8 weeks	1 mL/kg 30 min infusion 3x/wk	None	Non-GLP; age 8–15 months; 2♀
	IV (infusion)	48 hours	2 mL/kg 20 min infusion SD	None	0.9% saline; non-GLP; age 8–13 months; 2♂/2♀
	IV (infusion)	4 hours	2 mL/kg 20 min infusion SD	None	0.9% saline; non-GLP; age 8–9 months; 1♀
	Ocular	6 months	1 drop/eye TID	None	0.9% saline; GLP; age 6–7 months; 4♂/4♀
	PO	Single dose	0.282 mL/kg	Well tolerated	0.9%
	SC	1 month	0.025 mL	NOEL	0.9%
	SC	6 months	0.5 mL/kg total: 0.1 mL/kg at 5 separate locationsq21d	Occasional transient injection-site erythema	0.9% saline; GLP; age 5–6 months; 4♂/4♀

(Continued)

TABLE A.94 (*Continued*)

Sodium Chloride

Species	Route	Duration	Dose	Adverse Reactions/Toxicity	Notes
Minipig	SC	28 days	0.0225 mL/kg QD	None	0.9% saline; GLP; age 3 months; 4♂/4♀
	IM	49 days	0.5 mLq28d × 2 doses	None	0.9% saline; non-GLP; age 4 months; 3♂
	ID	49 days	0.5 mLq28d × 2 doses	None	0.9% saline; non-GLP; age 4 months; 3♂
Mouse	IM	30 doses		Well tolerated	0.9%
	IV	Single dose	10 mL/kg	Well tolerated	0.9%
	IV	2 days	10 mL/kg SD	None	0.9% saline and water to make 0.8% saline; GLP; age 4–7 weeks; 11♂/11♀
	SC	Single dose	10 mL/kg	Well tolerated	0.9%
	Topical	7 days	0.10 mL QD	None	0.9% saline; non-GLP; age 8 weeks; 28♂
Nonhuman Primate	SC	28 doses	0.67 mL/kg	Well tolerated	0.9%
	SC	56 doses	0.5 mL/kg/dose	Well tolerated	0.9%
	Slow Bolus	9 doses	10 mL/kg/dose	None	0.9%
Rabbit	IM	33 days	0.2 mLq14d × 3 doses	None	0.9% saline; GLP; age 6 months; 6♀
	IV	Single dose	0.1 mL/kg	Well tolerated	0.9%
	Perivascular	Single dose	0.1 mL/kg	Well tolerated	0.9%
	SC	Single dose	0.5 mL/kg	Well tolerated	0.9%

(*Continued*)

TABLE A.94 (*Continued*)

Sodium Chloride

Species	Route	Duration	Dose	Adverse Reactions/Toxicity	Notes
Rat	IP	90 days	10 mL/kg QD	None	0.9% saline; GLP; age 9 weeks; 10♂/10♀
	IV	6 hours	4 µL/g SD	None	0.9% saline; non-GLP; age 8–9 weeks; 2♂/2♀
	IV	Single dose	1 mL/kg	Well tolerated	0.9%
	IV	Single dose	2 mL/kg	None	0.9%
	IV	Single dose	10 mL/kg	Well tolerated	0.9%
	IV	3 doses	4 mL/kg	Well tolerated	0.9%
	IV	3 days	0.376 mL/kg QD	None	0.9% saline; GLP; age 4 months; 5♂
	IV	7 doses	1 mL/kg	None	0.9%
	IV	2 weeks	10 mL/kg/day	Well tolerated	0.9%
	IV infusion	13 weeks	10 mL/kg 30 min infusionq7d	None	0.9% saline; GLP; age 11 weeks; 20♂/20♀
	SC	Single dose	0.1-0.4 mL	Well tolerated	0.9%
	SC	14 days	0.5 mL/kg QD	None	0.9% saline; GLP; age 71–72 days; 15♂/15♀
	SC	28 doses	4 mL/kg	Well tolerated	0.9%
	SC	56 doses	2 mL/kg/dose	Well tolerated	0.9%
	Slow Bolus	Single dose	1 mL/kg	Well tolerated	0.9%
	Slow Bolus	Single dose	5 mL/kg	Well tolerated	0.9%
	Slow Bolus	Single dose	10 mL/kg	Well tolerated	0.9%
	Slow bolus	3 doses	2 mL/dose	Well tolerated	0.9%

TABLE A.95

Sodium Dihydrogen Phosphate Dihydrate (SDPD)

Species	Route	Duration	Dose	Adverse Reactions/Toxicity	Notes
Mouse	PO (gavage)	91 days	10 mL/kg QD	None	0.5M SDPD in DI water; age 5 weeks; ♂/♀
Nonhuman Primate	PO (gavage)	91 days	10 mL/kg QD	None	0.5M SDPD in DI water; age 2.5–3.5 years; ♂/♀

TABLE A.96

Sodium Metabisulfite

Species	Route	Duration	Dose	Adverse Reactions/Toxicity	Notes
Mouse	PO (gavage)	91 days	20 mL/kg QD	Well tolerated	10% in distilled water; age 6 weeks; ♂/♀
Nonhuman Primate	PO (gavage)	91 days	10 mL/kg QD	Well tolerated	10% in distilled water; age 2–4.5 years; ♂/♀
Rat	PO (gavage)	91 days	10 mL/kg QD	Well tolerated	10% in distilled water; age 6 weeks; ♂/♀

TABLE A.97

Sodium Phosphate Buffer

Species	Route	Duration	Dose	Adverse Reactions/Toxicity	Notes
Dog	PO	14 doses	10 mL/kg/dose	Well tolerated	70 mM
Mouse	PO (gavage)	90 days	10 mL/kg QD	Well tolerated	0.1 M; pH 7.0; age 6 weeks; ♂/♀
Nonhuman Primate	PO (gavage)	90 days	5 mL/kg QD	Well tolerated	0.1 M; pH 7.0; age 2–3 years; ♂/♀
	PO (gavage)	91 days	5 mL/kg QD	Well tolerated	0.1 M; pH 7.0; age 2–3 years; ♂/♀
Rat	PO	2 weeks	10 mL/kg QD	Well tolerated	70 mM
	PO (gavage)	90 days	10 mL/kg QD	None	0.1 M; pH 7.0; age 6 weeks; ♂/♀
	PO (gavage)	91 days	10 mL/kg QD	None	0.1 M; pH 7.0; age 6 weeks; ♂/♀

TABLE A.98

Sodium Sulfite

Species	Route	Duration	Dose	Adverse Reactions/Toxicity	Notes
Rabbit	Ocular (Topical)	28 days	50 mL/dose QID	None	10% in reverse osmosis DI water; age 7 months; ♂/♀

TABLE A.99

Solutol HS15™

Species	Route	Duration	Dose	Adverse Reactions/Toxicity	Notes
Dog	Any	1 dose	Varies	Poorly tolerated in significant amounts	Solutol® HS15/purified water
Mouse	IP	2 weeks	10 mL/kg 3x/wk	None	10% Solutol; non-GLP; age 4–5 weeks; 5♂/5♀
	IV	2 weeks	10 mL/kg 3x/wk	None	10% Solutol; non-GLP; age 4–5 weeks; 5♂/5♀
Rat	PO (gavage)	2 months	10 mL/kg/day	Well tolerated	10% Solutol HS15 in purified water; Sprague Dawley rats

TABLE A.100

Soybean Oil

Species	Route	Duration	Dose	Adverse Reactions/Toxicity	Notes
Dog	PO	Acute		Well tolerated	50%

(Continued)

TABLE A.100 (*Continued*)

Soybean Oil

Species	Route	Duration	Dose	Adverse Reactions/Toxicity	Notes
Rat	PO	13 weeks	2000 mg/kg	Well tolerated; reduced food consumption at 500 mg/kg and above	–
	PO	13 weeks	20% in diet	Well tolerated	–

TABLE A.101

Sulfobutylether-β-Cyclodextrin (SBECD) (see also Table A.20: Captisol™)

Species	Route	Duration	Dose	Adverse Reactions/Toxicity	Notes
Mouse	PO	7 days	10 mL/kg BID	None	10%; non-GLP; age 7 weeks; 3♂/3♀

TABLE A.102

Tartaric Acid

Species	Route	Duration	Dose	Adverse Reactions/Toxicity	Notes
Rabbit	PO (gavage)	Prelim. Segmt. II	3 mL/kg/day	Well tolerated	–
	PO (gavage)	Segmt. II	3 mL/kg/day	Well tolerated	–
Rat	PO (gavage)	39 weeks	0.5 mL/kg	Well tolerated	–
	PO (gavage)	Sys. nerv.	3 mL/kg/day	Well tolerated	–

TABLE A.103

Terbinafine HCL™ Placebo Nail Lacquer

Species	Route	Duration	Dose	Adverse Reactions/Toxicity	Notes
Pig	Dermal	1 month	600 µL/kg	Erythema, peeling or flaking skin	–

TABLE A.104

Transcutol ™

Species	Route	Duration	Dose	Adverse Reactions/Toxicity	Notes
Cat	IV	1 month	2 mL/kg single dose	Well tolerated	No evidence of hemolysis or hematotoxicity
	IV	Acute	1000 mg/kg	LD_{50}	–
Dog	IV	Acute	3000 mg/kg	LD_{50}	–
	PO	90 days	1500 mg/kg/day	NOAEL	–
	PO (gavage)	Acute (dose escalating)	500, 1000, 1500, 2000 mg/kg	MTD > 2000 mg/kg	Undiluted
	PO (gavage)	Subacute (7 days DRF)	0, 500, 1000, 2000 mg/kg/day	MTD> 2000 mg/kg/day	–
	PO (gavage)	Subchronic 13 weeks	0, 400, 1000, 2000/1500 mg/kg/day	NOAEL = 1000 mg/kg/day	–
Guinea Pig	PO	Acute	300 mg/kg	LD_{50}	–

(Continued)

TABLE A.104 ***(Continued)***

Transcutol ™

Species	Route	Duration	Dose	Adverse Reactions/Toxicity	Notes
Minipig	PO	90 days	0, 167, 500, 1500 mg/kg/day	NOAEL = 167 mg/kg/day	Uremea, death at 1500 mg/kg/day; high dose reduced to 1000 mg/kg/day after 21 days; histopath in doses > 500 mg/kg/day include hydropic degeneration of liver and proximal kidney tubules; at > 1000 mg/kg/day increased relative kidney weight and decreased RBC (males)
Mouse	IP	Acute	3900 mg/kg	LD_{50}	–
	IV	Acute	4300 mg/kg	LD_{50}	–
	IV (bolus, tail vein)	Acute (dose escalating)	25, 50, 100, 200, 400, 800, 1600, 6400, 3200 and 4800 mg/kg	MTD (IV): 3200 mg/kg	Males; physiological saline solution
	IV (bolus, tail vein)	Acute (dose escalating)	25, 50, 100, 200, 400, 800, 1600, 8000, 6400, 4800 and 3200 mg/kg	MTD (IV): 3200 mg/kg	Females; physiological saline solution
	PO	Acute	6.6 g/kg	Tested toxic	–
	PO	Acute	7250 mg/kg	LD_{50}	–
	PO (gavage)	Developmental (dosed GD 7–14, littered and reared to PND 3)	5500 mg/kg/day	No developmental toxicity	>99% pure; 50 mated CD1 mice; 14% maternal mortality, maternal weight gain decreased, no external malformations on pups

(Continued)

TABLE A.104 (*Continued*)

Transcutol ™

Species	Route	Duration	Dose	Adverse Reactions/Toxicity	Notes
	PO	90 days	300, 900, 2700, and 8100 mg/kg bw (0%, 0.2%, 0.6%, 1.8% 5.4% in diet respectively)	NOAEL = 850–1000 mg/kg bw (0.6% in diet)	At 8100 mg/kg bw intracellular edema of the kidney, increased organ weights, decreased RBC (males), liver cell enlargement, protein inclusions in bladder lumen (males), tubular degeneration and atrophy. At 2700 mg/kg bw increased relative kidney weights in males was seen.
	PO	Chronic (12 months)	850-1000 mg/kg	NOEL	–
	SC	Acute	5500 mg/kg	LD_{50}	–
Rabbit	Dermal	Skin irritation	0.5 mL over 2 cm2 area	Non-irritant	50%
	Dermal	28 days	0, 300, 1000, 3000 mg/kg/day	NOEL >1000 mg/kg/day	Undiluted
	IM	14 days	0, 0.62, 0.82, 1.6 mL/kg/day	NOAEL = 1.6 mL/kg/day	No treatment-related effects
	Ocular	Eye irritation	0.1 mL	Slight irritation	30%
	Ocular	Eye irritation	0.1 mL	Slight irritation	Undiluted
	PO	Acute	3620 mg/kg	LD_{50}	–
	SC	Acute	2000 mg/kg	LD_{50}	–
Rat	Inhalation (nasal)	28 days (6 h/day, 5 days/wk)	0, 16, 49, 200 ppm (0, 90, 270, 1100 mg/m^3 respectively)	NOAEL = 1100 mg/m^3	No systemic effects; mild local irritation; focal necrosis in larynx (males); 1100mg/m^3 was higher than max concentration, at which only vapor present

(*Continued*)

TABLE A.104 (*Continued*)

Transcutol ™

Species	Route	Duration	Dose	Adverse Reactions/Toxicity	Notes
	Inhalation (whole body)	Developmental (7 h/day from GD 7–15)	0, 102 ppm	No maternal or fetal toxicity	98–99.5% pure; Sprague Dawley rats
	IP	Acute	6300 mg/kg	LD_{50}	–
	IV	Acute	4000 mg/kg	LD_{50}	–
	PO	Acute	7500 mg/kg	LD_{50}	–
	PO	Acute	5.0 g/kg	LD_{50}> 5000 mg/kg	Undiluted
	PO	90 days	0%, 0.25%, 1% and 5%	NOEL = 1%	–
	PO (gavage)	90 days	250, 2500 mg/kg bw (0.5% and 5.0% in diet respectively)	NOAEL = 250 mg/kg bw	High dose group saw reduction in growth rate and food consumption, decreased hemoglobin, rbc decreased (females), oxalate crystals in urine (females), increased organ weights, calcification of renal cortex; CFE rats
	PO (gavage)	Fertility (Segment I)	0, 300, 1000, 2000 mg/kg/day	NOAEL (oral) = 2000 mg/kg/day	In sterile water
	PO (gavage)	6 weeks	1340, 2680, 5360 mg/kg/day	NOAEL = 1340 mg/kg/day	Death, hematological/clinical signs in intermediate and high dose groups; lethargy during first week; 10♂; Sprague Dawley rats

(*Continued*)

TABLE A.104 (*Continued*)

Transcutol ™

Species	Route	Duration	Dose	Adverse Reactions/Toxicity	Notes
	PO (gavage)	Embryo/fetal development study (Segment II)	0, 300, 1000, 2000 mg/kg/day	NOAEL(dev, mat) 1000 mg/kg/day	In sterile water
	PO	Fertility and embryotoxicity range-finding study	500, 1000, 2000, 4000 mg/kg/day	NOEL > 500 mg/kg/day	–
	SC	Acute	6000 mg/kg	LD_{50}	–

TABLE A.105

Trisodium Citrate Dihydrate

Species	Route	Duration	Dose	Adverse Reactions/Toxicity	Notes
Dog	PO (gavage)	52 weeks	10 mL/kg/day	Well tolerated	2.65% aqueous (pH 6.4)
Hamster	PO (gavage)	13 weeks	10 mL/kg/day	Well tolerated	2.65% aqueous (pH 6.4)
Mouse	PO (gavage)	13 weeks	10 mL/kg/day	Well tolerated	2.65% aqueous (pH 6.4)
Rat	PO (gavage)	4 weeks	10 mL/kg/day	Well tolerated	2.65% aqueous (pH 6.4)
	PO (gavage)	Segm. III	10 mL/kg/day	Well tolerated	2.65% aqueous (pH 6.4)
	PO (gavage)	39 weeks	10 mL/kg/day	Well tolerated	2.65% aqueous (pH 6.4)

TABLE A.106
Tween 20 ™

Species	Route	Duration	Dose	Adverse Reactions/Toxicity	Notes
Dog	IV, SC, IP	Single dose	–	Not tolerated in any significant amount	–
	PO	Single dose	–	Poorly tolerated in significant amounts	–
Mouse	PO	Acute	10 g/kg	Well tolerated	–
	PO	–	10 g/kg	Well tolerated	–
Nonhuman Primate	IV (slow bolus)	28 days	10 mL/kg 3x/wk	Red/back discoloration of the skin (anogenital region, and hindlimbs and forelimbs)	1.01% in sterile water for injection, USP; age 2–3 years; ♂/♀
Rat	PO	1 month	250 mg/kg	Well tolerated	–
	PO	90 days	500 g/kg	Diarrhea	–
	IV (slow bolus)	6 months	10 mL/kg 3x/wk	None	1.01% in sterile water for injection, USP; age 7–8 weeks; ♂/♀

TABLE A.107
Tween 80 ™

Species	Route	Duration	Dose	Adverse Reactions/Toxicity	Notes
Dog	IV, SC, IP	Single dose	Varies	Not tolerated in any significant amount; hypersensitivity	–
	PO	Single dose	Varies	Poorly tolerated in significant amounts	–
	PO (gavage)	ADME	5 mL/kg/day	Well tolerated	0.1%

(Continued)

TABLE A.107 (*Continued*)

Tween 80 ™

Species	Route	Duration	Dose	Adverse Reactions/Toxicity	Notes
	PO (gavage)	90 days	Dose vol 5 mL/kg, daily dose 10 mg/kg (0.2% w/v)	Well tolerated	Dose concentration 2 mg/mL
Mouse	Intranasal	3 days	10 µL/nostril	Well tolerated	0.2%
	IP	1 month	10 mL/kg	Well tolerated	2% solution
	PO (gavage)	35 days	10 mL/kg BID	Distended abdomen, skin cold to the touch, feces few/absent, limb function impaired, head tilt, swollen abdomen, death; vehicle was not tolerated following 35 days of daily administration	10% in distilled water; age 6 weeks; ♂/♀
	PO (gavage)	90 days	Dose vol 10 mL/kg, Daily dose 10 mg/kg (0.2% w/v)	Well tolerated	Dose concentration 1 mg/mL
Nonhuman Primate	PO (gavage)	Efficac.	5 mL/kg/day	Well tolerated	0.1%
	PO (gavage)	90 days	Dose vol 5 mL/kg, Daily dose 10 mg/kg (0.2% w/v)	Well tolerated	Dose concentration 2 mg/mL
Rat	IP	Efficac.	10 mL/kg/day	Well tolerated	0.2%
	IV	Acute	100 mg/kg	Well tolerated	–
	PO	Acute	350 mg/kg	Well tolerated	–
	PO	7 days	10 mL/kg	Well tolerated	1% solution

(*Continued*)

TABLE A.107 (*Continued*)

Tween 80 ™

Species	Route	Duration	Dose	Adverse Reactions/Toxicity	Notes
	PO (dietary)	–	2% (1 g/kg)	NOAEL	–
	PO (gavage)	4 weeks	5 mL/kg/day	Well tolerated	0.1%
	PO (gavage)	90 days	Dose vol 5 mL/kg, Daily dose 10 mg/kg (0.2% w/v)	Well tolerated	Dose concentration 2 mg/mL

TABLE A.108

Vitamin E TPGS

Species	Route	Duration	Dose	Adverse Reactions/Toxicity	Notes
Dog	PO (gavage)	32 days	5 mL/kg QD	None	20% in DI water; age 5–5.5 months; ♂/♀
Rat	PO (gavage)	QD for 4 days then off for 4 days (×5 cylces) for total of 32 days	10 mL/kg	None	20% in DI water; age 6 weeks; ♂/♀
	PO (gavage)	Single dose	10 mL/kg	None	10% Vitamin E TPGS in DI water; age 6 weeks; ♂

TABLE A.109

Water

Species	Route	Duration	Dose	Adverse Reactions/Toxicity	Notes
Dog	PO (gavage)	14 doses	10 mL/kg/dose	None	Distilled
	PO (gavage)	28 doses	5 mL/kg/day	None	Distilled
	PO (gavage)	30 doses	10 mL/kg	None	Distilled
	PO	9 months	0.052 mL/kg TID	None	GLP; age 5–6 months; 6♂/6♀
Minipig	SC & IA	q7d × 3 wks for 90 days	12 mL total: 2 mL/site	None	GLP; age 5–6 months; 4♂/4♀
Mouse	PO (gavage)	2 doses	20 mL/kg/day	None	Distilled
	PO (gavage)	28 doses	10 mL/kg	None	Deionized
Nonhuman Primate	PO (gavage)	14 doses	10 mL/kg	None	Sterile, USP
	PO (gavage)	28 doses	10 mL/kg	None	Deionized
Pig	Dermal	9 doses	10 mL/animal/day	None	Deionized
Rat	IV	11 doses	5 mL/kg	None	Sterile, USP
	IV (slow bolus)	4 doses	4 mL/kg	None	Sterile, USP
	PO (gavage)	Single dose	5 mL/kg	None	Sterile, USP
	PO (gavage)	Single dose	5 mL/kg	None	Deionized
	PO (gavage)	Single dose	10 mL/kg	None	Distilled
	PO (gavage)	Single dose	10 mL/kg	None	Deionized
	PO (gavage)	Single dose	10 mL/kg	None	Sterile, USP
	PO	Single dose	20 mL/kg	None	Sterile, USP

(*Continued*)

TABLE A.109 (*Continued*)

Water

Species	Route	Duration	Dose	Adverse Reactions/Toxicity	Notes
	PO (gavage)	5 doses	10 mL/kg	None	Deionized
	PO (gavage)	14 doses	5 mL/kg	None	Sterile, USP
	PO (gavage)	14 doses	10 mL/kg	None	*Distilled*
	PO (gavage)	14 doses	10 mL/kg	None	Sterile, USP
	PO	26 weeks	5 mL/kg TID	None	GLP; age 6 weeks; 25♂/25♀
	PO (gavage)	28 doses	5 mL/kg	None	Deionized
	PO (gavage)	28 doses	10 mL/kg	None	Deionized
	PO (gavage)	30 doses	10 mL/kg	None	Distilled

TABLE A.110

Xylitol

Species	Route	Duration	Dose	Adverse Reactions/Toxicity	Notes
Nonhuman Primate	Intranasal	1 month	Control and high dose of 1200 μL/day; intermediate dose of 400 μL/day; low dose of 200 μL/day	Well tolerated at 1200 μL/day	3.3% in water (w/v)

TABLE A.111

Combination Formulations

#	Formulation	Species	Route	Duration	Dose	Adverse Reactions/Toxicity	Notes
1	Acacia (10%)/Antifoam 1510-US™ (0.05%)/Water (purified)	Rat	PO (gavage)	14 days	10 mL/kg QD	Well tolerated	Age 9–11 weeks; ♂/♀
2	Acacia gum (10%)/DMSO (1%)	Nonhuman Primate	PO (gavage)	2 months (ADME)	5 mL/kg	Well tolerated	–
3	Acacia gum (10%)/DMSO (1%)	Rat	PO (gavage)	1 week (ADME)	5 mL/kg	Well tolerated	Sprague Dawley
4	Acacia gum (10%)/DMSO (1%)	Rat	PO (gavage)	1 month (ADME)	5 mL/kg	Well tolerated	Sprague Dawley
5	Acacia gum (10%)/DMSO (1%)	Rat	PO (gavage)	1 month (ADME)	5 mL/kg	Well tolerated	Sprague Dawley
6	Acacia gum (10%)/Tween 80™ (0.5%)	Rat	PO (gavage)	Prelim/ Segment II	5 mL/kg/ day	Well tolerated	–
7	Acacia gum (10%)/Tween 80™ (0.5%)	Rat	PO (gavage)	Segment II	5 mL/kg/ day	Well tolerated	–
8	Acetate buffer (pH 5)/Benzyl alcohol (1% pH 5.0) (acetate: 100 mM, citrate: 10 mM)	Nonhuman Primate	IV (into saphena vein)	2 weeks	0.92 mL/kg	Well tolerated	–
9	Acetate buffer pH 5 with 1.0% benzyl alcohol (acetate: 100 mM, citrate: 10 mM); pH 5.0	Rat	IV (into tail vein)	28 days	0.92 mL/kg	Well tolerated	Sprague Dawley
10	Acetic acid (0.01 M)/DMA (95/5)	Dog	30-minute infusion	1 day	3 mL/kg in single infusion	None	99% solution; beagle age 5 months; ♂/♀

(Continued)

TABLE A.111 (*Continued*)

Combination Formulations

#	Formulation	Species	Route	Duration	Dose	Adverse Reactions/Toxicity	Notes
11	Acetone/Cyclohexane (50/50)	Rat	Dermal	6h daily, 5d/wk over 14 days	1.5 mL/kg	None	100% solution; no sham group; age ~60 days; ♀
12	Acetonitrile/Acetic acid (99/1)	Dog	PO (Dietary)	ad libitum over 1 year	–	None	Age 7–8 months; ♂/♀
13	Benzyl alcohol: 60.2%/Citric acid: 0.1%/BHT: 0.1%	Cat	Topical	24 hours	0.3 mL SD	None	Non-GLP; age 1 year; 1♂/1♀
14	BHT/Benzyl alcohol/Isopropanol	Dog	Topical	q7d × 2 doses over 7 weeks	1.4 mL/kg divided into 3 doses 60 minutes apart	None	Non-GLP; age 7 weeks; 4♂/4♀
15	Capmul MCM™/Kolliphor EL™ (50/50)	Nonhuman Primate	PO (gavage)	Once weekly over 4 weeks	5 mL/kg	None	Age 3–6 years; ♂/♀
16	Capmul MCM NF™/Propylene glycol/Kolliphor EL™ in a ratio of 1:1:1 (by weight)	Rat	PO	13 weeks	3 mL/kg BD	Well tolerated	Wistar Hans (CRL) age ~7–9 weeks at study initiation; ♂/♀
17	Captisol™ (5.4%)/Dextrose (2.5%)/Water (pH 4)	Dog	IV (infusion)	3 days	2 mL/kg/hr, 24 hrs/day	None	Age 7–8 months; ♂/♀

(*Continued*)

TABLE A.111 (*Continued*)

Combination Formulations

#	Formulation	Species	Route	Duration	Dose	Adverse Reactions/Toxicity	Notes
18	Citrate buffer (50 mM containing 0.5% Methocel E50 Premium LV™ and 0.2% Tween 20™, pH 4 ± 0.1)	Rat	PO (gavage)	14 days	10 mL/kg QD	None	Age 8–9 weeks; ♂/♀
19	Citrate buffer (53.8–54.9 mM containing 0.5% Methocel E50 Premium LV™ and 0.2% Tween 20™, pH 4 ± 0.1)	Nonhuman Primate	PO (gavage)	1 day	5 mL/kg SD	None	Age 3–6 years; ♂
20	CMC (0.5%)/Tween 80™ (0.1% (w/w))	Rabbit	PO (gavage)	Prelim/ Segment III	5 mL/kg/ day	Well tolerated	–
21	CMC (1%)/Tween 80™ (0.5%)	Dog	PO	28 days	10 mL/kg	Well tolerated	–
22	CMC (1%)/Tween 80™ (0.5%)	Rat	PO	28 days	20 mL/kg	Well tolerated	–
23	CMC (high viscosity, 0.25%)/Tween 80™ (0.2%)/Water (sterile for injection, USP)	Mouse	PO (gavage)	102 weeks	10 mL/kg QD	None	Age 6 weeks; ♂/♀
24	CMC (low viscosity, 1%)/Tween 80™ (0.01%)/Water (distilled)	Rat	PO (gavage)	105 weeks	10 mL/kg QD	None	Age 6 weeks; ♂/♀
25	CMC (low viscosity, 1%)/Tween 80™ (0.01%)/Water (distilled)	Nonhuman Primate	PO (gavage)	≤ 365 days	10 mL/kg QD	None	Age 4.5–8 years; ♂/♀
26	CMC (low viscosity, 1%)/Tween 80™ (0.01%)/Water (distilled)	Mouse	PO (gavage)	91 weeks	10 mL/kg QD	None	Age 6 weeks; ♂/♀
27	CMC (medium viscosity, 0.5% w/v), Tween 80™ (0.1%, v/v)/ Water (sterile for injection, USP)	Mouse	PO (gavage)	13 weeks	10 mL/kg QD	None	Age 5 weeks; ♂/♀

(*Continued*)

TABLE A.111 (*Continued*)
Combination Formulations

#	Formulation	Species	Route	Duration	Dose	Adverse Reactions/Toxicity	Notes
28	CMC: 0.5%/Tween 80™: 0.05%/in PBS pH 7.4	Dog	IA	2 days	0.35 mL SD	1 day post-dosing, 1 animal had moderate hemorrhage of right stifle fat pad and mildly increased synovial fluid volume on necropsy, 2 had lameness 2 days post-injection, resolved within 24 hours	GLP; age 8 months; 6♂/6♀
29	CMC: 0.5%/Tween 80™: 0.05%/in PBS pH 7.4	Dog	IA	q30d over 3 months	350 μL	None	GLP; age 5–6 months; 5♂/5♀
30	CMC: 0.5%/Tween 80™: 0.05%/in PBS pH 7.4	Dog	IA	q30d over 9 months	350 μL	None	GLP; age 5–6 months; 5♂/5♀
31	Corn oil/Benzyl alcohol (99:1)	Rat	PO (gavage)	1 month	1 mL/kg/day	Well tolerated	Sprague Dawley; during the first 3 days of the study, the vehicle was di(ethyleneglycol) ethylether. From the 4th day it was replaced by corn oil/benzyl alcohol
32	Corn oil/EtOH 20%	Rat	PO (gavage)	1 month	2.5 mL/kg/day	Well tolerated	Sprague Dawley
33	Corn oil/EtOH 20%	Rat	PO (gavage)	1 month	5 mL/kg/day	Well tolerated	Sprague Dawley

(Continued)

TABLE A.111 (*Continued*)

Combination Formulations

#	Formulation	Species	Route	Duration	Dose	Adverse Reactions/Toxicity	Notes
34	Cyclodextrin/OraPlus™ suspension	Rat	PO (gavage)	28 doses	15 mL/kg	Fecal changes (soft, watery or mucoid); kidney lesions	–
35	Cyclodextrin/OraPlus™ suspension	Rat	PO (gavage)	Single dose	20 mL/kg	None	–
36	Cyclodextrin/OraPlus™ suspension	Rat	PO (gavage)	Single dose	20 mL/kg	None	–
37	Cyclodextrin/OraPlus™ suspension	Rat	PO (gavage)	5 doses		None	–
38	Dehydrated EtOH (200 proof): 86.249%/Hexylene glycol: 10.000%/Dimethiconol blend 20: 2.500%/Hydroxypropylcellulose: 1.250%/Anhydrous citric acid: 0.001%	Minipig	Topical	28 days	5 mL QD	None	GLP; age 3–5 months; 5♂/5♀
39	Dextrose/Trehalose (25%)	Rat	Bolus injection	28 doses	10 mL/kg	None	5% solution
40	Dextrose for injection (USP)/EtOH (5%)	Rat	PO (gavage)	Single dose	–	None	5% solution
41	Diethylacetamide/NaCl	Mouse	IV	–	MTD: 1.4 g/ kg; LD_{50}: 2.3-3.2 g/kg; NOEL: 468 mg/kg	–	CD-1 mice; 30% of a 5 mL/kg dose volume at MTD; (%v/v) in NaCl

(*Continued*)

TABLE A.111 (*Continued*)

Combination Formulations

#	Formulation	Species	Route	Duration	Dose	Adverse Reactions/Toxicity	Notes
42	Di-sodium hydrogen phosphate dihydrate (8 mM)/Sodium dihydrogen phosphate dihydrate (7 mM)/NaCl (50 mM)/Sucrose (146 mM)/Poloxamer 188™ (0.12 mM)/Water (for injection, pH 6.9 ± 0.4)	Nonhuman Primate	SC	26 weeks	2 mL/kg once weekly	None	Age 1–1.5 years; ♂/♀
43	DMSO (1.25%)/5% Mannitol (5%)/Kolliphor™ (1.25%)	Nonhuman Primate	PO (gavage)	ADME	1 mL/kg/day	Well tolerated	–
44	DMSO (2%)/PEG 400 (10%)	Nonhuman Primate	IV	1 month (ADME)	1.5 mL/kg	Well tolerated	–
45	DMSO (2%)/PEG 400 (10%)	Nonhuman Primate	IV	1 week (ADME)	1.5 mL/kg	Well tolerated	–
46	DMSO (2%)/PEG 400 (10%)	Nonhuman Primate	IV	2 months (ADME)	1.5 mL/kg	Well tolerated	–
47	DMSO (3.5%)/PEG 400	Nonhuman Primate	PO (gavage)	Prelim.	5 mL/kg/day	Soft /liquid feces	–
48	DMSO (5%)/Mannitol (4.5%)/Kolliphor™ (5%)	Nonhuman Primate	IV	2 weeks	2 mL/kg/day	Well tolerated	–

(*Continued*)

TABLE A.111 (*Continued*)

Combination Formulations

#	Formulation	Species	Route	Duration	Dose	Adverse Reactions/Toxicity	Notes
49	DMSO/NaCl	Mouse	IV	–	MTD: 2.2 g/kg; LD_{50}: 3.8-7.6 g/kg; NOEL: 1.6 g/kg	–	CD-1 mice; 40% of a 5 mL/kg dose volume at MTD; (%v/v) in NaCl
50	DMSO/PEG 400 (99%)	Rat	PO (gavage)	11 doses	–	None	1% solution
51	DMSO/PEG 400 (99%)	Rat	PO (gavage)	11 doses	2 mL/kg	None	1% solution
52	DMSO/PEG 400/Tris buffer	Nonhuman Primate	IV (bolus)	ADME	2 mL/kg	Well tolerated	–
53	DMSO/PEG 4000/Water (DI) (15/35/50)	Rat	IP	28 days	10 mL/kg	Well tolerated	–
54	DMSO/PEG 4000/Water (DI) (15/35/50)	Mouse	IP	28 days	10 mL/kg	Well tolerated	–
55	DMSO/Solutol HS15™/Water (pH 3–11) (20/5/75)	Rat	IV (bolus)	–	2 mL/kg	Well tolerated	–
56	DMSO/Tetraglycol (50/50)	Minipig	IV	1 day	0.4 mL/kg SD	Immediate cardiac arrest; study was cancelled after first dose	Age 3 months; 1♂
57	DMSO + 1% Tween 80 ™ at 0.12% in water for injection	Mouse	SC	RNA extraction, 7 days	4/10 mL/kg/injection	Well tolerated	–

(*Continued*)

TABLE A.111 (*Continued*)

Combination Formulations

#	Formulation	Species	Route	Duration	Dose	Adverse Reactions/Toxicity	Notes
58	EDTA (0.2 mg/mL)/Citric acid anhydrous (0.8 mg/mL)/NaCl (4.6 mg/mL, low iron)/Water (sterile for injection, USP, pH 3.9–4.0)	Rat	IV (slow bolus)	13 weeks	4.5 mL/kg QD	Red/blue skin discoloration and edema of the tail (non-adverse)	Age 6 weeks; ♂/♀
59	EDTA pH 7.3 (Phosphate 10 mM; NaCl 150 mM; EDTA 0.5 mM)	Rat	IV (into tail vein)	1 month	0.92 mL/kg	Well tolerated	Sprague Dawley
60	EDTA pH 7.3 (Phosphate 10 mM; NaCl 150 mM; EDTA 0.5 mM)	Nonhuman Primate	IV (into saphena vein)	2 weeks	0.92 mL/kg	Well tolerated	–
61	EtOH (190 Proof): 63.37%/Glycerol: 3%/Carbopol Ultrez™ 10: 2.5%/Tween 20 ™: 2%/Propylene glycol: 2%/Panthenol: 0.15%/Salicylic acid: 0.15%/EDTA: 0.05%/Water (DI): 26.78%	Minipig	Topical	92 days	1 g QD	None	GLP; age 3–4 months; 5♂/5♀
62	EtOH/Kolliphor EL™/Water for injection (10/5/85)	Nonhuman Primate	IV	ADME	0.4 mL/kg	Well tolerated	–
63	EtOH/Kolliphor EL™/Water for injection (10/5/85)	Rat	IV	ADME	2 mL/kg	Well tolerated	–

(*Continued*)

TABLE A.111 (*Continued*)

Combination Formulations

#	Formulation	Species	Route	Duration	Dose	Adverse Reactions/Toxicity	Notes
64	EtOH/NaCl	Mouse	IV		MTD: 986 mg/kg; LD_{50}: 1.6–4.3 g/kg; NOEL: 197 mg/kg	–	CD-1 mice; 25% of a 5 mL/kg dose volume at MTD; (%v/v) in NaCl
65	EtOH/Propylene glycol/Water (12.5/15.5/75, v/v/v)	Rat	PO (gavage)	2 weeks	5 mL/kg/day	Well tolerated	–
66	EtOH/Propylene glycol/Water (12.5/15.5/75, v/v/v)	Rat	SC	2 weeks	2 mL/kg/day	Well tolerated	–
67	EtOH/Propylene glycol/Water (30/10/60)	Dog	IV	–	5 mL/kg at a rate of 0.3 mL/kg	Hemolyis *in vitro* dog blood	–
68	EtOH/Propylene glycol/Water (30/20/50)	Dog	IV	–	5 mL/kg at a rate of 0.3 mL/kg	Hemolyis *in vitro* dog blood	–
69	EtOH/Propylene glycol/Water (40/10/50)	Dog	IV	–	5 mL/kg at a rate of 0.3 mL/kg	Hemolyis *in vitro* dog blood	–
70	EtOH/Solutol HS15™/Water	Nonhuman Primate	PO (gavage)	39 weeks	1 mL/kg/day	Well tolerated	–
71	EtOH/Solutol HS15™/Water	Nonhuman Primate	PO (gavage)	4 weeks	1 mL/kg/day	Well tolerated	–

(*Continued*)

TABLE A.111 (*Continued*)

Combination Formulations

#	Formulation	Species	Route	Duration	Dose	Adverse Reactions/Toxicity	Notes
72	EtOH/Water (30/70)	Rat	IV	–	5 mL/kg at a rate of 0.3 mL/kg	Hematuria	–
73	EtOH: 50%/Propylene glycol: 50%/with BHA: 0.05%/BHT: 0.05%	Minipig	Topical	16 weeks	1 mL TID	All males had increased AST at end of study, up to 500 IU/L (ref: 19-263 IU/L)	GLP; age 3–4 months; 6♂/6♀
74	Gelatin (0.5% m/v)/Mannitol (5% m/v)/Water for injection	Nonhuman Primate	PO (gavage)	Prelim.	5 mL/kg/day	Well tolerated	–
75	Gelucire™/PEG 400/NMP/Transcutol HP™ (50/30/10/10)	Minipig	PO	4 days	5 mL/kg QD	None	Non-GLP; age 4–6 months; 3♀
76	Gelucire 44/14™/PEG 400/NMP/Transcutol HP™ (50/30/10/10)	Minipig	PO	3 days	5 mL/kg QD	Mild transient diarrhea 24–30 hours post-first dose	Non-GLP; age 7–8 months; 3♀
77	Gelucire 44/14™/PEG 400/NMP/Transcutol HP™ (50/30/10/10)	Mouse	PO	5 days	5 mL/kg QD	None	Non-GLP; age 9–14 weeks; 6♂/6♀
78	Histidine (20 mM, pH 6.5)/Sucrose (8.8%)	Rat	IV (slow bolus)	14 days	5 mL/kg QD	None	Age 6 weeks; ♂/♀
79	Histidine (20 mM, pH 6.5)/Sucrose (8.8%)	Dog	SC	14 days	1 mL/kg QD	None	Age 5–6 months; ♂/♀
80	Histidine buffered solution (10 mM, pH 6.5)/NaCl (130 mM)/Water (sterile)	Rat	IV (bolus)	1 day	7.14 ml/kg Single dose	None	Age 5 months; ♂

(Continued)

TABLE A.111 (*Continued*)

Combination Formulations

#	Formulation	Species	Route	Duration	Dose	Adverse Reactions/Toxicity	Notes
81	HPMC (0.5%)/Tween 80™ (0.1%)/ Water (DI)	Nonhuman Primate	PO (gavage)	28 days	5 mL/kg QD	Soft/watery feces	Age 2–4 years; ♂/♀
82	HPMC (0.5%)/Tween 80 ™ (5%)	Rat	PO (gavage)	2 weeks	5/1 mL/kg/ day	Well tolerated	–
83	HPMC (1%)/Poloxamer 188™ (1%)	Dog	PO	2 weeks	5 mL/kg/ day	Well tolerated	Beagle
84	HPMC (1%)/Poloxamer 188™ (1%)	Dog	PO	2 weeks	10 or 20 mL/ kg/day	Well tolerated	Beagle
85	HPMC (1%)/Poloxamer 188™ (1%)	Dog	PO	2 months	10 mL/kg/ day	Well tolerated	Beagle
86	HPMC (1%)/Tween 80 ™ (0.25%)/ Water (purified)	Rat	PO (gavage)	11 days	10 mL/kg QD	None	Age 8–10 weeks; ♀
87	HPMC/Sodium lauryl sulfate (2%)/Water (distilled)	Rat	PO (gavage)	Single dose	10 mL/kg	None	0.5% solution
88	HPMC acetate succinate (90 mg/mL)/ Methylcellulose (0.5%)	Mouse	PO	14 days	5 mL/kg EOD	None	GLP; age 61–63 days; 20♂/20♀
89	HPMC acetate succinate (90 mg/ mL)/Methylcellulose (0.5%)	Minipig	PO	14 days	6 mL/kg EOD	None	GLP; age 3–5 months; 5♂/5♀
90	HPMC: 1%/Fumaric acid: 0.5%/ Tween 80 ™: 0.1%/Glycerol: 0.1%/in water	Dog	PO	14 days	5 mL/kg QD	None	Non-GLP; age 7–9 months; 1♂/1♀
91	HPβCD (20% solution)/ DMSO (99/1)	Dog	IV	ADME	4 mL/kg/ day	Well tolerated	–

(*Continued*)

TABLE A.111 (*Continued*)

Combination Formulations

#	Formulation	Species	Route	Duration	Dose	Adverse Reactions/Toxicity	Notes
92	HPβCD (20%)/Sodium acetate (25 mM, pH 4)	Rat	PO (gavage)	28 days	5 mL/kg QD	None	Age 6 weeks; ♂/♀
93	HPβCD (20%)/Sodium acetate (25 mM, pH 4)	Nonhuman Primate	PO (gavage)	28 days	2.5 mL/kg/dose BID	None	Age 2–3.5 years; ♂/♀
94	HPβCD (30%)/DMSO (5%)/Water (purified) (acidic solution, w/w)	Nonhuman Primate	IV	4 weeks	1 mL/kg/day	Well tolerated	–
95	HPβCD/Citric acid solution (0.05 M)	Dog	PO (gavage)	14 doses	3 mL/kg	None	20% solution
96	HPβCD/NaCl solution (0.9%)	Dog	IV	Single dose	1 mL/kg	None	10% solution
97	HPβCD/NaCl solution (0.9%)	Rat	IV	Single dose	1 mL/kg	None	10% solution
98	HPβCD: 12%/EtOH: 8%/Propylene glycol: 2%/Water for injection	Dog	IV (1-hr infusion)	q7d × 3 doses over 16 days	2 mL/kg	May cause temporary mild diarrhea	Non-GLP; age 1–4 years; 2♂/2♀
99	HPβCD: 12%/EtOH: 8%/Propylene glycol: 2%/Water for injection	Dog	IV (1-hr infusion)	q7d × 3 doses over 16 days	1.5 mL/kg	May cause diarrhea	Non=GLP; age 1–4 years; 2♂/2♀
100	Hydroxyethylcellulose (1%)/Tween 80™ (0.25%)/Antifoam™ (0.05%)/Water (purified)	Rat	PO (gavage)	1 day	10 mL/kg Single dose	None	Age 6 weeks; ♂

(Continued)

TABLE A.111 (*Continued*)

Combination Formulations

#	Formulation	Species	Route	Duration	Dose	Adverse Reactions/Toxicity	Notes
101	Hydroxyethylcellulose (1%, w/v)/Tween 80™ (0.25%)/Antifoam 1510-US™ (0.05%)/Water (purified)	Rat	PO (gavage)	4 days	10 mL/kg QD	None	Age 8–9 weeks; ♀
102	Hymetellose (1%)/Poloxamer 188™ (0.1%)	Rabbit	PO (gavage)	1 month	3 mL/kg/day	Well tolerated	–
103	Imwitor 742™/Tween 80™ (1:1, w/w)	Rat	PO (gavage)	7 days	10 mL/kg QD	None	Age 7 weeks; ♂/♀
104	Imwitor 742™/Tween 80™ (1:1, w/w)	Nonhuman Primate	PO (gavage)	7 days	5 mL/kg QD	Soft/watery feces	Age 2–3 years; ♂/♀
105	Imwitor 742™/Tween 80™ (1:1, w/w)	Hamster	PO (gavage)	7 days	5 mL/kg QD	None	Age 5 weeks; ♂/♀
106	Kolliphor EL™/PEG 300/(4:1, w/w)	Mouse	PO (gavage)	104 weeks	3 mL/kg/day	Well tolerated	–
107	Kolliphor RH40™: 41.5%/PEG 400: 20%/TPGS: 20%/Propylene glycol: 10%/Tween 80 ™: 8.5%	Rat	PO (gavage)	92 days	10 mL/kg QD	None	Age 10–23 weeks; ♂/♀
108	Kolliphor EL™/10% 190 proof EtOH/80% 5% dextrose in water (D5W)	Rabbit	IV	Single dose	5 mL/kg	None	10 % solution
109	Kolliphor EL™/10% 190 proof EtOH/80% 5% dextrose in water (D5W)	Rabbit	IV	Single dose	5 mL/kg	Local irritation, very slight to well-defined erythema	10 % solution

(Continued)

TABLE A.111 (*Continued*)

Combination Formulations

#	Formulation	Species	Route	Duration	Dose	Adverse Reactions/Toxicity	Notes
110	Kolliphor EL™/10% 190 proof EtOH/ 80% 5% dextrose in water (D5W)	Rabbit	Perivascular Injection	Single dose	0.5 mL/kg	None	10 % solution
111	Kolliphor EL™/10% 190 proof EtOH/80% 5% dextrose in water (D5W)	Rabbit	Perivascular Injection	Single dose	0.5 mL/kg	Erythema	10 % solution
112	Labrafil™/Tween 80™ (0.1%)	Rat	PO (gavage)	2 doses	2 mL/kg/dose	None	–
113	Labrasol™/Kplliphor HS15™/Transcutol HP™ (60/30/10)	Minipig	PO (gavage)	Once weekly over 4 weeks	3.75 mL/kg	None	Age 6–7 months; ♂/♀
114	Labrasol™/Labrafil™/Transcutol™	Rat	PO	4 weeks	0, 5, 10, or 20 mL/k/day	Tolerated at 5 mL/kg/day	Changes in appearance and behavior at 10 mL/kg/day; lethality and renal and hepatic effects at 20 mL/kg/day; Wistar rats
115	Labrasol™/PEG-400 (60/40)	Rat	PO	7 days	5 mL/kg BD	Well tolerated	Sprague Dawley rats (Harlan); age ~8–10 weeks; ♂/♀

(*Continued*)

TABLE A.111 (*Continued*)

Combination Formulations

#	Formulation	Species	Route	Duration	Dose	Adverse Reactions/Toxicity	Notes
116	Lactated Ringer's injection (USP)	Nonhuman Primate	IV	28 days	2 mL/kg QD	None	–
117	L-Ascorbic Acid/Isotonic NaCl	Rat	PO	90 days	500 g/kg	Hematologic changes, weight loss	–
118	Mannitol (47 mg/mL)/Succinic acid (1.181 mg/mL)/Water (sterile, USP)	Dog	SC	28 days	0.1 mL/kg QD	None	Age 5 months; ♂/♀
119	Mannitol (5%)/Acetate buffer pH4 (4:6)	Minipig	IM	Tolerance	0.8/1.2 mL/kg/day	Well tolerated	–
120	Mannitol (5%)/Gelatin (0.5%)/Tween 80™ (0.2%) (aqueous solution, % m/v)	Nonhuman Primate	PO (gavage)	2 weeks	2 mL/kg/day	Well tolerated	–
121	Mannitol: 120 mg/Na2HPO4: 12 mg/NaCl: 10.5 mg/Water: 4 mL	Dog	IV	One week (placebo) then one-day (with the test item)	0.15 mL/kg	Well tolerated	Beagle dogs
122	Mannitol: 250 mM/Sodium succinate: 25 mM/Water pH 4.9	Rat	SC	28 days	4 mL/kg	Well tolerated	–
123	Mannitol: 250 mM/Sodium succinate: 25 mM/Water pH 4.9	Mouse	SC	28 days	20 mL/kg	Well tolerated	–
124	Methane sulfonic acid (5%)/EtOH (5%)/Water	Rat	PO	28 days	10 mL/kg	Well tolerated	–

(Continued)

TABLE A.111 (*Continued*)

Combination Formulations

#	Formulation	Species	Route	Duration	Dose	Adverse Reactions/Toxicity	Notes
125	Methocel™ (0.5%)/Tween 80™ (0.1%)/Water (reverse osmosis)	Rat	PO (gavage)	182 days	10 mL/kg QD	None	Age 6 weeks; ♀
126	Methocel™ (A4M Premium, 0.5%)/Simethicone (0.1%)/TPGS (10%)/Citrate buffer (17 mM, pH 4 ± 0.05)	Rat	PO (gavage)	28 days	10 mL/kg QD	None	Age 8 weeks; ♂/♀
127	Methyl methacrylate/Glycol dimethacrylate cross polymer, propylene glycol dicaprylate/Dicaprate, BHT	Rat	Topical	90 days	2.4 mL/kg × 44–46 d, then 0.75 mL/kg QD	None	GLP; Age 8 weeks; 15♂/15♀
128	Methyl Paraben, Propyl Paraben	Minipig	Topical	91 days	0.05 mL/cm² TID	None	Age 2–3 months; 6♂/6♀
129	Methylcellulose (0.5% w/v)/Tween 80™ (0.1%)	Guinea Pig	PO	28 days	10 mL/kg	Well tolerated	–
130	Methylcellulose (aqueous, 0.5% w/w)/Tween 80™ (0.5% w/w)	Rat	PO (gavage)	Prelim/Segment III	5 mL/kg/day	Well tolerated	–
131	Methylcellulose/PEG 200 (5%)	Rat	PO (gavage)	Single dose	10 mL/kg	None	0.5% solution
132	Methylcellulose/Tween 80™ (0.1% v/v)	Rat	PO (gavage)	Single dose	7.5 mL/kg	None	0.5% solution
133	Methylcellulose/Tween 80™ (0.1% v/v)	Rat	PO (gavage)	Single dose	5 mL/kg	None	0.5% solution
134	Methylcellulose/Tween 80™ (0.1% v/v)	Rat	PO (gavage)	28 doses	5 mL/kg/dose	None	0.5% solution

(Continued)

TABLE A.111 (*Continued*)

Combination Formulations

#	Formulation	Species	Route	Duration	Dose	Adverse Reactions/Toxicity	Notes
135	Methylcellulose/Tween 80™ (0.1% v/v)	Dog	PO (gavage)	28 doses	1 mL/kg/dose	None	0.5% solution
136	Methylcellulose/Tween 80™ (0.1% v/v)	Dog	PO (gavage)	Single dose	5 mL/kg/dose	None	0.5% solution
137	Methylcellulose/Tween 80™ (0.1%)	Rat	PO (gavage)	Single dose	10 mL/kg	None	0.5% solution
138	Methylcellulose 1500 cps (0.5%)/Tween 80™ (0.1%)/Acetate buffer (10 mM)/Water (distilled, pH 4.5 ± 0.1)	Mouse	PO (gavage)	28 days	10 mL/kg QD	None	Age 6 weeks; ♂/♀
139	Methylcellulose 1500 cps (0.5%)/Tween 80™ (0.1%)/Acetate buffer (10 mM)/Water (distilled, pH 4.5 ± 0.1)	Mouse	PO (gavage)	13 weeks	10 mL/kg QD	None	Age 6 weeks; ♂/♀
140	Methylcellulose 1500 cps (0.5%)/Tween 80™ (0.1%)/Acetate buffer (10 mM)/Water (distilled, pH 4.5 ± 0.1)	Nonhuman Primate	PO (gavage)	91 days	5 mL/kg QD	None	Age 2–4 years; ♂/♀
141	Methylcellulose 1500 cps (0.5%)/Tween 80™ (0.1%)/Acetate buffer (10 mM)/Water (distilled, pH 4.5 ± 0.1)	Rat	PO (gavage)	91 days	10 mL/kg QD	None	Age 6 weeks; ♂/♀
142	Methylcellulose 400 cps (0.5%)/Sodium lauryl sulfate (0.5%)	Rat	PO (gavage)	26 weeks	10 mL/kg QD	None	Age 6 weeks; ♂/♀

(*Continued*)

TABLE A.111 (*Continued*)
Combination Formulations

#	Formulation	Species	Route	Duration	Dose	Adverse Reactions/Toxicity	Notes
143	Methylcellulose 400 cps (0.5%)/ Sodium lauryl sulfate (0.5%)/ Simethicone (0.01%)/Water (DI) (% w/v)	Rat	PO (gavage)	13–26 weeks	5 mL/kg QD	None	Age 8 weeks; ♂/♀
144	Methylcellulose 400 cps (0.5%)/ Sodium lauryl sulfate (0.5%)/ Water (DI)	Dog	PO (gavage)	26 weeks	5 mL/kg QD	Soft/watery feces (non-adverse)	Age 4–5 months; ♂/♀
145	Methylcellulose 400 cps (0.5%)/ Sodium lauryl sulfate (0.5%)/ Water (DI)	Dog	PO (gavage)	52 weeks	5 mL/kg QD	Soft/watery feces (non-adverse)	Age 4–5 months; ♂/♀
146	Methylcellulose 400 cps (0.5%)/ Sodium lauryl sulfate (0.5%)/ Water (DI)	Rat	PO (gavage)	28 days	10 mL/kg QD	None	Age 8 weeks; ♂/♀
147	Methylparaben (0.17%)/ Propylparaben (0.03%)/Acetyl cysteine (0.5%)/in citrate buffer (100 mM, pH 6.5)/Sodium hydroxide (10%) added to pH 6.0	Minipig	SC	56 days	0.138 mL/kg QD	Temporary dose-site irritation post-injection, resolved within a few minutes	GLP; age 4–7 months; 3♂/3♀
148	NaCl (0.9 5)/Propylene glycol/ EtOH (50/40/10)	Dog	IV	–	5 mL/kg at a rate of 0.3 mL/kg	Hemolysis *in vitro* dog blood	–
149	NaCl (0.9%)/EtOH (60/40)	Rat	IV	–	5 mL/kg at a rate of 0.3 mL/kg	Hematuria	–

(*Continued*)

TABLE A.111 (*Continued*)
Combination Formulations

#	Formulation	Species	Route	Duration	Dose	Adverse Reactions/Toxicity	Notes
150	NaCl (0.9%)/EtOH (70/30)	Rat	IV	–	5 mL/kg at a rate of 0.3 mL/kg	Hematuria	–
151	NaCl (0.9%)/EtOH/PEG 400 (50/30/20)	Dog	IV	–	5 mL/kg at a rate of 0.3 mL/kg	Partial hemolysis *in vitro* dog blood, RBC discolored	–
152	NaCl (0.9%)/EtOH/PEG 400 (50/40/10)	Dog	IV	–	5 mL/kg at a rate of 0.3 mL/kg	Partial hemolysis *in vitro* dog blood, RBC discolored	–
153	NaCl (0.9%)/EtOH/Propylene glycol (60/30/10)	Dog	IV	–	5 mL/kg at a rate of 0.3 mL/kg	Hemolysis *in vitro* dog blood	–
154	NaCl (0.9%)/EtOH/Propylene glycol (50/30/20)	Dog	IV	–	5 mL/kg at a rate of 0.3 mL/kg	Hemolysis *in vitro* dog blood	–
155	NaCl (0.9%)/EtOH/Propylene glycol (50/40/10)	Dog	IV	–	5 mL/kg at a rate of 0.3 mL/kg	Hemolysis *in vitro* dog blood	–
156	NaCl (0.9%)/Propylene glycol/EtOH (50/30/20)	Dog	IV	–	5 mL/kg at a rate of 0.3 mL/kg	Hemolysis *in vitro* dog blood	–
157	NaCl/DMSO/Tetraglycol (90/5/5)	Rat	IV	14 days	4.3 mL/kg SD	None	GLP; age 8 weeks; 5♂/5♀
158	NaCl/DMSO/Tetraglycol (90/5/5)	Minipig	IV	16 days	2.6 mL/kg SD	None	GLP; age 2–4 months; 1♂/1♀

(*Continued*)

TABLE A.111 (*Continued*)

Combination Formulations

#	Formulation	Species	Route	Duration	Dose	Adverse Reactions/Toxicity	Notes
159	NaCl for injection (USP)/Mannitol (20 mg/mL)/Tween™ 80 (4 mg/mL), sterile filtered	Rat	IV (infusion)	Single dose	6.67 mL/kg/dose	None	0.9% solution
160	NaCl for Injection (USP)/10% EtOH and 0.9%	Dog	IV (bolus)	Single dose	5 mL/kg	None	10% solution
161	NaCl for injection/10% EtOH	Rat	IV	Single dose	5 mL/kg	None	10% solution
162	NaCl for injection/20% PET in 0.9%	Rat	IV (slow bolus)	7 doses	0.6 mL/kg	None	20% solution
163	NaCl USP: 15.00%/Potassium chloride: 7.50%/L-Arginine HCl USP: 7.50%/Glyceryl stearate SE: 7.00%/Cetyl alcohol NF: 7.00%/Propylene glycol: 5.00%/Squalene NF: 4.00%/Tween 20™ NF: 2.00%/Sodium hydroxide: 1.30%/Oleic acid NF: 1.00%/Isopropyl myristate: 1.00%/Keltrol RD: 0.50% (Xanthan Gum)/Keltrol BT™: 0.30% (Xanthan Gum)/Water (Purified): 40.90%	Minipig	Topical	7 days	3.0 g/kg BID	Severe dose-site erythema in 2/6 animals	GLP; age 3–5 months; 3♂/3♀
164	Neobee 1053 Oil™/EtOH/BHT (94.95/5/0.05)	Mouse	IV	28 days	10 mL/kg	Well tolerated	–
165	Neobee 1053 Oil™/EtOH/BHT (94.95/5/0.05)	Rat	PO	28 days	5 mL/kg	Well tolerated	–

(*Continued*)

TABLE A.111 (*Continued*)

Combination Formulations

#	Formulation	Species	Route	Duration	Dose	Adverse Reactions/Toxicity	Notes
166	NMP/NaCl	Mouse	IV	–	MTD: 1.3 g/kg; LD_{50}: 54-3600 mg/kg; NOEL: 257 mg/kg	–	CD-1 mice; 25% of a 5 mL/kg dose volume at MTD; (%v/v) in NaCl
167	Octoxynol-40, Vitamin E	Dog	Ocular (topical)	q1h × 8 doses over 14 days	0.35 µL (one drop)/eye	None	GLP; age 8 months; 4♂/6♀
168	Octoxynol-40, Vitamin E	Rabbit	Ocular (topical)	q1h × 8 doses over 14 days	0.35 µL (one drop)/eye	None	GLP; age 5–6 months; 5♂/5♀
169	Octoxynol-40, Vitamin E	Rabbit	Ocular (topical)	q1h × 8 doses over 13 weeks	0.35 µL (one drop)/eye	None	GLP; age 6 months; 6♂/6♀
170	Oleic Acid/PEG 400/Kolliphor EL™ (80/10/10, w/w)	Dog	PO (capsule)	9 months	0.6 mL/kg QD	None	Age 6 months; ♂/♀
171	Oleic Acid/PEG 400/Kolliphor EL™ (80/10/10, w/w)	Rat	PO (gavage)	104 weeks	2 mL/kg QD	Decreases in body weight gain (non-adverse)	Age 4 weeks: ♂/♀
172	Oleic Acid/PEG 400/Kolliphor EL™ (80/10/10, w/w)	Mouse	PO (gavage)	13 weeks	2 mL/kg QD	None	Age 6 weeks; ♂/♀

(*Continued*)

TABLE A.111 (*Continued*)

Combination Formulations

#	Formulation	Species	Route	Duration	Dose	Adverse Reactions/Toxicity	Notes
173	Oleic Acid/PEG 400/Kolliphor EL™ (80/10/10, w/w)	Nonhuman Primate	PO (gavage)	14 days	2 mL/kg/dose BID	Emesis and fecal changes	Age 2–3 years; ♂/♀
174	Olive Oil: 28.25%/Tween 80™: 11.25–13.5%/Oleyl alcohol NF: 10.00%/Lanolin alcohol NF: 8.00% /Cyclomethicone NF: 3.00%/Cetyl acetate: 1.5–3.75%/Shea butter: 2.00%/0.50% Sorbitan tristearate/acetylated lanolin alcohol: 0.15–0.75%/Methylparaben NF: 0.20%/Propylparaben NF: 0.05%/Water (USP purified): 33.00%	Minipig	Topical	14 days	0.32 mL/kg QD	4/6 experienced persistent mild (Draize score 1/4) dose-site erythema or miliary (disseminated in surrounding cutis) erythema	GLP; age 3–5 months; 5♂/5♀
175	Peanut oil/EtOH 100 (8:1)	Dog	SC	–	0.33 mL/kg	Well tolerated	Beagle
176	Peceol™ (Gattefossé)/Tween 80™/PEG 400/Vitamin E (TPPE) (50/40/10/0.2)	Rat	PO (gavage)	91 days	10 mL/kg/dose BID	None	Age 6 weeks; ♂/♀
177	Peceol™ (Gattefossé)/Tween 80™/PEG 400/Vitamin E (TPPE) (50/40/10/0.2)	Dog	PO (gavage)	91 days	5 mL/kg/dose BID	None	Age 7–8 months; ♂/♀
178	PEG (5%)/Methylcellulose (0.5%)	Pig	PO (gavage)	28 doses	5 mL/kg	None	5% solution
179	PEG/DAM (70/30, v/v)	Dog	IV	2 weeks	0.32 mL/kg for single IV injection	Well tolerated	–

(*Continued*)

TABLE A.111 (*Continued*)

Combination Formulations

#	Formulation	Species	Route	Duration	Dose	Adverse Reactions/Toxicity	Notes
180	PEG/DAM (70/30, v/v)	Rat	IV	3 weeks	Bolus 0.8–1.07 mL/kg infusion 0.266–0.356 mL/kg IV injection (into tail vein) followed by an IV injection for 6 h	Well tolerated	–
181	PEG/DAM (70/30, v/v)	Dog	IV (into cephalic of saphenous vein) followed by an IV injection for 6 hrs	2 weeks	Bolus 0.24–0.33 mL/kg infusion 0.08–0.11 mL/kg/hr	Well tolerated	–

(*Continued*)

TABLE A.111 (*Continued*)
Combination Formulations

#	Formulation	Species	Route	Duration	Dose	Adverse Reactions/Toxicity	Notes
182	PEG/DAM (70/30, v/v)	Rat	IV	3 weeks	Bolus 0.8–1.07 mL/kg infusion 0.266–0.356 mL/kg IV injection (into tail vein) followed by an IV injection for 6 hrs	Well tolerated	–
183	PEG 200/95% Methylcellulose (0.5%)	Rat	PO	3 doses	5 mL/kg of body weight	None	5% solution
184	PEG 200/EtOH/Dextrose (5%) (70/15/15, v/v/v)	Minipig	IV	2 weeks	1 mL/kg	Well tolerated	–
185	PEG 300 (40%)/Cavisol W7™ (25/75 v/v)	Dog	PO	28 days	10 mL/kg	Well tolerated	–
186	PEG 300 (40%)/Cavisol W7™ (25/75 v/v)	Rat	PO	28 days	10 mL/kg	Well tolerated	–
187	PEG 300/DMA (90/10)	Rat	PO (gavage)	14 days	2.5 mL/kg QD	None	Age 6 weeks; ♂/♀
188	PEG 300/DMA (90/10)	Dog	PO (gavage)	14 days	2.5 mL/kg QD	Body weight loss (>20%) (adverse)	Age 5–6 months; ♂/♀

(*Continued*)

TABLE A.111 (*Continued*)

Combination Formulations

#	Formulation	Species	Route	Duration	Dose	Adverse Reactions/Toxicity	Notes
189	PEG 300/NaCl (0.9%) (40/60, v/v)	Rat	PO (gavage)	ADME	5 mL/kg/day	Well tolerated	–
190	PEG 300/Propylene glycol/Water (DI) (55/25/20)	Rat	PO (gavage)	7 days	10 mL/kg QD	None	Age 9–10 weeks; ♂/♀
191	PEG 400/Captisol™/EtOH/Water (pH 3) (45/7/5/43)	Dog	IV (Infusion)	–	1 mL/kg	Well tolerated	–
192	PEG 400/DMA (50/50)	Dog	IV (Infusion)	–	0.1 mL/kg	Well tolerated	–
193	PEG 400/DMSO (20%)	Mouse	IV (into tail vein)	Acute	1128 mg/kg (dose vol 5 mL/kg)	NOEL	20% PEG 400
194	PEG 400/DMSO (20%)	Mouse	IV (into tail vein)	Acute	3948 mg/kg (dose vol 5 mL/kg)	MTD; ventral recombency, ataxia, tremors and hypoactivity shortly after dosing. Tremors for up to 4 min, hypoactivity for up to 10 min is typical	70% PEG 400
195	PEG 400/DMSO (95/5)	Rabbit	PO (gavage)	12 doses	0.33 mL/kg	None	5% solution
196	PEG 400/EtOH (10%)/DMSO (10%)	Mouse	IV (into tail vein)	Acute	1692 mg/kg (dose vol 5 mL/kg)	NOEL	30% PEG 400
197	PEG 400/EtOH (10%)/DMSO (10%)	Mouse	IV (into tail vein)	Acute	2820 mg/kg (dose vol 5 mL/kg)	MTD; ventral recumbency, tremors, ataxia and hypoactivity shortly after dosing, recovery by 10 min is typical	50% PEG 400

(*Continued*)

TABLE A.111 (*Continued*)

Combination Formulations

#	Formulation	Species	Route	Duration	Dose	Adverse Reactions/Toxicity	Notes
198	PEG 400/EtOH (20%)	Mouse	IV (into tail vein)	Acute	3384 mg/kg (dose vol 5 mL/kg)	MTD; vocalization and struggling at dosing, ventral recumbency, rapid breathing, tremors, and ataxia shortly after dosing, recovery by 5 min is typical	60% PEG 400
199	PEG 400/EtOH (200 proof) (95/5, v/v)	Rat	PO	14 days	10 mL/kg BD	Abnormal clinical observations included anogenital or urogenital staining, soft feces/watery diarrhea, stained body surface, apparent dehydration, staining around mouth or nose/nares, and wet body surface	Wistar Han (CRL) rats; age ~8–10 weeks at study initiation; ♂/♀
200	PEG 400/EtOH/Propylene glycol/Water (sterile) (30/20/20/30)	Rat	IP	7 days	2 mL/kg QD	None	Non-GLP; age 7–8 weeks; 6♂
201	PEG 400/EtOH/Water (pH 3–11) (45/5/50)	Rat	IV (bolus)	–	2 mL/kg	Well tolerated	–
202	PEG 400/EtOH/Water (pH 3–11) (45/5/50)	Rat	IV (infusion)	–	5 mL/kg	Well tolerated	–
203	PEG 400/EtOH/Water (pH 3–11) (45/5/50)	Dog	IV (bolus)	–	1 mL/kg	Well tolerated	–
204	PEG 400/EtOH/Water (pH 3–11) (45/5/50)	Dog	IV (Infusion)	–	2 mL/kg	Well tolerated	–

(*Continued*)

TABLE A.111 (*Continued*)

Combination Formulations

#	Formulation	Species	Route	Duration	Dose	Adverse Reactions/Toxicity	Notes
205	PEG 400/EtOH/Water (sterile) (1:1:1)	Guinea Pig	IV	28 days	2 mL/kg	Well tolerated	–
206	PEG 400/Kolliphor RH40™ (70/30)	Rat	PO (gavage)	91 days	10 mL/kg/dose	Sporadic incidences of fecal changes (soft and/or loose/watery); slight brown/orange staining around anus; fluid contents in the cecum; 10% decrease in mean body weights (male rats); lowered food consumption; mildly increased serum urea; minimally decreased serum sodium and chloride values (male rats); minimally increased total serum cholesterol values (females); alterations in urine electrolytes; organ weight changes; minimal, focal or multifocal coagulative hepatocellular necrosis (in 3 females and 1 male)	6♂/6♀
207	PEG 400/Kolliphor RH40™ (90/10)	Dog	PO (gavage)	28 days	2 mL/kg/dose	Emesis; administration associated with minimal lamina propria hemorrhage in gastric glandular mucosa in 1/3 dogs	Beagle dogs; ♂

(*Continued*)

TABLE A.111 (*Continued*)

Combination Formulations

#	Formulation	Species	Route	Duration	Dose	Adverse Reactions/Toxicity	Notes
208	PEG 400/Kolliphor RH40™ (90/10)	Dog	PO (gavage)	28 days	5 mL/kg/dose	Fecal alterations (loose/watery, mucoid or red) present beginning on day one; emesis; administration associated with minimal lamina propria hemorrhage in gastric glandular mucosa in 2/3 dogs; single/multifocal red areas in stomach; minimal increase in group mean serum urea	Beagle dogs; ♂
209	PEG 400/Labrasol™/Kolliphor EL™ (50/30/20)	Mouse	PO (gavage)	182 days	10 mL/kg QD	Unkempt appearance (potential effect)	Age 6 weeks; ♂/♀
210	PEG 400/NaCl	Mouse	IV	–	MTD: 4.5 g/kg; LD_{50}: 8.6–9.7; NOEL: 1.7 g/kg	Over tested range, expect hypoactivity, tremors, mild ataxia with increasing duration with dose	CD-1 mice; 80% of a 5 mL/kg dose volume at MTD; (%v/v) in NaCl
211	PEG 400/NaCl (0.9%)/EtOH (50/40/10)	Dog	IV	–	5 mL/kg at a rate of 0.3 mL/kg	Partial hemolysis *in vitro* dog blood, RBC discolored	–
212	PEG 400/Propylene glycol/EtOH/Water (pH 3) (40/20/5/35)	Dog	IV (infusion)	–	2.5 mL/kg	Well tolerated	–
213	PEG 400/Propylene glycol/Tween 80 ™/Water (25/15/6/54)	Rat	IV	28 days	10 mL/kg	Well tolerated	–

(*Continued*)

TABLE A.111 (*Continued*)

Combination Formulations

#	Formulation	Species	Route	Duration	Dose	Adverse Reactions/Toxicity	Notes
214	PEG 400/Propylene glycol/Tween 80™/Water (25/15/6/54)	Guinea Pig	IV	28 days	10 mL/kg	Well tolerated	–
215	PEG 400/Propylene glycol/Tween 80™/Water (25/15/6/54)	Mouse	IV	28 days	10 mL/kg	Well tolerated	–
216	PEG 400/PVP K30/TPGS (90/5/5)	Dog	PO (gavage)	28 doses	2.5 mL/kg/day	None	90% solution
217	PEG 400/PVP K30/TPGS (90/5/5)	Rat	PO (gavage)	28 doses	5 mL/kg/day	None	90% solution
218	PEG 400/Solutol HS15™ (70/30)	Rat	PO (gavage)	91 days	10 mL/kg	Males: Sporadic fecal changes (soft and/or loose/watery); slight brown/orange staining around anus; 6% decrease in mean body weight; lower food consumption; organ weight changes; increased urine volume; alterations in urine electrolytes (both genders); fluid contents in the cecum (both genders)	6♂/6♀
219	PEG 400/Solutol HS15™ (70/30)	Dog	PO (gavage)	28 days	2 mL/kg/dose	Intermittent loose/watery feces; sporadic emesis starting on day one; 1/3 animals had minimal mucus cell hypertrophy of the ileal mucosa (direct effect of vehicle or effect of loose stools?)	Beagle; 3♀

(Continued)

TABLE A.111 (*Continued*)

Combination Formulations

#	Formulation	Species	Route	Duration	Dose	Adverse Reactions/Toxicity	Notes
220	PEG 400/Solutol HS15™ (70/30)	Dog	PO (gavage)	28 days	5 mL/kg/dose	Consistent incidence of loose/watery feces starting on day one; sporadic emesis starting on day one; 2/3 animals had minimal mucus cell hypertrophy of the ileal mucosa (direct effect of vehicle or effect of loose stools?); increase in RBC mass; decrease in urine volume	Beagle; 3♀
221	PEG 400/Solutol HS15™ (90/10)	Dog	PO (gavage)	28 days	2 mL/kg/dose	Intermittent loose/watery feces; sporadic emesis starting on day one	Beagle; 3♀
222	PEG 400/Solutol HS15™ (90/10)	Dog	PO (gavage)	28 days	5 mL/kg/dose	Consistent incidence of loose/watery feces starting on day one; sporadic emesis starting on day one; all had minimal mucus cell hypertrophy of the ileal mucosa (direct effect of vehicle or effect of loose stools?)	Beagle; 3♀
223	PEG 400/TPGS/PVP VA 64/EtOH (80/10/5/5)	Rat	PO (gavage)	182 days	2 mL/kg QD	White feces (non-adverse)	Age 8 weeks; ♂/♀
224	PEG 400/TPGS/PVP VA 64/EtOH (80/10/5/5)	Dog	PO (gavage)	9 months	1 mL/kg QD	None	Age 10–11 months; ♂/♀

(*Continued*)

TABLE A.111 (*Continued*)

Combination Formulations

#	Formulation	Species	Route	Duration	Dose	Adverse Reactions/Toxicity	Notes
225	PEG 400/Tween 20™/Poloxamer 124™ (70/20/10)	Rat	PO (gavage)	26 weeks	2 mL/kg/dose BID	None	Age 6 weeks; ♂/♀
226	PEG 400/Tween 20™/Poloxamer 124™ (70/20/10)	Dog	PO (gavage)	26 weeks	0.5 mL/kg/dose BID	None	Age 6.5–7.5 months; ♂/♀
227	PEG 400/Tween 20™/TPGS/Poloxamer 124™ (50/20/20/10)	Mouse	PO (gavage)	91 days	2 mL/kg QD	None	Age 6 weeks; ♂/♀
228	PEG 400/Tween 20™/TPGS/Poloxamer 124™ (50/20/20/10)	Dog	PO (capsule)	91 days	≤ 5 mL of vehicle QD	Soft/watery feces (non-adverse)	Age 6–7 months; ♂/♀
229	PEG 400/Tween 80™ (95/5)	Mouse	PO	2-year carc	5 mL/kg BD	During the early stages of the study, a number of mice receiving the vehicle developed gastrointestinal atony resulting in severe bloating and sometimes death.	CD-1 (Harlan) mice age ~7 weeks at study initiation; ♂/♀
230	PEG 400/Tween 80™ (95/5)	Rat	PO	2-year carc	5 mL/kg BD	Well tolerated	Vacuolation of the tubular epithelium in the kidney associated with PEG-linked proteins; Sprague Dawley (Harlan) rats age ~6 weeks at study initiation; ♂/♀

(*Continued*)

TABLE A.111 ***(Continued)***

Combination Formulations

#	Formulation	Species	Route	Duration	Dose	Adverse Reactions/Toxicity	Notes
231	Phosal 53 MCT™/PEG 400/Poloxamer 124™/Kolliphor RH40™ (40/20/20/20)	Rat	PO (gavage)	99 weeks	2 mL/kg QD	None	Age 6–110 weeks; ♂/♀
232	Phosphate (50 mM)/NaCl (100mM)/Tween 80™ (0.01%)	Nonhuman Primate	SC	Prelim/2 week	1 mL/kg/injection	Well tolerated	–
233	Phosphate buffer 0.5 M at pH 7.5/0.4% Mannitol™	Rat	SC	Segment I	0.5 mL/animal/injection	Well tolerated	–
234	Polawax™: 4.80%/Alcohol denatured SDA 40–2 (190 proof): 4.25%/Propylene glycol: 4.00%/Isopropyl myristate: 2.50%/Sodium hydroxide solution (10.0): 1.20%/Phenoxyethanol: 1.00%/Carbomer 974P™: 0.55%/Water (purified): 81.7%	Minipig	Topical	210 days	0.4 g/kg QD	None	GLP; age 3 months; 5♂/5♀
235	Polawax™: 4.80%/Propylene glycol: 4.00%/EtOH 200 proof: 4.00%/Ispropyl myristate: 2.50%/Sodium hydroxide 10% solution in purified water: 1.20%/Phenoxyethanol: 1.00%/Carbomer 974P™: 0.55%	Mouse	Topical	90 days	3.4 mL/kg QD	None	GLP; age 7–8 weeks; 18♂/18♀
236	Poloxamer 188™ (0.5%)/NaCl for injection (USP, 0.9%)	Rat	SC	Single dose	2.5 mL/kg	None	Age 6 weeks; ♂/♀

(Continued)

TABLE A.111 (*Continued*)

Combination Formulations

#	Formulation	Species	Route	Duration	Dose	Adverse Reactions/Toxicity	Notes
237	Poloxamer 188™ (0.5%)/NaCl for injection (USP, 0.9%)	Minipig	SC		1 mL/kg	None	Gottingen Minipigs age 5 months; ♂/♀
238	Poloxamer 188™ (1%, w/v)/ Citrate buffer (100 mM, pH 3)	Rat	PO (gavage)	14 days	10 mL/kg QD	None	Age 6 weeks; ♂/♀
239	Propylene glycol (USP)/Glycerol (USP)/EtOH (200 Proof, USP) (65/25/10, w/w/w)	Rabbit	SC	2x/wk over 3 weeks	1 mL/kg	None	Age 6.5 months; ♂/♀
240	Propylene glycol (USP)/Glycerol (USP)/EtOH (200 Proof, USP) (65/25/10, w/w/w)	Nonhuman Primate	SC	1x/wk over 3 weeks	0.5 mL/kg	Scratching and red skin discoloration at the dose site (non-adverse)	Age 3–6 years; ♂/♀
241	Propylene glycol/Capmul PG8™/ EtOH/Water (75/12.5/10/2.5)	Dog	PO (gavage)	7 days	5 mL/kg QD	None	Age 5 months; ♂/♀
242	Propylene glycol/Capmul PG8™/ EtOH/Water (75/12.5/10/2.5)	Rat	PO (gavage)	21 days	10 mL/kg QD	Aspiration, salivation, material around the mouth/nose, audible breathing, stereotypical behavior (scratching in the cage following dosing), death	Consistency of vehicle considered to have contributed to aspiration risk & related observations in the rat; age 6–9 weeks; ♂/♀
243	Propylene glycol/EtOH/Water (20/5/75)	Rat	IV (bolus)	–	1 mL/kg	Well tolerated	–

(*Continued*)

TABLE A.111 (*Continued*)

Combination Formulations

#	Formulation	Species	Route	Duration	Dose	Adverse Reactions/Toxicity	Notes
244	Propylene glycol/EtOH/Water (30/20/50)	Dog	IV	–	5 mL/kg at a rate of 0.3 mL/kg	Hemolyis *in vitro* dog blood.	–
245	Propylene glycol/EtOH/Water (50/10/40)	Dog	IV	14 days	4 mL/kg/day at a rate of 6 mL/min	Frank red urine after first dose, this was occasionally observed throughout the 2-week period, decreases in hematocrit, hemoglobin, and erythrocyte count. Urinalyses were positive for occult blood, bilirubin, ketones and proteins. Swelling at injection site.	–
246	Propylene glycol/EtOH/Water (60/20/20)	Dog	IV (bolus)	–	0.5 mL/kg	Well tolerated	–
247	Propylene glycol/EtOH/Water (pH 3–11) (40/5/55)	Rat	IV (bolus)	–	2 mL/kg	Well tolerated	–
248	Propylene glycol/EtOH/Water (pH 3–11) (40/5/55)	Rat	IV (infusion)	–	5 mL/kg	Well tolerated	–
249	Propylene glycol/EtOH/Water (pH 3–11) (40/5/55)	Dog	IV (bolus)	–	1 mL/kg	Well tolerated	–
250	Propylene glycol/EtOH/Water (pH 3–11) (40/5/55)	Dog	IV (infusion)	–	2 mL/kg	Well tolerated	–
251	Propylene glycol/NaCl	Rat	PO (gavage)	4 weeks	2 mL/kg/day	Well tolerated	–

(*Continued*)

TABLE A.111 (*Continued*)

Combination Formulations

#	Formulation	Species	Route	Duration	Dose	Adverse Reactions/Toxicity	Notes
252	Propylene glycol/NaCl	Mouse	IV	–	MTD: 1.5 g/kg; LD_{50}: 5.0–8.6 g/kg; NOEL: 1 g/kg	–	CD-1 mice; 30% of a 5 mL/kg dose volume at MTD; (%v/v) in NaCl
253	Propylene glycol/NaCl (0.9%)/EtOH (50/40/10)	Dog	IV	–	5 mL/kg at a rate of 0.3 mL/kg	Hemolysis *in vitro* dog blood	–
254	Propylene glycol/PEG 400/Water/EtOH (40/25/25/10)	Nonhuman Primate	PO (gavage)	4 weeks	2 mL/kg/day	Well tolerated	–
255	Propylene glycol/TPGS/Capmul MCM NF™ (5:5:2 by weight)	Dog	PO	13 weeks	5 mL/kg BD	Administration of the vehicle was associated with emesis and abnormal fecal quality throughout the 13 weeks of the study	Beagle (Marshall) age ~6 months at study initiation; ♂/♀
256	PVP K30 (10%)/Sodium citrate buffer (50 mM, pH 5)	Rat	PO (gavage)	1 day	10 mL/kg Single dose	None	Age 6 weeks; ♂/♀
257	PVP K30 (10%)/Sodium citrate buffer (50 mM, pH 5)	Nonhuman Primate	PO (gavage)	4 weeks	5 mL/kg once weekly	None	Age 5–7 years; ♂/♀
258	PVP K30 (10%)/Sodium citrate buffer (50 mM, pH 5)	Nonhuman Primate	PO (gavage)	28 days	5 mL/kg QD	None	Age 2–3 years; ♂/♀
259	Sesame oil/EtOH (96/4)	Dog	PO	28 days	2 mL/kg	Well tolerated	–
260	Sesame oil/EtOH (96/4)	Rat	PO	28 days	2 mL/kg	Well tolerated	–
261	Sodium acetate (25 mM, USP)/Lactose (70 mg/mL)/Water (sterile for injection, USP, pH 4.5)	Dog	SC	14 doses	1 mL/kg	Red discoloration and swelling at injection sites; chronic active inflammation and hemorrhage	–

(*Continued*)

TABLE A.111 (*Continued*)
Combination Formulations

#	Formulation	Species	Route	Duration	Dose	Adverse Reactions/Toxicity	Notes
262	Sodium acetate (25 mM, USP)/ Lactose (70 mg/mL)/Water (sterile for injection, USP, pH 4.5)	Rat	SC	14 doses	10 mL/kg	Scabbing at injection site	–
263	Sodium acetate (25 mM, USP)/ Lactose (70 mg/mL)/Water (sterile for injection, USP, pH 4.5)	Rat	SC	14 doses	4 mL/kg	None	–
264	Sodium acetate (USP, 25 mM)/ Lactose (USP/EP, 70 mg/mL)/ Water (pH 4.5)	Rat	SC	28 days	10 mL/kg	Well tolerated	–
265	Sodium acetate in NaCl, 5 mM	Rat	IV	1 month	1 mL/kg	Well tolerated	–
266	Sodium acetate trihydrate buffer (50 mM)/Tween 80™ (1%)	Mouse	PO (gavage)	26 weeks	10 mL/kg/ day	Well tolerated	–
267	Sodium acetate trihydrate buffer (50 mM) /Tween 80™ (1%)	Mouse	PO (gavage)	7 days	10 mL/kg/ day	Well tolerated	–
268	Sodium citrate/NaCl buffer	Rat	IV	2 weeks	4 mL/kg bolus, 2/10min	Well tolerated	–
269	Sodium CMC (0.1%)/ Methylparaben sodium (0.1%)/ Propylparaben sodium (0.02%)/ Water (purified) (w/v)	Mouse	PO (gavage)	105 weeks	10 mL/kg/ dose BID	None	Age 7 weeks; ♂/♀

(*Continued*)

TABLE A.111 (*Continued*)

Combination Formulations

#	Formulation	Species	Route	Duration	Dose	Adverse Reactions/Toxicity	Notes
270	Sodium CMC (0.1%)/ Methylparaben sodium (0.1%)/ Propylparaben sodium (0.02%)/ Water (purified) (w/v)	Rat	PO (gavage)	104 weeks	10 mL/kg/ dose BID	None	Age 6 weeks; ♂/♀
271	Sodium hydroxide (0.1 M)/NaCl for injection, USP (0.9%)	Nonhuman Primate	IV (30 minute infusion)	14 days	10 mL/kg QD	None	Age 2–4.5 years; ♂/♀
272	Sodium hydroxide (10% solution): 2%/Phenoxyethanol: 1%/ Carbomer 974P NF™: 1%/in purified water	Mouse	Topical	28 days	3.4 mL/kg QD	None	GLP; age 7 weeks; 12♂/12♀
273	Sodium phosphate (10 mM)/NaCl (0.8%)/Tween 20™ (0.05%)/Water (sterile for injection, USP, pH 6.0 ± 0.3)	Rat	SC	26 weeks	1500 μL/kg twice weekly	None	Age 6 weeks; ♂/♀
274	Sodium phosphate (20 mM)/ Sucrose (1%)/Mannitol (4%)/ Water (for injection)	Rat	SC	26 weeks	1.38 mL/kg QD	None	Age 7 weeks; ♂/♀
275	Sodium phosphate (20 mM)/ Sucrose (1%)/Mannitol (4%)/ Water (for injection)	Rabbit	SC	39 weeks	0.58 mL/kg QD	None	Age 5–6 months; ♂/♀

(*Continued*)

TABLE A.111 (*Continued*)

Combination Formulations

#	Formulation	Species	Route	Duration	Dose	Adverse Reactions/Toxicity	Notes
276	Sodium phosphate buffer (0.3 M)/PEG 400, pH:8 (70:30, w/w)	Nonhuman Primate	PO (gavage)	4 weeks	2 mL/kg/day (0.4 mL/min/kg)	Well tolerated	–
277	Sodium phosphate buffer (20 mM)/Dextrose (4%)/Sodium hydroxide (pH 7.9–8.1)	Rat	Infusion	4 days	2 mL/kg/hr	None	Age 8 weeks; ♀
278	Sodium succinate (25 mM)/Lactose (45 mg/mL)/NaCl (0.45%)/Water (sterile for injection, USP)	Dog	IV	7 doses	1 mL/kg	None	–
279	Sodium succinate (25 mM)/Lactose (45 mg/mL)/NaCl (0.45%)/Water (sterile for injection, USP)	Dog	Sc	14 doses	1 mL/kg	None	–
280	Sodium succinate (25 mM)/Lactose (45 mg/mL)/NaCl (0.45%)/Water (sterile for injection, USP)	Rat	IV	7 doses	1 mL/kg	None	–
281	Sodium succinate (25 mM)/Lactose (45 mg/mL)/NaCl (0.45%)/Water (sterile for injection, USP)	Rat	SC	14 doses	1 mL/kg	None	–
282	Sodium succinate (25 mM)/Mannitol™ (250 mM)/Water (sterile for injection, pH 4.6)	Rat	SC	Single dose	4 mL/kg	None	–
283	Solutol HS 15™/EtOH/Water (40/10/50, v/v/v)	Dog	PO	q7d over 8 weeks	1.5 mL/kg	None	Non-GLP; age 1–4 years; 3♂/3♀

(Continued)

TABLE A.111 (*Continued*)

Combination Formulations

#	Formulation	Species	Route	Duration	Dose	Adverse Reactions/Toxicity	Notes
284	Solutol HS 15™: 15%/EtOH: 5%/PBS	Dog	PO	5 days	20 mL/kg QD	May cause vomiting, loose stool	Non-GLP; age 1–2 years; 2♂
285	Solutol™: 10%/in NaCl	Rat	IV	14 days	6 mL/kg SD	None	Non-GLP; age 10 weeks; 3♂/3♀
286	Solutol HS15™/EtOH/Water(40/10/50)	Nonhuman Primate	PO (gavage)	9 months	3 mL/kg/day	Well tolerated	–
287	Sorbitol (5%)/Histidine (10 mM)/Tween 80™ (0.01%)/Water (sterile for injection, USP, pH 5.8)	Dog	IV (bolus)	5 weeks	4.98 mL/kg once weekly	None	Age 2–3 years; ♂/♀
288	Soybean oil: 50.00%/Coconut oil: 23.60%/Mineral oil: 5.80%/Cyclomethicone: 5.00%/Cetostearyl alcohol: 3.50%/Stearic acid: 3.00%/Myristyl alcohol: 2.50%/Hydrogenated castor oil: 2.00%/White wax (beeswax): 2.00%/Stearyl alcohol: 1.50%/Docosanol: 1.10%	Minipig	Topical	3 weeks	0.25 mL/kg QD	None	GLP; age 3–4 months; 5♂/5♀
289	Sucrose (1%)/NaCl (100 mM)/L-arginine hydrochloride (25 mM)/Sodium phosphate (25 mM, pH 6.3)/Water (for injection, USP)	Nonhuman Primate	SC	4 weeks	2 mL/kg twice weekly	Soft/watery feces	Age 2.5–3.5 years; ♂/♀
290	Sucrose acetate isobutyrate/EtOH/PEG 300 (90/5/5)	Cat	Oral mucosa	8 hours	0.1 mL SD	None	Non-GLP; age >6 months; 3♀

(*Continued*)

TABLE A.111 (*Continued*)

Combination Formulations

#	Formulation	Species	Route	Duration	Dose	Adverse Reactions/Toxicity	Notes
291	TPGS (2%)/HPMC acetate succinate (1% HF grade)/PVP K30 (0.25%)/Water (DI)	Rat	PO (gavage)	91 days	10 mL/kg QD	None	Age 6 weeks; ♂/♀
292	TPGS (2%)/HPMC acetate succinate (1% HF grade)/PVP K30 (0.25%)/Water (DI)	Dog	PO (gavage)	91 days	10 mL/kg QD	None	Age 6–7 months; ♂/♀
293	TPGS (2%)/HPMC acetate succinate (1% HF grade)/PVP K30 (0.25%)/Water (DI)	Rabbit	PO (gavage)	14 days	10 mL/kg QD	None	Age 5–8 months; ♂/♀
294	TPGS (2%)/HPMC acetate succinate (1.5% HF grade)/PVP VA 64 (1.5%)/Sodium citrate (50 mM pH 5)/Water (DI)	Rat	PO (gavage)	28 days	10 mL/kg QD	None	Age 6–8 weeks; ♂/♀
295	TPGS (5%)/Methylcellulose 400 cps (0.5%)/Water (DI)	Rat	PO (gavage)	1 day	10 mL/kg single dose	None	Age 6 weeks; ♂/♀
296	Trehalose (9%)/Lactic acid (10mM)	Dog	IA	q28d × 4 doses over 85 days	0.4 mL	None	GLP; age 9–11 months; 5♂/5♀
297	Trehalose (9%)/Lactic acid (10mM)	Dog	IV	2×/ wk × 8 doses over 28 days	0.5 mL/kg	None	GLP; age 1 year; 3♀
298	Trehalose (9%)/Lactic acid (10mM)	Rat	IV	14 days	0.5–5 mL/kg SD	None	GLP; age 16–20 weeks; 6♂/6♀

(*Continued*)

TABLE A.111 ***(Continued)***

Combination Formulations

#	Formulation	Species	Route	Duration	Dose	Adverse Reactions/Toxicity	Notes
299	Tween 20™ (0.01%)/Sodium acetate (10 mM)/Sorbitol (5%, pH 5)	Nonhuman Primate	IV (bolus)	12 weeks	5 mL/kg once weekly	None	Age 2–4 years; ♂/♀
300	Tween 20™ (0.01%)/Sodium acetate (10 mM)/Sorbitol (5%, pH 5)	Rat	IV (bolus)	12 weeks	5 mL/kg once weekly	None	Age 7 weeks; ♂/♀
301	Tween 80™/CMC/Dimethicone (0.01 %)(Ratio PS80/CMC of 1:1, 0.2 % de PS80 and CMC)	Rabbit	PO (gavage)	1 month	3 mL/kg/day	Well tolerated	–
302	Tween 80™ (10%)/Citric acid (10.5 mg/mL)/Water (sterile solution)	Rat	IV	2 doses	3 mL/kg	None	–
303	Vitamin E (20%)/Sodium citrate buffer (50 mM, pH 5)	Rat	PO (gavage)	28 days	10 mL/kg QD	None	Age 8 weeks; ♂/♀
304	Water (sterile for injection, USP)/Sodium hydroxide	Rat	SC	Single dose	1 mL/kg	Necrosis of the subcutaneous muscle panniculus carnosus, inflamed injection	–
305	Xanthan gum NF (aka: Xantural 180™)(0.2%, w/v)/Tween 80™ NF (0.255%, w/v)/Water (sterile for injection, USP)	Rat	PO (gavage)	5 days	10 mL/kg/dose BID	None	Age 8–10 weeks; ♀

References

Aerpio technical document. (2015) Type C Meeting Information Packet. May 26, 2015.

Ahmad SK, Brinch DS, Friis EP, Pedersen PB. (2004) Toxicological studies on lactose oxidase from microdochium nivale expressed in fusarium venenatum. *Regul Toxicol Pharmacol*, 39(3):256–70.

Albers E, Muller BW. (1995) Cyclodextrin derivatives in pharmaceutics. *Crit Rev Ther Drug Carrier Syst*, 12(4):311–337.

Ali BH. (2001) Dimethyl sulfoxide: Recent pharmacological and toxicological research. *Vet Hum Toxicol*, 43(4):228–231.

Allen B, Gentry R, Shipp A, Van Landingham C. (1998) Calculation of benchmark doses for reproductive and developmental toxicity observed after exposure to isopropanol. *Regul Toxicol Pharmacol*, 28(1):38–44.

American Lecithin (2007) *Phosal® 53 MCT*. American Lecithin Website. http://www.americanlecithin.com/TDS/TDS_53MCT.PDF, February 2007. Accessed August 7, 2015.

Anderson DM. (1986) Evidence for the safety of gum Arabic (Acacia senegal (L.) Willd.) as a food additive – A brief review. *Food Addit Contam*, 3(3):225–230.

Anderson DM. (1989) Evidence for the safety of gum tragacanth (Asiatic Astragalus spp.) and modern criteria for the evaluation of food additives. *Food Addit Contam*, 6(1):1–12.

Anderson RC, Harris PN, Chen KK. (1950) Toxicological studies on synthetic glycerin. *J Am Pharm Ass, Sci Ed*, 39(10):583–585.

Augustine KA, Zhang Q, Welsh MJ. (2000) A study of vehicles for dosing cultured rodent embryos with non-aqueous soluble compounds. *Toxicologist*, 54(1):297.

Bachmann E, Weber E, Post M, Zbinden G. (1978) Biochemical effects of gum arabic, gum tragacanth, methylcellulose and carboxymethylcellulose-na in rat heart and liver. *Pharmacology*, 17(1):39–49.

Baldrick P, Bamford DG. (1997) A toxicological review of lactose to support clinical administration by inhalation. *Food Chem Toxicol*, 35(7):719–733.

Bar A, Til HP, Timonen M. (1995) Subchronic oral toxicity study with regular and enzymatically depolymerized sodium carboxymethylcellulose in rats. *Food Chem Toxicol*, 33(11):909–917.

Barrie MD, Dahlstrom DL, Goswami E, Kaetzel R. (2012) Sodium chloride. In: Bingham E and B Cohrssen (Eds.) *Patty's Toxicology*, 6th Ed., Vol. 1. New York, NY: John Wiley and Sons, pp. 1073–1074.

Bartsch W, Sponer G, Dietmann K, Fuchas G. (1976) Acute toxicity of various solvents in the mouse and rat. LD50 of ethanol, diethylacetamide, dimethylformamide, dimethylsulfoxide, glycerine, n-methylpyrrolidone, polyethylene glycol 400, 1, 2-propanediol and Tween 20. *Arzneimittelforschung*, 26(8):1581–1583.

Beckwith-Hall BM, Holmes E, Lindon JC, et al. (2002) NMR-Based metabonomic studies on the biochemical effects of commonly used drug carrier vehicles in the rat. *Chem Res Toxicol*, 15(9):1136–1141.

Beji S, Kaaroud H, Ben Moussa F, et al. (2006) Acute renal failure following mucosal administration of povidone iodine. *Presse Med*, 35(1 Pt 1):61–63.

Bellantone R, Bossola M, Carriero C, et al. (1999) Structured versus long-chain triglycerides: A safety, tolerance and efficacy randomized study in colorectal surgical patients. *J Parenter Enteral Nutr*, 23(3):123–127.

Bendich A, Cohen M. (1990) Ascorbic acid safety: Analysis of factors affecting iron absorption. *Toxicol Lett*, 51(2):189–201.

Benita S. (1999) Prevention of topical and ocular oxidative stress by positively charged submicron emulsion. *Biomed Pharmacother*, 53(4):193–206.

Bevan C. (2012a) Ethanol. In: Bingham E and B Cohrssen (Eds.) *Patty's Toxicology*, 6th Ed., Vol. 3. New York, NY: John Wiley & Sons, pp. 926–934.

Bevan C. (2012b) Isopropanol. In: Bingham E and B Cohrssen (Eds.) *Patty's Toxicology*, 6th Ed., Vol. 3. New York, NY: John Wiley & Sons, pp. 937–943.

Bevan C. (2012c) C16 Alcohols. In: Bingham E and B Cohrssen (Eds.) *Patty's Toxicology*, 6th Ed., Vol. 4. New York, NY: John Wiley & Sons, p. 21.

Briggs D, Lok E, Nera EA, Karpinski K, Clayson DB. (1989) Short-term effects of butylated hydroxytoluene on the Wistar rat liver, urinary bladder and thyroid gland. *Cancer Lett*, 46(1):31–36.

Buard A, Carlton BD, Floch F, Simon GS. (2003) Subchronic toxicity, mutagenicity and allergenicity studies of a cultured dextrose food product. *Food Chem Toxicol*, 41(5):689–694.

Budden R, Kühl UG, Buschmann G. (1978) Studies on the pharmacodynamic activity of several drug solvents. 1st communication: Diethyleneglycol monoethylether, n, n-diethylacetamide, dimethylsufoxide (author's transl). *Arzneimittelforschung*, 28(9):1571–1579.

Burleigh-Flayer H, Gill M, Hurley N, et al. (1998) Motor activity effects in female Fischer 344 rats exposed to isopropanol for 90 days. *J Appl Toxicol*, 18(5):373–381.

Cambridge MedChem Consulting. (2012) *Formulation*. Cambridge MedChem Consulting website. http://www.cambridgemedchemconsulting.com/resources/formulation.html Cambridge, UK. Updated February 17, 2012. Accessed October 3, 2015.

Campbell RL, Bruce RD. (1981) Comparative dermatotoxicology. I. Direct comparison of rabbit and human primary skin irritation responses to isopropylmyristate. *Toxicol Appl Pharmacol*, 59(3):555–563.

Caraccio TR, McGuigan MA. (2004) Over-the-counter products. In: Dart RC, et al. (Eds.) *Medical Toxicology*, 3rd Ed. Philadelphia, PA: Lippincott Williams & Wilkins, p. 1051.

Carlton WW, Boitnott JK, Dungworth DL, et al. (2001) Assessment of the morphology and significance of the lymph nodal and hepatic lesions produced in rats by the feeding of certain mineral oils and waxes. Proceedings of a pathology workshop held at the Fraunhofer Institute of Toxicology and Aerosol Research Hannover, Germany, May 7–9, 2001. *Exp Toxicol Pathol*, 53(4):247–255.

Carpenter CP, Shaffer CB. (1952) A study of the polyethylene glycols as vehicle for intramuscular and subcutaneous injection. *J Am Pharm Assoc*, 41:27–29.

Caujolle F, Canh PH, Dat-Xuong N, Azum-Gelade MC. (1970) Toxicological studies upon acetamide and its n-methyl and n-ethyl derivatives. *Arzneimittelforschung*, 20(9):1242–1246.

Cavender FL. (2012a) Propylene glycol. In: Bingham E and B Cohrssen (Eds.) *Patty's Toxicology*, 6th Ed., Vol. 4. New York, NY: John Wiley & Sons, pp. 611–616.

Cavender FL. (2012b) Polyethylene glycols. In: Bingham E and B Cohrssen (Eds.) *Patty's Toxicology*, 6th Ed., Vol. 4. New York, NY: John Wiley & Sons, pp. 608–611.

Cavender FL. (2012c) Hydroxypropyl cellulose. In: Bingham E and B Cohrssen (Eds.) *Patty's Toxicology*, 6th Ed., Vol. 4. New York, NY: John Wiley & Sons, pp. 945–946.

Cavender FL. (2012d) Sodium carboxymethyl cellulose. In: Bingham E and Cohrssen B (Eds.) *Patty's Toxicology*, 6th Ed., Vol. 4. New York, NY: John Wiley & Sons, p. 943.

Cavender FL. (2012e) Ethylenediaminetetraacetic acid. In: Bingham E and B Cohrssen (Eds.) *Patty's Toxicology*, 6th Ed., Vol. 2. New York, NY: Wiley & Sons, pp. 483–484.

Challa R, Ahuja A, Ali J, Khar RK. (2005) Cyclodextrins in drug delivery: An updated review. *AAPS PharmSciTech*, 6(2):E329–E357.

Cho YA, Gwak HS. (2004) Transdermal delivery of ketorolactromethamine: Effects of vehicles and penetration enhancers. *Drug Deg Ind Pharm*, 30:557–564.

Church AS, Witting MD. (1997) Laboratory testing in ethanol, methanol, ethylene glycol, and isopropanol toxicities. *J Emerg Med*, 15(5):687–692.

Clark DP, Hanke CW, Swanson NA. (1989) Dermal implants: Safety of products injected for soft tissue augmentation. *J Am Acad Dermatol*, 5(1):992–998.

Coon JS, Knudson W, Clodfelter K, Lu B, Weinstein RS. (1991) Solutol®HS15, nontoxic polyethylene esters of 12-hydroxystearic acid, reverses multidrug resistance. *Cancer Res*, 51:897–902.

Cosmetic Ingredient Review. (1999) Final report on the safety assessment of propylene glycol (PG) dicaprylate, PG dicaprylate/dicaprate, PG dicocoate, PG dipelargonate, PG isostearate, PG laurate, PG myristate, PG oleate, PG oleate SE, PG dioleate, PG dicaprate, PG diisostearate, and PG dilaurate. *Int J Toxicol*, 18:35–52.

Cosmetic Ingredient Review. (2001) Final report on the safety assessment of peanut (arachis hypogaea) oil, hydrogenated peanut oil, peanut acid, peanut glycerides, and peanut (arachis hypogaea) flour. *Int J Toxicol*, 20(Suppl 2):65–77.

Cosmetic Ingredient Review. (2004) Final report on the amended safety assessment of glyceryl laurate, glyceryl laurate SE, glyceryl laurate/oleate etc. *Int J Toxicol*, 23(2):55–94.

Coussement W, Van Cauteren H, Vandenberghe J, et al. (1990) Toxicological profile of hydroxypropyl b-cyclodextrin (HP B-CD) in laboratory animals. In: Duchene D (Ed.) *Minutes 5th International Symposium on Cyclodextrins*. Paris: Editions de Sante, pp 552–613.

Curry DJ, Wright DA, Lee RC, Kang UJ, Frim DM. (2004) Poloxamer 188 volumetrically decreases neuronal loss in the rat in a time-dependent manner. *Neurosurgery*, 55(4):943–949.

Daher CF, Baroody GM, Howland RJ. (2003) Effect of a surfactant, Tween 80, on the formation and secretion of chylomicrons in the rat. *Food Chem Toxicol*, 41(4):575–582.

Dalbey WE and Biles RW. (2003) Respiratory toxicology of mineral oils in laboratory animals. *Appl Occup Environ Hyg*, 18(11):921–929.

David RM, Bachman AN, Butala JH, Piper JT, Shelp CJ. (2012) Methyl benzoate. In: Bingham E and B Cohrssen (Eds.) *Patty's Toxicology*, 6th Ed., Vol. 4. New York, NY: John Wiley & Sons, pp. 152–156.

DeWitt JC, Meyer EB, Henshel DS. (2005) Environmental toxicity studies using chickens as surrogates for wildlife: Effects of vehicle volume. *Arch Environ Contam Toxicol*, 48(2):260–269.

Diethylacetamide. (2015) ChemIDPlus website. http://chem.sis.nlm.nih.gov/.chemidplus/name/diethylacetamide, Rockville, MD. Accessed August 7, 2015.

Dordunoos SK, Ford JL, Rubinstein MH. (1996) Solidification studies of polyethylene glycols, Gelucire 44/14 or their dispersions with triamterene of Temazepam. *J Pharm Pharmacol*, 48(8):782–789.

Dow Corning. (2015a) *Antifoam 1510-US. MSDS*. Down Corning Website. http://www.dowcorning.com/DataFiles/090277018224ba08.pdf, updated April 2015. Accessed August 7, 2015.

Dow Corning. (2015b) *Dimethiconol Blend* 20. MSDS. Dow Corning Website. http://www .dowcorning.com/DataFiles/090277018213b935.pdf, updated April 2015. Accessed August 7, 2015.

Dupont J, White PJ, Carpenter MP, et al. (1990) Food uses and health effects of corn oil. *J Am Coll Nutr*, 9(5):438–470.

Dykes MHM, Meier P. (1975) Ascorbic acid and the common cold: Evaluation of its efficacy and toxicity. *JAMA*, 231(10):1073–1079.

Earl LK, Baldrick P, Hepburn PA. (2002) A 13-week feeding study in the rat with shea oleine and hardened shea oleine. *Int J Toxicol*, 21:13–22.

Evangelista CM, Antunes LM, Francescato HD, Bianchi ML. (2004) Effects of the olive, extra virgin olive and canola oils on cisplatin-induced clastogenesis in Wistar rats. *Food Chem Toxicol*, 42(8):1291–1297.

Farber TM, Ritter DL, Weinberger MA, et al. (1976) The toxicity of brominated sesame oil and brominated soybean oil in miniature swine. *Toxicology*, 5(3):319–336.

Feitoza AB, Gostout CJ, Burgart LJ, et al. (2003) Hydroxypropyl methylcellulose: A better submucosal fluid cushion for endoscopic mucosal resection. *Gastrointest Endosc*, 57(1):41–7.

Fini A, Moyano JR, Ginés JM, Perez-Martinez JI, Rabasco AM. (2005) Diclofenac salts, II. Solid dispersions in PEG6000 and Gelucire 50/13. *Eur J Pharm Biopharm*, 60(1):99–111.

Fisherman EW, Cohen GN. (1974) Tween 80 exacerbated intrinsic rhinitis and asthma in non-aspirin-sensitive patients. *Ann Allergy*, 32(6)307–320.

Fort FL, Heyman IA, Kesterson JW. (1984) Hemolysis study of aqueous polyethylene glycol 400, propylene glycol and ethanol combinations in vivo and in vitro. *J Parenter Sci Technol*, 38(2):82–87.

Freeman C, Weiner ML, Kotkoskie LA, Borzelleca J, Butt M. (2003) Subchronic and developmental toxicity studies in rats with Ac-Di-Sol croscarmellose sodium. *Int J Toxicol*, 22(3):149–157.

Frim DM, Wright DA, Curry DJ, et al. (2004) The surfactant Poloxamer-188 protects against glutamate toxicity in the rat brain. *Neuroport*, 15(1):171–174.

Gad SE. (2014) Cyclohexane. In: Wexler P, et al. (Eds.) *Encyclopedia of Toxicology*, 3rd Ed., Vol. 1. Oxford: Elsevier, pp. 1106–1108.

Gattefossé technical document. Gelucire 50/13. October 25, 2004.

Gattefossé technical document. Labrafil M1944 CS. February 24, 2003a.

Gattefossé technical document. Plurol Oleique CC497. December 16, 2003b.

Geerling G, Daniels JT, Dart JK, Cree IA, Khaw PT. (2001) Toxicity of natural tear substitutes in a fully defined culture model of human corneal epithelial cells. *Invest Opthamol Vis Sci*, 42(5):948–956.

Gelderblom H, Verweij J, Nooter K, Sparreboom A. (2001) Cremophor EL: The drawbacks and advantages for drug formulation. *Eur J Cancer*, 37(13):1590–1598.

Gelperina SE, Khanansky AS, Skidan IN, et al. (2002) Toxicological studies of doxorubicin bound to polysorbate 80-coated poly (butyl cyanoacrylate) nanoparticles in healthy rats and rats with intracranial glioblastoma. *Toxicol Lett*, 126(2):131–141.

Genovese RF, Newman DB, Gordon KA, Brewer TG. (1999) Acute high dose artemether toxicity in rats. *Neurotoxicology*, 20(5):851–859.

Gerloczy A, Hoshino T, Pitha J. (1994) Safety of oral cyclodextrins: Effects of hydroxypropyl cyclodextrins, cyclodextrins sulfates and cationic cyclodextrins on steroid balance in rats. *J Pharm Sci*, 83(2):193–196.

Gould S, Scott RC. (2005) 2-Hydroxypropyl-beta-cyclodextrin (HP-beta-CD): A toxicology review. *Food Chem Toxicol*, 43(10):1451–1459.

Grindel JM, Jaworski T, Piraner O, Emanuele RM, Balasubramanian M. (2002) Distribution, metabolism, and excretion of a novel surface-active agent, purified Poloxamer 188, in rats, dogs, and humans. *J Pharm Sci*, 91(9):1936–1947.

Gupta U, Beaulieu J, Chapin Hopper J, Hagler AR, Hills-Perry P. (1996) Teratogenic evaluation of alternative vehicles: PEG 400, cremophor, carboxy-methylcellulose; Comparisons with methylcellulose. *Teratology*, 53(2):111.

Gutiérrez-Cabano CA. (2000) Protection by ingastric polyethylene glycol 400 in rat stomach against ethanol damage involves α2-adrenoceptors. *Dig Dis Sci*, 45(1):105–109.

Hagiwara A, Boonyaphiphat P, Kawabe M, et al. (1992) Lack of carcinogenicity of tragacanth gum in B6C3F1 mice. *Food Chem Toxicol*, 30(8):673–679.

Hardin BD, Goad PT, Burg JR. (1984) Developmental toxicity of four glycol ethers applied cutaneously to rats. *Environ Health Perspect*, 57:69–74.

Hardin BD. (1983) Reproductive toxicity of the glycol ethers. *Toxicology*, 27(2):91–102.

Heimbach J, Rieth S, Mohamedshah F, et al. (2000) Safety assessment of iron EDTA: Summary of toxicological, fortification and exposure data. *Food Chem Toxicol*, 38(1):99–111.

Hermansky SJ, Neptun DA, Loughran KA, Leung HW. (1995) Effects of polyethylene glycol 400 (PEG 400) following 13 weeks of gavage treatment in Fischer-344 rats. *Food Chem Toxicol*, 33(2):139–149.

Hironishi M, Kordek R, Yanagihara R, and Garruto RM. (1996) Maltol (3-hydroxy-2-methyl-4-pyrone) toxicity in neuroblastoma cell lines and primary murine fetal hippocampal neuronal cultures. *Neurodegeneration*, 5(4):325–329.

Horvath IP, Somafi-Relle S, Hegedus L, Jarman M. (1982) Toxicity, antitumor and haematological effects of 1,2-anhydro-6-bromogalactitol and d-mannitol: A comparison with the related dibromo- and dianhydro-derivatives. *Eur J Cancer Clin Oncol*, 18(6):573–577.

Hu Z, Prasad Yv R, Tawa R, et al. (2002) Diethyl ether fraction of Labrasol having a stronger absorption enhancing effect on gentamicin than Labrasol itself. *Int J Pharm*, 234(1–2):223–235.

Hu Z, Tawa R, Konishi T, Shibata N, Takada K. (2001) A novel emulsifier, Labrasol, enhances gastrointestinalabsorption of gentamicin. *Life Sci*, 69(24):2899–2910.

JECFA. (1997) *Alginic Acid*. 49th JECFA FNP 52 Addendum 5. Food & Agriculture Organization Website. http://www.fao.org/ag/agn/jecfa-additives/specs/Monograph1/Additive-009.pdf. Accessed August 6, 2015.

Jefferson R. (2012) Trisodium phosphate. In: Bingham E and B Cohrssen (Eds.) *Patty's Toxicology*, 6th Ed., Vol. 1. New York, NY: John Wiley & Sons, pp. 942–943.

Jockovich M, Murray T, Excalona-Benz E, Hernandez E, Feuer W. (2006) Anecortave acetate as a single and adjuvant therapy in the treatment of retinal tumors of LDbetaTag mice. *Invest Opthal Visual Sci*, 47:1264–1268.

Kärber G. (1951) Über sogenanntes Oxydwachs. *Arch Exp Pathol Pharmakol*, 212:509–521.

Kawakami K, Miyoshi K, Ida Y. (2004) Solubilization behavior of poorly soluble drugs with combined use of Gelucire 44/14 and cosolvent. *J Pharm Sci*, 93(6):1471–1479.

Kawashima H, Toyoda-Ono Y, Suwa Y, et al. (2009) Subchronic (13-week) oral toxicity study of dihomo-y-linolenic acid (DGLA) oil in rats. *Food Chem Toxicol*, 47(6):1280–1286.

Kennedy GL Jr. (2012) N-Methyl-2-Pyrrolidinone. In: Bingham E and B Cohrssen (Eds.) *Patty's Toxicology*, 6th Ed. Vol. 2. New York, NY: John Wiley & Sons, pp. 727–731.
Kim Y, Oksanen DA, Massefski W Jr, et al. (1998) Inclusion complexation of ziprasidone mesylate with beta-cyclodextrin sulobutyl ether. *J Pharm Sci*, 87(12):1560–1567.
Kligman AM. (1983) Lanolin allergy: Crisis or comedy. *Contact Dermatitis*, 9(2):99–107.
Komatsu H, Asaba K, Suzuki M. (1979) Some biochemical effects of isopropyl myristate and squalane on rabbit skin. *J Soc Cosmet Chem*, 30:263–278.
Kreckmann KH, Baldwin JK, Roberts LG, et al. (2000) Inhalation developmental toxicity and reproduction studies with cyclohexane. *Drug Chem Toxicol*, 23:555–573.
Lanigan RS, Yamarik TA. (2002a) Final report on the safety assessment of BHT(1). *Int J Toxicol*, 21(Suppl 2):19–94.
Lanigan RS, Yamarik TA. (2002b) Final report on the safety assessment of sorbitan caprylate, sorbitan cocoate, sorbitan diisostearate, sorbitan dioleate, sorbitan distearate, sorbitan isostearate, sorbitan olivate, sorbitan sesquiisostearate, sorbitan sesquistearate, and sorbitan triisostearate. *Int J Toxicol*, 21(1):93–112.
Lanigan RS, Yamarik TA. (2002c) Final report on the safety assessment of EDTA, calcium disodium EDTA, diammonium EDTA, dipotassium EDTA, disodium EDTA, TEA-EDTA, tetrasodium EDTA, tripotassium EDTA, trisodium EDTA, HEDTA, and trisodium HEDTA. *Int J Toxicol*, 2:95–142.
Leadscope Data Portal. (2008a) LS-711. Modified May 19, 2008.
Leadscope Data Portal. (2008b) LS-55950. Modified May 20, 2008.
Lee KP, Chromey NC, Culik R, Barnes JR, Schneider PW. (1987) Toxicity of n-methyl-2-pyrrolidone (NMP): Teratogenic, subchronic, and two-year inhalation studies. *Fundam Appl Toxicol*, 9(2):222–235.
Lemieux P, Guerin N, Paradis G, et al. (2000) A combination of poloxamers increases gene expression of plasmid DNA in skeletal muscle. *Gene Ther*, 7(11):986–991.
Li B-Q, Dong X, Fang S-H, et al. (2011) Systemic toxicity and toxicokinetics of a high dose of polyethylene glycol 400 in dogs following intravenous injection. *Drug ChemToxicol*, 34(2):208–212.
Li P, Ghosh A, Wagner RF, et al. (2005) Effect of combined use of nonionic surfactant on formation of oil-in-water microemulsions. *Int J Pharm*, 288:27–34.
Lina BA, Bos-Kuijpers MH, Til HP, Bar A. Chronic toxicity and carcinogenicity study of erythritol in rats. *Regul Toxicol Pharmacol*, 24(2 Pt 2):S264–S279.
Liu H, Li S, Wang Y, Yao H, Zhang Y. (2006a) Effect of vehicles and enhancers on the topical delivery of cyclosporin A. *Int J Pharm*, 311:182–186.
Liu Z, Li J, Nie S, Guo H, Pan W. (2006b) Effects of Transcutol on the corneal permeability of drugs and evaluation of its ocular irritation of rabbit eyes. *J Pharm Pharmacol*, 58(1):45–50.
Lorenz W, Reimann HJ, Schmal A, et al. (1977) Histamine release in dogs by Cremophor EL and its derivatives: Oxethylated oleic acid is the most effective constituent. *Agents Actions*, 7(1):63–67.
Maki KC, Davidson MH, Torri S, et al. (2000) High-molecular-weight hydroxypropylmethylcellulose taken with or between meals is hypocholesterolmic in adult men. *Amer Soc Nutr Sci*, 130(7):1705–1710.
Malley LA, Bamberger JR, Stadler JC, et al. (2000) Subchronic toxicity of cyclohexane in rats and mice by inhalation exposure. *Drug Chem Toxicol*, 23(4):513–537.
Marttin E, Verhoef JC, Merkus FW. (1998) Efficacy, safety and mechanism of cyclodextrins as absorption enhancers in nasal delivery of peptide and protein drugs. *J Drug Target*, 6(1):17–36.

McCarthy DM, Haas M, Thuluvath PJ, et al. (2002) Pulmonary embolization of bovine collagen. *Arch Pathol Lab Med*, 127(2):67–69.

Mehlman MA. (2012a) Carboxymethylcellulose. In: Bingham E and B Cohrssen (Eds.) *Patty's Toxicology*, 6th Ed., Vol. 3. New York, NY: John Wiley & Sons, pp. 644–645.

Mehlman MA. (2012b) Hydroxypropyl Methylcellulose. In: Bingham E and B Cohrssen (Eds.) *Patty's Toxicology*, 6th Ed., Vol. 3. New York, NY: John Wiley & Sons, pp. 643–644.

Mehlman MA. (2012c) Methylcellulose. In: Bingham E and Cohrssen B (Eds.) *Patty's Toxicology*, 6th Ed., Vol. 3. New York, NY: John Wiley & Sons, pp. 642–643.

Meneely GR, Ball CO. (1958) Experimental epidemiology of chronic sodium chloride toxicity and the protective effect of potassium chloride. *Am J Med*, 25:713–725.

Meneely GR, Tucker RG, Darby WJ, Auerbach SH. (1953) Chronic sodium chloride toxicity in the albino rat: II. *JEM*, 98:71–80.

Modderman JP. (1993) Safety assessment of hydrogenated starch hydrolysates. *Regul Toxicol Pharmacol*, 18:80–114.

Moore GL, Boswell GW, Ledford ME, et al. (1988) Toxicity and clearance of sodium phosphate intravenously injected into rabbits. *Mil Med*, 153(4):203–206.

Moorman M, Safrit C, Richardson J, Gwaltney-Brant S. (Speakers) (2001) *Winter Hazards Part II - Antifreeze Toxicity*. National Animal Poison Control Center, Nanette Walker Smith, N. (Moderator). December 10, 2001. Web. http://www.vspn.org/library/rounds/vspn_vspn011210.htm. Last accessed June 16, 2015.

Murakami K, Ishida K, Watakabe K, et al. (2006) Maltol/iron-mediated apoptosis of HL60 cells: Participation of reactive oxygen species. *Toxicol Lett*, 161(2):102–107.

Nair B. (2001) Final report on the safety assessment of benzyl alcohol, benzoic acid, and sodium benzoate. *Int J Toxicol*, 20(Suppl 3):23–50.

Nakagawa Y, Tayama K, Nakao T, Hiraga K. (1984) On the mechanism of butylated hydroxytoluene-induced hepatic toxicity in rats. *Biochem Pharma*, 33(16):2669–2674.

Nash JF, Gettings SD, Diembeck W, Chudowski M, Kraus AL. (1996) A toxicological review of topical exposure to white mineral oils. *Food Chem Toxicol*, 34(2):213–225.

National Toxicology Program. (1992) NTP toxicology and carcinogenesis studies of polysorbate 80 (CAS No. 9005-65-6) in F344/N rats and B6C3F1 mice (feed studies). *National Toxicology Program Tech Rep Ser*, 415:1–225.

National Toxicology Program. (2001) Toxicology and carcinogenesis studies of coconut oil acid diethanolamine condensate (CAS No. 68603-42-9 in F344/N rats and B6C3F1 mice (dermal studies). *Natl Toxicol Program Tech Rep Ser*, 479:5–226.

O'Sullivan SM, Woods JA, O'Brien NM. (2004) Use of Tween 40 and Tween 80 to deliver a mixture of phytochemicals to human colonic adenocarcinoma cell (CaCo-2) monolayers. *Brit J Nutr*, 91(5):757–764.

Obara S, Muto H, Kokubo H, et al. (1992) Studies on single-dose toxicity of hydrophobically modified hydroxypropyl methylcellulose in rats. *J Toxicol Sci*, 17(1):13–19.

Oz M, Spivak CE, Lupica CR. (2004) The solubilizing detergents, Tween 80 and Triton X-100 non-competitively inhibit alpha 7-nicotinic acetylcholine receptor function in xenopus oocytes. *J Neurosci Methods*, 137(2):167–173.

Paddock Laboratories (2015) *Ora-Plus Oral Suspending Vehicle*. Paddock Laboratories Website. http://medicalfs.com.au/portal/images/stories/Orarange/Ora-Plus-Sell-Sheet.pdf. Copyright 2010. Accessed August 7, 2015.

Passerini N, Perissutti B, Moneghini M, et al. (2002) Charcterization of carbamazepine-Gelucire 50/13 microparticle prepared by a spray-congealing process using ultrasounds. *J Pharm Sci*, 91(3):699–707.

Patel M, Patel G, Patel R, et al. (2005) PEGylation: An innovative approach for protein delivery. *Drug Delivery Tech*, 5(6):48–56.

Pestel S, Martin HJ, Maier GM, Guth B. (2006) Effect of commonly used vehicles on gastrointestinal, renal and liver function in rats. *J Pharmacol Toxicol Methods*, 54(2):200–214.

Pfordte K. (1971) Polyäthyleneglykol, ein wenig toxisches lösemittel zur intravenösen applikation wasserunlöslicher substanzen. *Zentralbl Plarm*, 110:449–451.

PharmPK Discussion (2002a) Cosolvents/surfactants in IV formulations for preclinical studies, 2 Oct 2002. Web. http://www.pharmpk.com/PK02/PK2002075.html. Last accessed March 31, 2015.

PharmPK Discussion (2002b) Cremophor EL toxicity, October 3, 2002, Fiona Starvos Response.

Prasamthi K, Muralidhara, Rajini PS. (2005) Fenvalerate-induced oxidative damage in rat tissues and its attenuation by dietary sesame oil. *Food Chem Toxicol*, 43(2):299–306.

Quadbeck G. (1950) Äthylenglykol und verwandte verbindungen als pharmazeutische lösungmittel. *Chem Z*, 74:91–94.

Rajewski RA, Taiger G, Bresnahan J, et al. (1995) Preliminary safety evaluation of parenterally administered sulfoalkyl ether beta-cyclodextrin derivatives. *J Pharm Sci*, 84(8):927–932.

Ramadan LA, El-Habit OH, Arafa H, Sayed-Ahmad MM. (2001) Effect of Cremophor-EL on cisplatin-induced organ toxicity in normal rat. *J Egyptian Nat Cancer Inst*, 13(2):139–145.

Ratsimbazafy V, Bourret E, Duclos R, Brossard C. (1999) Rheological behavior of drug suspensions in Gelucire mixtures and proxyphylline release from matrix hard gelatin capsules. *Eur J Pharm Biopharm*, 48:247–252.

Registry of Toxic Effects of Chemical Substances (RTECS) – Methyl sulfoxide. Website. http://www.cdc.gov/niosh-rtecs/pv5ec1d0.html. Published 2002. Updated May, 2009. Accessed August 7, 2015.

Robertson RP, Harmon J, Tran PO, Tanaka Y, Takahashi H. (2003) Glucose toxicity in B-cells: Type 2 diabetes, good radicals, one bad, and the glutathione connection. *Diabetes*, 52:581–587.

Rosen PA, Brooks AMV, Ramsay RJ, et al. (1987) Efficacy and safety of hypromellose in ocular implant surgery. *Aust NZ J Opthalmol*, 15:193–199.

Rowe RC, Sheskey PJ, Cook WG, Fenton ME. (Eds.) (2012) *Handbook of Pharmaceutical Excipients*, 7th Ed. London: Pharmaceutical Press, London.

Rowe VK, Wolf MA. (1982) Polyethylene glycols. In: Clayton GD and FE Clayton (Eds.) *Patty's Industrial Hygiene and Toxicology*, Vol. 2C. New York, NY: Wiley & Sons, pp. 3844–3852, 3902, 3904–3905.

Ruble GR, Giardino OZ, Fosscedo, SL, et al. (2006) The effects of commonly used vehicles on canine hematology and clinical chemistry values. *J Am Assoc Lab Anim Sci*, 45(1):25–29.

Ruchatz E. (1998) *Applications of Solutol® HS15–A Potent Solubilizer with a Low Toxicity.* Ludwigshafen: BASF.

Schonwald S. (2004a) Irrigating solutions. In: Dart RC, et al. (Eds.) *Medical Toxicology*, 3rd Ed. Philadelphia, PA: Lippincott Williams & Wilkins, pp. 918–920.

Schonwald S. (2004b) Acidifying and alkalinizing agents. In: Dart RC, et al. (Eds.) *Medical Toxicology*, 3rd Ed. Philadelphia, PA: Lippincott Williams & Wilkins, pp. 911–913.

Sellers RS, Antman M, Phillips J, Khan KN, Furst SM. (2005) Effects of miglyol 812 on rats after 4 weeks of gavage as compared with methylcellulose/Tween 80. *Drug Chem Toxicol*, 28:423–432.

Serbest G, Horwitz J, Barbee K. (2005) The effect of Poloxamer-188 on neuronal cell recovery from mechanical injury. *J Neurotrauma*, 22(1):119–132.

Shadnia S, Rahimi M, Pajoumand A, Rasouli MH, Abdollahi M. (2005) Successful treatment of acute aluminium phosphide poisoning: Possible benefit of coconut oil. *Hum Exp Toxicol*, 24(4):215–218.

Sharma PK, Chaudhari PD, Badagale KD, Kulkarni PA, Barhate NS. (2006) Current trends in solid dispersions techniques. *Pharmainfo.net*.

Shertzer HG. (2012) Methanesulfonic acid. In: Bingham E and B Cohrssen (Eds.) *Patty's Toxicology*, 6th Ed., Vol. 4. New York, NY: John Wiley & Sons, p. 1063.

Shideman FE, Procita L. (1951) Some pharmacological actions of polypropylene glycols of average molecular weight 400, 750, 1200, and 2000. *J Pharmacol Exp Ther*, 103:293–305.

Sivilotti MLA. (2004) Ethanol, isopropanol, and methanol. In: Dart RC, et al. (Eds.) *Medical Toxicology*, 3rd Ed. Philadelphia, PA: Lippincott Williams & Wilkins, pp. 1215–1216.

Smyth HF Jr, Carpenter CP, Shaffer CB. (1945) The subacute toxicity and irritation of polyethylene glycols of approximate molecular weights of 200, 300, 400. *J Am Pharm Assoc*, 34:172–174.

Smyth HF Jr, Carpenter CP, Shaffer CB. (1947) The toxicity of high molecular weight polyethylene glycols; Chronic oral and parenteral administration. *J Am Pharm Assoc*, 36:157–160.

Smyth HF Jr, Carpenter CP, Shaffer CB, Seaton J, Fischer L. (1942) Some pharmacological properties of polyethylene glycols of high molecular weight ("carbowax" compounds). *J Ind Hyg Toxicol*, 24:281–284.

Smyth HF Jr, Carpenter CP, Weil CS. (1950) The toxicology of polyethylene glycols. *J Am Pharm Assoc*, 39:349–354.

Smyth HF Jr, Carpenter CP, Weil CS. (1955) The chronic oral toxicity of the polyethylene glycols. *J Am Pharm Assoc*, 44:27–30.

Smyth HF Jr, Weil CS, Woodside MD, et al. (1970) Experimental toxicity of high molecular weight poly(ethylene oxide). *Toxicol Appl Pharmacol*, 16:442–445.

Sodicoff M, Lamperti A, Ziskin MC. (1990) Transdermal absorption of radioprotectors using permeation-enhancing vehicles. *Radiat Res*, 121(2):212–219.

Solomon GM, Morse EP, Garbo MJ, Milton DK. (1996) Stillbirth after occupational exposure to n-methyl-2-pyrrolidine: A case report and review of the literature. *J Occup Env Med*, 38(7):705–713.

Sourkes TL, Koppanyi T. (1950) Correlation between the acute toxicity and the rate of elimination of tartaric acid and certain of its esters. *J Am Pharm Assoc*, 39(5):275–276.

Stella VJ, He Q. (2008) Cyclodextrins. *Toxicol Pathol*, 36:30–42.

Stokes AH, Kemp DC, Faiola B, et al. (2013) Effects of Solutol (Kolliphor) and Cremophor in polyethylene glycol 400 vehicle formulations in Sprague-Dawley rats and beagle dogs. *Int J Toxicol*, 32(3)189–197.

Strickley RG. (2008) Formulation in drug discovery. In: Macor JE (Ed.) *Annual Reports in Medicinal Chemistry*, Vol. 43. London: Academic Press, pp. 419–451.

Sullivan DW Jr, Gad SC, Julien M. (2014) A review of the nonclinical safety of Transcutol®, a highly purified form of diethylene glycol monoethyl ether (DEGEE) used as a pharmaceutical excipient. *Food Chem Toxicol*, 72:40–50.
Susananta W, Craid DQ, Newton JM. (1995) An evaluation of the mechanisms of drug release from glyceride bases. *J Pharm Pharmacol*, 47(3):182–187.
Szilagyi M. (2012a) Acetic acid. In: Bingham E and B Cohrssen (Eds.) *Patty's Toxicology*, 6th Ed., Vol. 3. New York, NY: John Wiley & Sons, pp. 477–480.
Szilagyi M. (2012b) Citric acid. In: Bingham E and B Cohrssen (Eds.) *Patty's Toxicology*, 6th Ed., Vol. 3. New York, NY: John Wiley & Sons, pp. 519–521.
Szilagyi M. (2012c) Succinic acid. In: Bingham E and B Cohrssen (Eds.) *Patty's Toxicology*, 6th Ed., Vol. 3. New York, NY: John Wiley & Sons, pp. 513–514.
Szilagyi M. (2012d) Tartaric acid. In: Bingham E and B Cohrssen (Eds.) *Patty's Toxicology*, 6th Ed., Vol. 3. New York, NY: John Wiley & Sons, pp. 516–517.
Takahashi KK, Onodera, Akiba Y. (1999) Effect of dietary Xylitol on growth and inflammatory responses in immune stimulated chickens. *Br Poult Sci*, 40:552–554.
Talsma J. (2004) Anecortave acetate proving helpful for inhibiting CNV growth. *Opthamology Times*, 29(6):34.
Temple BR. (2004) Vitamin toxicity. In: Dart RC, et al. (Eds.) *Medical Toxicology*, 3rd Ed. Philadelphia, PA: Lippincott Williams & Wilkins, pp. 1021–1023.
Thackaberry EA. (2014) Solvent-based formulations for intravenous mouse pharmacokinetic studies: Tolerability and recommended solvent dose limits. *Xenobiotica*, 44(3):235–241.
Thackaberry EA, Kopytek S, Sherratt P, Trouba K, McIntyre B. (2009) Comprehensive investigation of hydroxypropyl methylcellulose, propylene glycol, polysorbate 80, and hydroxypropyl-beta-cyclodextrin for use in general toxicology studies. *Toxicol Sci*, 117(2):485–492.
TOXNET (2015a) *Arabic Gum*. TOXNET Website. http://toxnet.nlm.nih.gov/. Accessed August 7, 2015.
TOXNET (2015aa) *Cetyl Alcohol*. TOXNET Website. http://toxnet.nlm.nih.gov/. Accessed August 7, 2015..
TOXNET (2015ab) *Coconut Oil*. TOXNET Website. http://toxnet.nlm.nih.gov/. Accessed August 7, 2015.
TOXNET (2015bs) *Lactose*. TOXNET Website. http://toxnet.nlm.nih.gov/. Accessed August 7, 2015.
TOXNET (2015ac) *Corn Oil*. TOXNET Website. http://toxnet.nlm.nih.gov/. Accessed August 7, 2015.
TOXNET (2015ad) *Cottonseed Oil*. TOXNET Website. http://toxnet.nlm.nih.gov/. Accessed August 7, 2015.
TOXNET (2015ae) *Cyclodextrin*. TOXNET Website. http://toxnet.nlm.nih.gov/. Accessed August 7, 2015.
TOXNET (2015af) *Cyclohexane*. TOXNET Website. http://toxnet.nlm.nih.gov/. Accessed August 7, 2015.
TOXNET (2015ag) *Cyclomethicone NF*. TOXNET Website. http://toxnet.nlm.nih.gov/. Accessed August 7, 2015.
TOXNET (2015ah) *Diacetylmonoxime*. TOXNET Website. http://toxnet.nlm.nih.gov/. Accessed August 7, 2015.

TOXNET (2015ai) *Dextrose*. TOXNET Website. http://toxnet.nlm.nih.gov/. Accessed August 7, 2015.

TOXNET (2015aj) *Dichlorvos*. TOXNET Website. http://toxnet.nlm.nih.gov/. Accessed August 7, 2015.

TOXNET (2015ak) *N, N-Diethylacetamide*. TOXNET Website. http://toxnet.nlm.nih.gov/. Accessed August 7, 2015.

TOXNET (2015al) *Diethylene Glycol Monoethyl Ether*. TOXNET Website. http://toxnet.nlm.nih.gov/. Accessed August 7, 2015.

TOXNET (2015am) *Dimethicone*. TOXNET Website. http://toxnet.nlm.nih.gov/. Accessed August 7, 2015.

TOXNET (2015an) *N, N-Dimethylacetamide*. TOXNET Website. http://toxnet.nlm.nih.gov/. Accessed August 7, 2015.

TOXNET (2015ao) *Dimethylsulfoxide*. TOXNET Website. http://toxnet.nlm.nih.gov/. Accessed August 7, 2015.

TOXNET (2015ap) *Di-Sodium Hydrogen Phosphate Dehydrate*. TOXNET Website. http://toxnet.nlm.nih.gov/. Accessed August 7, 2015.

TOXNET (2015aq) *Docosanol*. TOXNET Website. http://toxnet.nlm.nih.gov/. Accessed August 7, 2015.

TOXNET (2015ar) *Edetic Acid*. TOXNET Website. http://toxnet.nlm.nih.gov/. Accessed August 7, 2015.

TOXNET (2015as) *Ethyl Alcohol*. TOXNET Website. http://toxnet.nlm.nih.gov/. Accessed August 7, 2015.

TOXNET (2015at) *Fumaric Acid*. TOXNET Website. http://toxnet.nlm.nih.gov/. Accessed August 7, 2015.

TOXNET (2015au) *Gelatin*. TOXNET Website. http://toxnet.nlm.nih.gov/. Accessed August 7, 2015.

TOXNET (2015av) *Gelucire 44/14*. TOXNET Website. http://toxnet.nlm.nih.gov/. Accessed August 7, 2015.

TOXNET (2015aw) *Gelucire 50/13*. TOXNET Website. http://toxnet.nlm.nih.gov/. Accessed August 7, 2015.

TOXNET (2015ax) *Gluconic Acid*. TOXNET Website. http://toxnet.nlm.nih.gov/. Accessed August 7, 2015.

TOXNET (2015ay) *Glycerol*. TOXNET Website. http://toxnet.nlm.nih.gov/. Accessed August 7, 2015.

TOXNET (2015az) *Glycofurol*. TOXNET Website. http://toxnet.nlm.nih.gov/. Accessed August 7, 2015.

TOXNET (2015b) *Acetate Ion*. TOXNET Website. http://toxnet.nlm.nih.gov/. Accessed August 7, 2015.

TOXNET (2015ba) *Tragacanth*. TOXNET Website. http://toxnet.nlm.nih.gov/. Accessed August 7, 2015.

TOXNET (2015bb) *Xanthan Gum*. TOXNET Website. http://toxnet.nlm.nih.gov/. Accessed August 7, 2015.

TOXNET (2015bc) *Histidine*. TOXNET Website. http://toxnet.nlm.nih.gov/. Accessed August 7, 2015.

TOXNET (2015bd) *Hydrochloric Acid*. TOXNET Website. http://toxnet.nlm.nih.gov/. Accessed August 7, 2015.

TOXNET (2015be) *Hydrogenated Castor Oil*. TOXNET Website. http://toxnet.nlm.nih.gov/. Accessed August 7, 2015.

TOXNET (2015bf) *Hydroxyethylcellulose.* TOXNET Website. http://toxnet.nlm.nih.gov/. Accessed August 7, 2015.

TOXNET (2015bg) *Hydroxypropyl Betadex.* TOXNET Website. http://toxnet.nlm.nih.gov/. Accessed August 7, 2015.

TOXNET (2015bh) *Hydroxypropyl Cellulose.* TOXNET Website. http://toxnet.nlm.nih.gov/. Accessed August 7, 2015.

TOXNET (2015bi) *Hydroxypropyl Methylcellulose.* TOXNET Website. http://toxnet.nlm.nih.gov/. Accessed August 7, 2015.

TOXNET (2015bj) *Hypromellose Acetate Succinate.* TOXNET Website. http://toxnet.nlm.nih.gov/. Accessed August 7, 2015.

TOXNET (2015bk) *Hymetellose.* TOXNET Website. http://toxnet.nlm.nih.gov/. Accessed August 7, 2015.

TOXNET (2015bl) *Isopropyl Alcohol.* TOXNET Website. http://toxnet.nlm.nih.gov/. Accessed August 7, 2015.

TOXNET (2015bm) *Isopropyl Myristate.* TOXNET Website. http://toxnet.nlm.nih.gov/. Accessed August 7, 2015.

TOXNET (2015bn) *Polyoxyl 35 Castor Oil.* TOXNET Website. http://toxnet.nlm.nih.gov/. Accessed August 7, 2015.

TOXNET (2015bo) *Polyoxyl 40 Hydrogenated Castor Oil.* TOXNET Website. http://toxnet.nlm.nih.gov/. Accessed August 7, 2015.

TOXNET (2015bp) *Labrafil.* TOXNET Website. http://toxnet.nlm.nih.gov/. Accessed August 7, 2015.

TOXNET (2015bq) *Ringer's Lactate Solution.* TOXNET Website. http://toxnet.nlm.nih.gov/. Accessed August 7, 2015.

TOXNET (2015br) *Lactic Acid.* TOXNET Website. http://toxnet.nlm.nih.gov/. Accessed August 7, 2015.

TOXNET (2015bt) *Lanolin.* TOXNET Website. http://toxnet.nlm.nih.gov/. Accessed August 7, 2015.

TOXNET (2015bu) *Lanolin Alcohols.* TOXNET Website. http://toxnet.nlm.nih.gov/. Accessed August 7, 2015.

TOXNET (2015bv) *L-Arginine Hydrochloride.* TOXNET Website. http://toxnet.nlm.nih.gov/. Accessed August 7, 2015.

TOXNET (2015bw) *Ascorbic Acid.* TOXNET Website. http://toxnet.nlm.nih.gov/. Accessed August 7, 2015.

TOXNET (2015bx) *Propylene Glycol.* TOXNET Website. http://toxnet.nlm.nih.gov/. Accessed August 7, 2015.

TOXNET (2015by) *Lycasin.* TOXNET Website. http://toxnet.nlm.nih.gov/. Accessed August 7, 2015.

TOXNET (2015bz) *Maltol.* TOXNET Website. http://toxnet.nlm.nih.gov/. Accessed August 7, 2015.

TOXNET (2015c) *Acetic Acid.* TOXNET Website. http://toxnet.nlm.nih.gov/. Accessed August 7, 2015.

TOXNET (2015ca) *Mannitol, D-* TOXNET Website. http://toxnet.nlm.nih.gov/. Accessed August 7, 2015.

TOXNET (2015cb) *Methanesulfonic Acid.* TOXNET Website. http://toxnet.nlm.nih.gov/. Accessed August 7, 2015.

TOXNET (2015cc) *Methyl Methacrylate.* TOXNET Website. http://toxnet.nlm.nih.gov/. Accessed August 7, 2015.

TOXNET (2015cd) *Methylcellulose*. TOXNET Website. http://toxnet.nlm.nih.gov/. Accessed August 7, 2015.

TOXNET (2015ce) *Methylparaben*. TOXNET Website. http://toxnet.nlm.nih.gov/. Accessed August 7, 2015.

TOXNET (2015cf) *Methylpyrrolidone*. TOXNET Website. http://toxnet.nlm.nih.gov/. Accessed August 7, 2015.

TOXNET (2015cg) *Mineral Oil*. TOXNET Website. http://toxnet.nlm.nih.gov/. Accessed August 7, 2015.

TOXNET (2015ch) *Myristyl Alcohol*. TOXNET Website. http://toxnet.nlm.nih.gov/. Accessed August 7, 2015.

TOXNET (2015ci) *Octoxynol*. TOXNET Website. http://toxnet.nlm.nih.gov/. Accessed August 7, 2015.

TOXNET (2015cj) *Oleic Acid*. TOXNET Website. http://toxnet.nlm.nih.gov/. Accessed August 7, 2015.

TOXNET (2015ck) *Oleyl Alcohol*. TOXNET Website. http://toxnet.nlm.nih.gov/. Accessed August 7, 2015.

TOXNET (2015cl) *Olive Oil*. TOXNET Website. http://toxnet.nlm.nih.gov/. Accessed August 7, 2015.

TOXNET (2015cm) *Panthenol*. TOXNET Website. http://toxnet.nlm.nih.gov/. Accessed August 7, 2015.

TOXNET (2015cn) *Peanut Oil*. TOXNET Website. http://toxnet.nlm.nih.gov/. Accessed August 7, 2015.

TOXNET (2015co) *Glyceryl Monooleate*. TOXNET Website. http://toxnet.nlm.nih.gov/. Accessed August 7, 2015.

TOXNET (2015cp) *Macrogol*. TOXNET Website. http://toxnet.nlm.nih.gov/. Accessed August 7, 2015.

TOXNET (2015cq) *Petrolatum*. TOXNET Website. http://toxnet.nlm.nih.gov/. Accessed August 7, 2015.

TOXNET (2015cr) *Phenoxyethynol*. TOXNET Website. http://toxnet.nlm.nih.gov/. Accessed August 7, 2015.

TOXNET (2015cs) *Phosphate Ion*. TOXNET Website. http://toxnet.nlm.nih.gov/. Accessed August 7, 2015.

TOXNET (2015ct) *Poloxalene*. TOXNET Website. http://toxnet.nlm.nih.gov/. Accessed August 7, 2015.

TOXNET (2015cu) *Decaglyceryl Monooleate*. TOXNET Website. http://toxnet.nlm.nih.gov/. Accessed August 7, 2015.

TOXNET (2015cv) *Povidone*. TOXNET Website. http://toxnet.nlm.nih.gov/. Accessed August 7, 2015.

TOXNET (2015cw) *Potassium Chloride*. TOXNET Website. http://toxnet.nlm.nih.gov/. Accessed August 7, 2015.

TOXNET (2015cx) *Propylene Glycol Dicaprylate/Dicaprate*. TOXNET Website. http://toxnet.nlm.nih.gov/. Accessed August 7, 2015.

TOXNET (2015cy) *Propylparaben*. TOXNET Website. http://toxnet.nlm.nih.gov/. Accessed August 7, 2015.

TOXNET (2015cz) *Copovidone*. TOXNET Website. http://toxnet.nlm.nih.gov/. Accessed August 7, 2015.

TOXNET (2015d) *Acetone*. TOXNET Website. http://toxnet.nlm.nih.gov/. Accessed August 7, 2015.

TOXNET (2015da) *Safflower Oil*. TOXNET Website. http://toxnet.nlm.nih.gov/. Accessed August 7, 2015.
TOXNET (2015db) *Salicylic Acid*. TOXNET Website. http://toxnet.nlm.nih.gov/. Accessed August 7, 2015.
TOXNET (2015dc) *Sesame Oil*. TOXNET Website. http://toxnet.nlm.nih.gov/. Accessed August 7, 2015.
TOXNET (2015dd) *Shea Butter*. TOXNET Website. http://toxnet.nlm.nih.gov/. Accessed August 7, 2015.
TOXNET (2015de) *Simethicone*. TOXNET Website. http://toxnet.nlm.nih.gov/. Accessed August 7, 2015.
TOXNET (2015df) *Sodium Acetate*. TOXNET Website. http://toxnet.nlm.nih.gov/. Accessed August 7, 2015.
TOXNET (2015dg) *Sodium Citrate*. TOXNET Website. http://toxnet.nlm.nih.gov/. Accessed August 7, 2015.
TOXNET (2015dh) *Sodium Phosphate, Monobasic, Dihydrate*. TOXNET Website. http://toxnet.nlm.nih.gov/. Accessed August 7, 2015.
TOXNET (2015di) *Sodium Hydroxide*. TOXNET Website. http://toxnet.nlm.nih.gov/. Accessed August 7, 2015.
TOXNET (2015dj) *Sodium Lauryl Sulfate*. TOXNET Website. http://toxnet.nlm.nih.gov/. Accessed August 7, 2015.
TOXNET (2015dk) *Sodium Metabisulfite*. TOXNET Website. http://toxnet.nlm.nih.gov/. Accessed August 7, 2015.
TOXNET (2015dl) *Methylparaben Sodium*. TOXNET Website. http://toxnet.nlm.nih.gov/. Accessed August 7, 2015.
TOXNET (2015dm) *Sodium Phosphate*. TOXNET Website. http://toxnet.nlm.nih.gov/. Accessed August 7, 2015.
TOXNET (2015dn) *Propylparaben Sodium*. TOXNET Website. http://toxnet.nlm.nih.gov/. Accessed August 7, 2015.
TOXNET (2015do) *Butanedioic Acid, Disodium Salt*. TOXNET Website. http://toxnet.nlm.nih.gov/. Accessed August 7, 2015.
TOXNET (2015dp) *Sodium Sulfite*. TOXNET Website. http://toxnet.nlm.nih.gov/. Accessed August 7, 2015.
TOXNET (2015dq) *Solutol HS15*. TOXNET Website. http://toxnet.nlm.nih.gov/. Accessed August 7, 2015.
TOXNET (2015dr) *Sorbitan Tristearate*. TOXNET Website. http://toxnet.nlm.nih.gov/. Accessed August 7, 2015.
TOXNET (2015ds) *Sorbitol*. TOXNET Website. http://toxnet.nlm.nih.gov/. Accessed August 7, 2015.
TOXNET (2015dt) *Soybean Oil*. TOXNET Website. http://toxnet.nlm.nih.gov/. Accessed August 7, 2015.
TOXNET (2015du) *Squalene*. TOXNET Website. http://toxnet.nlm.nih.gov/. Accessed August 7, 2015.
TOXNET (2015dv) *Stearic Acid*. TOXNET Website. http://toxnet.nlm.nih.gov/. Accessed August 7, 2015.
TOXNET (2015dw) *Stearyl Alcohol*. TOXNET Website. http://toxnet.nlm.nih.gov/. Accessed August 7, 2015.
TOXNET (2015dx) *Sucrose*. TOXNET Website. http://toxnet.nlm.nih.gov/. Accessed August 7, 2015.

TOXNET (2015dy) *Sucrose Acetate Isobutyrate*. TOXNET Website. http://toxnet.nlm.nih.gov/. Accessed August 7, 2015.

TOXNET (2015dz) *Sulfobutylether Beta Cyclodextrin*. TOXNET Website. Accessed August 7, 2015.

TOXNET (2015e) *Acetonitrile*. TOXNET Website. http://toxnet.nlm.nih.gov/. Accessed August 7, 2015.

TOXNET (2015ea) *Tartaric Acid*. TOXNET Website. http://toxnet.nlm.nih.gov/. Accessed August 7, 2015.

TOXNET (2015eb) *Terbinafine Hydrochloride*. TOXNET Website. http://toxnet.nlm.nih.gov/. Accessed August 7, 2015.

TOXNET (2015ec) *Tetraglycol*. TOXNET Website. http://toxnet.nlm.nih.gov/. Accessed August 7, 2015.

TOXNET (2015ed) *Trehalose*. TOXNET Website. http://toxnet.nlm.nih.gov/. Accessed August 7, 2015.

TOXNET (2015ee) *Tromethamine*. TOXNET Website. http://toxnet.nlm.nih.gov/. Accessed August 7, 2015.

TOXNET (2015ef) *Polysorbate 20*. TOXNET Website. http://toxnet.nlm.nih.gov/. Accessed August 7, 2015.

TOXNET (2015eg) *Polysorbate 80*. TOXNET Website. http://toxnet.nlm.nih.gov/. Accessed August 7, 2015.

TOXNET (2015eh) *Alpha-tocopherol*. TOXNET Website. http://toxnet.nlm.nih.gov/. Accessed August 7, 2015.

TOXNET (2015ei) *Tocophersolan*. TOXNET Website. http://toxnet.nlm.nih.gov/. Accessed August 7, 2015.

TOXNET (2015ej) *White Wax*. TOXNET Website. http://toxnet.nlm.nih.gov/. Accessed August 7, 2015.

TOXNET (2015ek) *Xylitol*. TOXNET Website. http://toxnet.nlm.nih.gov/. Accessed August 7, 2015.

TOXNET (2015f) *Acetylated Lanolin Alcohol*. TOXNET Website. http://toxnet.nlm.nih.gov/. Accessed August 7, 2015.

TOXNET (2015g) *N-Methylacetamide*. TOXNET Website. http://toxnet.nlm.nih.gov/. Accessed August 7, 2015.

TOXNET (2015h) *Alginic Acid*. TOXNET Website. http://toxnet.nlm.nih.gov/. Accessed August 7, 2015.

TOXNET (2015i) *Anecortave Acetate*. TOXNET Website. http://toxnet.nlm.nih.gov/. Accessed August 7, 2015.

TOXNET (2015j) *Cellulose, Microcrystalline*. TOXNET Website. http://toxnet.nlm.nih.gov/. Accessed August 7, 2015.

TOXNET (2015k) *Benzoic Acid*. TOXNET Website. http://toxnet.nlm.nih.gov/. Accessed August 7, 2015.

TOXNET (2015l) *Benzyl Alcohol*. TOXNET Website. http://toxnet.nlm.nih.gov/. Accessed August 7, 2015.

TOXNET (2015m) *beta-Cyclodextrin*. TOXNET Website. http://toxnet.nlm.nih.gov/. Accessed August 7, 2015.

TOXNET (2015n) *Butylated Hydroxyanisole*. TOXNET Website. http://toxnet.nlm.nih.gov/. Accessed August 7, 2015.

TOXNET (2015o) *Butylated Hydroxytoluene*. TOXNET Website. http://toxnet.nlm.nih.gov/. Accessed August 7, 2015.

TOXNET (2015p) *Bicarbonate Ion*. TOXNET Website. http://toxnet.nlm.nih.gov/. Accessed August 7, 2015.
TOXNET (2015q) *Calcium Chloride*. TOXNET Website. http://toxnet.nlm.nih.gov/. Accessed August 7, 2015.
TOXNET (2015r) *Canola Oil*. TOXNET Website. http://toxnet.nlm.nih.gov/. Accessed August 7, 2015.
TOXNET (2015s) *Beta-cyclodextrin Sulfobutyl Ether Sodium Salt*. TOXNET Website. http://toxnet.nlm.nih.gov/. Accessed August 7, 2015.
TOXNET (2015t) *Carbomer Homopolymer Type B (allyl pentaerythritol crosslinked)*. TOXNET Website. http://toxnet.nlm.nih.gov/. Accessed August 7, 2015.
TOXNET (2015u) *Carbopol Ultrez 10*. TOXNET Website. http://toxnet.nlm.nih.gov/. Accessed August 7, 2015.
TOXNET (2015v) *Carboxymethylcellulose*. TOXNET Website. http://toxnet.nlm.nih.gov/. Accessed August 7, 2015.
TOXNET (2015w) *Carboxymethylcellulose Calcium*. TOXNET Website. http://toxnet.nlm.nih.gov/. Accessed August 7, 2015.
TOXNET (2015x) *Carboxymethylcellulose Sodium*. TOXNET Website. http://toxnet.nlm.nih.gov/. Accessed August 7, 2015.
TOXNET (2015y) *Cetosteryl Alcohol*. TOXNET Website. http://toxnet.nlm.nih.gov/. Accessed August 7, 2015.
TOXNET (2015z) *Cetyl Acetate*. TOXNET Website. http://toxnet.nlm.nih.gov/. Accessed August 7, 2015.
Toyoda K, Shoda T, Uneyama C, Takada K, Takahashi M. (1997) Carcinogenicity study of beta-cyclodextrin in F344 rats. *Food Chem Toxicol*, 35(3–4):331–336.
Traul KA, Driedger A, Ingle DL, Nakhasi D. (2000) Review of the toxicologic properties of medium-chain triglycerides. *Food Chem Toxicol*, 38(1):79–98.
Trimmer GW, Freeman JJ, Priston RA, Urbanus J. (2004) Results of chronic dietary toxicity studies of high viscosity (P70H and P100H) white mineral oils in Fischer 344 rats. *Toxicol Pathol*, 32(4):439–447.
Tyl RW, Masten LW, Marr MC, et al. (1994) Developmental toxicity evaluation of isopropanol by gavage in rats and rabbits. *Fund Appl Toxicol*, 22(1):139–151.
Ueda H, Ou D, Endo T, et al. (1998) Evaluation of a sulfobutyl ether beta-cyclodextrin as solubilizing agent for several drugs. *Drug Dev Ind Pharm*, 24(9):863–867.
Walker R, El Harith EA. (1978) Nutritional and toxicological properties of some raw and modified starches. *Ann Nutr Aliment*, 32(2–3):671–679.
Waner T, Borelli G, Cadel S, Privman I, Nyska A. (1995) Investigation of potential oncogenetic effects of beta-cyclodextrin in the rat and mouse. *Life Sci Res Israel*, 69(9):631–639.
White CW, Rodriguez J, Marrs GE Jr. (1984) *Acute Oral Toxicity of DMSO (dimethyl sulfoxide) Process Stream Samples in Male and Female Rats*. Govt Reports Announcements & Index (GRA&I), Issue 9, National Technical Information Service (NTIS), Springfield, VA.
Wieland TM, Lin X, Odle J. (1993) Utilization of medium-chain triglycerides by neonatal pigs: Effects of emulsification and dose delivered. *J Anim Sci*, 71(7):1863–1868.
Wood DC, Weber FS, Palmquist MA. (1971) Continued studies in the toxicology of dimethyl sulfoxide (DMSO). *J Pharmacol Exp Ther*, 177(3):520–527.

Working PK, Newman MS, Johnson J, Cornacoff JB. (1997) Safety of poly (ethylene glycol) and poly (ethylene glycol) derivatives. *ACS Symp Ser*, 680(4):45–57.

Wu B, Iwakiri R, Ootani A, et al. (2004) Dietary corn oil promotes colon cancer by inhibiting mitochonria-dependent apoptosis in azoxymethane-treated rats. *Exp Bio and Med*, 229:1017–1025.

Xcess Biosciences, Inc. Vehicle Guidance for Preclinical Studies, San Diego, CA. Web. http://xcessbiosciences.com/uploads/tmp/in%20vivo%20study%20formulation%20guidance.pdf. Accessed June 5, 2015.

Index